The Role of Insectivorous Birds in Forest Ecosystems

Proceedings of a Symposium: The Role of Insectivorous Birds
in Forest Ecosystems, Held July 13 and 14, 1978 in
Nacogdoches, Texas

THE ROLE OF INSECTIVOROUS BIRDS IN FOREST ECOSYSTEMS

Edited by

James G. Dickson   Richard N. Conner

USDA Forest Service
Southern Forest Experiment Station
Wildlife Habitat and Silviculture Laboratory
Nacogdoches, Texas

Robert R. Fleet   James C. Kroll

Stephen F. Austin State University
School of Forestry
Nacogdoches, Texas

Jerome A. Jackson

Mississippi State University
Department of Biological Sciences
Mississippi State, Mississippi

ACADEMIC PRESS

New York   San Francisco   London   1979

A Subsidiary of Harcourt Brace Jovanovich, Publishers

ACADEMIC PRESS, INC.
111 Fifth Avenue, New York, New York 10003

United Kingdom Edition published by
ACADEMIC PRESS, INC. (LONDON) LTD.
24/28 Oval Road, London NW1 7DX

Library of Congress Cataloging in Publication Data

Main entry under title:

Proceedings of the symposium, the role of insectivorous birds in forest ecosystems.

The symposium held in Nacogdoches, July 13-14,
1978, was sponsored by the School of Forestry,
Stephen F. Austin State University, and the USDA
Forest Service, Southern Forest Experiment Station,
Wildlife Habitat and Silviculture Laboratory, Nacog-
doches, Texas.
1. Birds—Food—Congresses. 2. Birds—Ecology—
Congresses. 3. Forest ecology—Congresses.
4. Forest insects—Congresses. I. Dickson, James G.
II. Stephen F. Austin State University. School
of Forestry. III. Wildlife Habitat and Silviculture
Laboratory. IV. Title: Insectivorous birds in
forest ecosystems.
QL698.3.P76   574.5'264   79-12111
ISBN 0-12-215350-2

PRINTED IN THE UNITED STATES OF AMERICA

79 80 81 82   9 8 7 6 5 4 3 2 1

577.3
Rol

CONTENTS

CONCLUSIONS

CONTRIBUTORS AND PARTICIPANTS

Numbers in parentheses indicate the pages on which authors' contributions begin.

Stanley H. Anderson (203, 375), Migratory Bird and Habitat Research Laboratory, U.S. Fish and Wildlife Service, Laurel, Maryland 20811

Keith A. Arnold, Department of Wildlife and Fisheries Sciences, Texas A&M University, College Station, Texas 77843

Richard N. Conner [1]*(95),* Department of Biology, Virginia Polytechnic Institute and State University, Blacksburg, Virginia 24061

William A. Copper (217), Division of Biological Control, University of California, Berkeley, California 94720

Robert Coulson (53), Department of Entomology, Texas A&M University, College Station, Texas 77843

Donald L. Dahlsten (217), Division of Biological Control, University of California, Berkeley, California 94720

James G. Dickson (261), USDA Forest Service, Southern Forest Experiment Station, Nacogdoches, Texas 75962

Lewis Edson (53), Department of Entomology, Texas A&M University, College Station, Texas 77843

Charles D. Fisher, Department of Biology, Stephen F. Austin State University, Nacogdoches, Texas 75962

Robert R. Fleet (269), School of Forestry, Stephen F. Austin State University, Nacogdoches, Texas 75962

[1]Present address: USDA Forest Service, Southern Forest Experiment Station, Nacogdoches, Texas 75962

Leon J. Folse, Jr. (9), Department of Wildlife and Fisheries Sciences, Texas A&M University, College Station, Texas 77843

Edward O. Garton (107), University of Idaho, College of Forestry, Wildlife and Range Sciences, Moscow, Idaho 83843

Thomas C. Grubb, Jr. (119), Department of Zoology, The Ohio State University, Columbus, Ohio 43210

Lowell K. Halls, USDA Forest Service, Southern Forest Experiment Station, Nacogdoches, Texas 75962

Jerome A. Jackson (1, 69), Department of Biological Sciences, Mississippi State University, Mississippi State, Mississippi 39762

Ross D. James (137), Department of Ornithology, Royal Ontario Museum, 100 Queen's Park, Toronto, Canada M5S 2C6

James C. Kroll (269), School of Forestry, Stephen F. Austin State University, Nacogdoches, Texas 75962

Dan W. Lay, Texas Parks and Wildlife Department, Nacogdoches, Texas 75962

B. Riley McClelland (283), School of Forestry, University of Montana, Missoula, Montana 59812

Shaun M. McEllin (301), Department of Science and Mathematics, New Mexico Highlands University, Las Vegas, New Mexico 87701

Imre S. Otvos (341), Newfoundland Forest Research Centre, Canadian Forestry Service, Department of Fisheries and Environment, St. Johns, Newfoundland A1C 5X8

Benedict C. Pinkowski (165), Ecological Sciences Division, NUS Corporation, Pittsburgh, Pennsylvania 15220

Paul Pulley (53), Data Processing Center, Texas A&M University, College Station, Texas 77843

William M. Shields[2]*(23),* Department of Zoology, Ohio State University, Columbus, Ohio 43210

Herman H. Shugart, Jr. (203), Environmental Sciences Division, Oak Ridge National Laboratory, Oak Ridge, Tennessee 37830

Thomas M. Smith (203), Environmental Sciences Division, Oak Ridge National Laboratory, Oak Ridge, Tennessee 37830

Jerry W. Via (191), Department of Biology, Virginia Polytechnic Institute and State University, Blacksburg, Virginia 24061

R. Montague Whiting, Jr. (331), School of Forestry, Stephen F. Austin State University, Nacogdoches, Texas 75962

[2]Present address: Department of Environmental and Forest Biology, State University of New York, College of Environmental Science and Forestry, Syracuse, New York 13210

PREFACE

This publication presents the proceedings of a symposium, The Role of Insectivorous Birds in Forest Ecosystems, held on July 13 – 14, 1978, in Nacogdoches, Texas. The symposium was sponsored by the School of Forestry, Stephen F. Austin State University, and the USDA Forest Service, Southern Forest Experiment Station, Wildlife Habitat and Silviculture Laboratory, both in Nacogdoches.

The proceedings discuss the ecology of insectivorous birds and acknowledge their roles in forest ecosystems. The papers presented here should enhance understanding of this subject, which should be helpful to forest managers in managing for insectivorous birds; as well as that point out areas where information is lacking. Currently, many mixed forests are being converted to pure forest stands that are vulnerable to insect attacks; there are problems with insect epidemics and there are hazards and restrictions in insectidal control of forest insects. The discussions herein should prove especially valuable in formulating biological control strategies for use in forests.

The volume includes an introduction to and history of insectivorous birds and their roles in forest ecosystems, discussions of sampling methods for bird and insect populations, bird foraging strategies, ecology of insectivorous bird species and communities, and an overview and conclusion.

For their assistance in conducting the symposium, we thank Jim Anderson, Tom Carbone, Greg Hiser, Brian Locke, Diana Montgomery, Gary Reed, Monty Whiting, Howard Williamson, and Bob Zaiglin.

INSECTIVOROUS BIRDS
AND NORTH AMERICAN FOREST ECOSYSTEMS

Jerome A. Jackson

Department of Biological Sciences
Mississippi State University
Mississippi State, Mississippi

During the past 200 years the face of North America has undergone a change that was accomplished in temperate Europe only over a period of thousands of years. Forests have been replaced by our cities and towns, our agricultural fields, our highways, and other artifacts of our civilization. The pattern of decrease in forest lands is well established and not unique to our own culture. At first the people were few and the forests relatively unlimited. Forests were a place to fear because they offered hiding places for enemies of both humans and livestock. Clearing of the forests around settlements served the multiple purposes of providing raw resources for homes and fuel, providing open land for crops, and increasing visibility of approaches. Birds and other wildlife too seemed limitless. There is little mention in the early history of our continent of concern for destruction of the forests by man. There is also little mention of concern for destruction of the forests by insect pests. Concern over man's effects on forests arose only as the resource became limited. Then too forest insect pests were recognized and the science of economic entomology was born. This was in the latter half of the last century. With the flood of human immigrants to this continent and the rise of American trade among nations, a host of insect immigrants arrived and also became established. Most were adventives, brought in with human commodities and unrecognized until it was too late. The Gypsy Moth (Porthetria dispar) was brought in on purpose. The result was the same. Dwindling forest resources under increased human pressure were subjected to the ravages of exotic as well as native pests. Simultaneous with growing concern for effects of harmful insects on our forests, the U.S. Biological Survey sponsored studies of the food habits of North American birds.

1

F.E.L. Beal published numerous accounts in the 1890's and early 1900's detailing food habits of our birds. Economic ornithology was born with these studies. William McAtee's (1926) study of "The Relation of Birds to Woodlots in New York State" and Edward Forbush's (1913) book on "Useful Birds and Their Protection," in addition to Beal's work, offered firm evidence of the positive role of insectivorous birds in control of forest insects.

While early studies pointed the way, they did not provide a quantitative assessment of effects of birds on insect populations nor did they document effects insect populations had on birds. Such studies began to appear in the 1940's. Kendeigh (1947) for example, demonstrated both functional and numerical responses of bird populations during an outbreak of the Spruce Budworm (Choristoneura fumiferana). McFarlane (1976) provides a good review of several quantitative studies dealing with impact of birds on forest insect pests. Basically, we know that some bird species can drastically influence populations of some pests but that in other cases impact of avian predators on pests seems negligible. As an example of the former, Knight (1958) reported that woodpeckers reduced Engelmann Spruce Beetle (Dendroctonus engelmanni) populations by 45-98%. Other studies relating individual pest species to individual bird species or groups of species offer sufficient, well-documented evidence that birds can and should be considered as an important component of modern integrated pest management schemes. A few studies, such as that of Morris et al. (1958) have begun to examine simultaneous changes in populations of several bird species during an insect outbreak. Before we can truly employ birds in our pest management schemes, we must know more than just the relationship between pest and predator species. We must have an understanding of the dynamics of the forest ecosystem to which these species belong.

In more than magnitude, forest ecosystems of North America are different from those of two hundred years ago. Yet, thanks to the foresight of men like Gifford Pinchot and Theodore Roosevelt, we have maintained some of the forest diversity and integrity that has been lost in Europe. Only 4% of Great Britain remains forested in contrast to over 30% of North America (Platt, 1965). Changes that have taken place in North America include tremendous forest fragmentation and isolation of small forest areas, reduction in forest age by increased harvesting and change in tree species composition as a result of selective harvesting, planting, control of undesirable species, and losses to exotic diseases such as chestnut blight and Dutch elm disease. Increased mechanization of harvesting techniques to meet increasing demands for forest products is producing new forest ecosystems. In creating these systems little thought has been given to effects of changes in tree species composition and diversity, and forest age structure on the other endemic plants and animals. As a result of human manipulation, forests of today's North

America are being stressed by insect pests that two centuries ago
might have been kept in check by natural predators.

If we are to attempt to restore some balance to our forest
ecosystems, we must make effective use of natural controls for
forest insect pests. To do so we need to recognize the potential
predators and provide for their needs - all year long, not just
during visible stages of forest destruction. We must also
recognize the dynamics of predation during all stages in pest
life cycles. Even without the increased costs and environmental
hazards of chemical controls for insect pests, I feel that
management to enhance forest bird populations could be economically advantageous. In evaluating roles of insectivorous birds
in forest ecosystems, we must first be cognizant of those birds.
To give you an indication of the group we might be talking about,
I considered the habitat requirements, foraging behavior, and
food habits of each species of bird known to breed in North
America. This list includes approximately 640 bird species. Of
these, approximately 44% include insects as 10% or more of their
diet during at least one season and forage in or above forests
or in association with trees at the forest edge (Table I). Data
on each species were gleaned from Robbins et al. (1966), Martin
et al. (1951), and various volumes of Bent's "Life Histories of
North American ..." birds which were published as U.S. National
Museum Bulletins. Of these 283 species, 30% are residents, 70%
are migrants, 23% are cavity nesters, 42% forage on the ground,
67% forage in the shrub layer, 55% forage in trees, and 28%
forage in the air.

Before considering the birds further, I would like to turn
briefly to some of the more economically important forest insect
pests. I compiled a list in Table I from a variety of forest
entomology text books and technical sources, but used Anderson
(1960) as a primary source. While I may have excluded some
serious pests, I feel that those listed exemplify the worst
problem species. Note first of all that these include members
of only three insect orders: Lepidoptera, Coleoptera and Hymen-
optera. The hordes of warblers, vireos, and other songbirds
that migrate to North America to nest in the spring have an
important impact on adults and larvae of Lepidoptera when their
populations are beginning to build up in the spring and summer.
Impact of early arrivals such as the Blue-gray Gnatcatcher
(Polioptila caerulea) and Northern Parula (Parula americana) must
be greater than that of later arrivals since the former likely
get some overwintering insects before they have an opportunity to
lay eggs. In order for these migrant species to take up resi-
dence, however, there must be more than an abundant food supply
for them. They must also have suitable singing perches and nest
sites. For example, cuckoos are among the most voracious eaters
of caterpillars, but to allow a build-up of their population in
an area there must be at least patches of thicket in the understory. For this is the type of nest site they select. During

TABLE I. Characteristics of North American Insectivorous Forest Birds.

Family	Period of insectivory		Migratory status		Cavity nesting	Foraging Site			
	All year	Breeding season	Resident	Migrant		Ground	Shrub	Tree	Air
Anatidae		2		2	2	2			
Accipitridae	4			4		2			2
Falconidae	1			1	1	1			
Meleagrididae		1	1			1			
Cracidae	1		1			1			
Tetraonidae		3	3			3			
Phasianidae	1	4	5			5			
Gruidae	1			1		1			
Scolopacidae	2			2		2			
Cuculidae	5		3	2		2	5		3
Strigidae	8		4	4	8	8	8	8	
Caprimulgidae	5			5		1			5
Apodidae	4		4						4
Picidae	17	1	14	6	18	6	17	2	18
Cotingidae	1			1				1	1
Tyrannidae	31			31	4				31
Hirundinidae	8			8	2				8
Corvidae	15		15	1		15	15	15	
Paridae	14		14		11		14	14	
Pycnonotidae	1		1			1	1	1	
Sittidae	4		3	1	4			4	
Certhiidae	1			1	1			1	
Troglodytidae	7		2	5	7	7	7	7	
Mimidae	6		3	3		6	6		
Turdidae	11			11	3	11	11	3	3
Sylviidae	5		1	4			5	5	2
Bombycillidae		2		2					2
Laniidae	2			2		2			2
Sturnidae	2		2	1	2	2			
Vireonidae	12		1	11			12	12	
Parulidae	53		4	52	1	4	53	49	3
Ploceidae		2	2			2	2	2	
Icteridae	11			11		1	10	10	
Thraupidae	4			4				4	
Fringillidae	14	17	7	21		34	24	3	

TABLE II. Major Insect Pests of North American Forests, the Parts of Trees They Attack, and Their Overwintering Site and Developmental Stage During Overwintering. Avian Predators on These Species Vary with Site of Injury, and Overwintering Site.

Species	Site of injury		Overwintering site & stage	
	Bark	Foliage	On tree	On ground
Pine Butterfly		X	Eggs	
Pandora Moth		X		Pupa
Tussock Moth		X	Eggs	
Gypsy Moth		X	Eggs	
Tent Caterpillar		X	Eggs	
Cankerworm		X		Pupa
Pine Tip Moth		X	Larva	Pupa
Spruce Budworm		X		Pupa
Jack Pine Budworm		X		Pupa
Larch Case Bearer		X	Larva	
White Pine Weevil	X			Pupa
Fir Engraver Beetle	X		All	
Dendroctonus spp.	X		All	
Turpentine Beetle	X		All	
Ips spp.	X		All	
Larch Sawfly		X		Larva

the spring and summer we need also to be aware that a host of bird species that are typically seed and fruit eaters turn insectivorous as they begin to feed their young. We need also to recognize that these, and even typically insectivorous species such as the woodpeckers, bring different food to young of different ages. At first prey are smaller and softer bodied; as the young near fledging, larger, harder bodied larvae and adults may be brought to the nest. In other words, the impact of a bird community on an arthropod population will vary with nesting phenology.

During the fall, winter, and early spring when migrants are gone, arthropod populations are also less obvious, but significant predation on overwintering arthropods can be an important factor in determining extent of arthropod populations during the coming year. What species prey on overwintering insects? Woodpeckers, nuthatches, creepers, chickadees and wrens are among the most important predators. For perhaps 5-6 months these are the major avian insectivores in the forest. These species are also important during the warmer months. In particular, woodpeckers are the only species that can make significant impacts on subsurface bark beetles.

If we are to make maximum use of these species in controlling our forest insect pests, we must be aware of and provide for the needs of these resident species. All of those mentioned: woodpeckers, nuthatches, creepers, chickadees, and wrens, are cavity nesting, cavity roosting species. Their populations are often likely limited by availability of nest and roost cavities or suitable sites to excavate them. Competition for cavities is often severe (e.g., Jackson, 1978). European foresters learned the value of tits decades ago and have systematically provided nest boxes for them throughout their well-manicured forests. But they, like us, have often removed the dead trees as a part of "forest sanitation," thus depriving woodpeckers and other species that don't readily use bird houses as a place to live. Woodpeckers in Europe are in serious difficulty. Birdwatchers in England consider many species rare.

In recognition of the role that insectivorous forest birds play in forest ecosystems, an integrated pest management program needs to include leaving some lightning struck trees as future nest trees for cavity nesters. We know that such trees quickly attract insects and that is why we remove them. We should consider, however, that this concentration of arthropod food also serves as an immediate attractant for avian predators. Perhaps such trees play a significant role in maintenance of insectivorous birds during times of stress. Perhaps they can make the difference in number of birds an area can support. During bad weather many of our wintering insectivorous birds and early arriving migrants turn to other resources such as buds and fruit. Do our forests provide the diversity of resources to allow us to realize maximum benefit from natural controls of forest insects? Perhaps slightly modifying forest management practices such as tailoring cuts more to terrain and leaving fingers of forest within a cut, thus increasing edge and cover for birds, will help. In some cases more drastic modifications of management practices may be needed to save endangered species. There are lots of questions concerning the role of insectivorous birds in forest ecosystems. That's what this symposium is all about. The key word is "Ecosystem." We must look at the whole system and the subtleties of our effects on it.

REFERENCES

Anderson, R. (1960). "Forest and Shade Tree Entomology." John Wiley & Sons, New York.

Forbush, E. (1913). "Useful Birds and Their Protection." 4th ed. The Massachusetts State Board of Agriculture, Boston.

Jackson, J. (1978). Competition for Cavities and Red-cockaded Woodpecker Management. Proc. Symp. on Management Techniques for Preserving Endangered Birds, p. 103. Univ. of Wisconsin, Madison, 17-20 Aug. 1977.

Kendeigh, S. (1947). "Bird Population Studies in the Coniferous Forest Biome During a Spruce Budworm Outbreak." Ontario Dept. Lands Forests, Biol. Bull. 1.

Knight, F. (1958). J. Econ. Entomol. 51, 603.

Martin, A., Zim, H., and Nelson, A. (1951). "American Wildlife & Plants, A Guide to Wildlife Food Habits." McGraw-Hill Book Co., Inc., New York.

McAtee, W. (1926). "The Relation of Birds to Woodlots in New York State." Roosevelt Wild Life Bull. 4, 1.

McFarlane, R. (1976). The Biologist 58, 123.

Morris, R., Cheshire, W., Miller, C., and Mott, D. (1958). Ecology 39, 487.

Platt, R. (1965). "The Great American Forest." Prentice-Hall, Inc., Englewood Cliffs, N.J.

Robbins, C., Bruun, B., and Zim, H. (1966). "A Guide to Field Identification, Birds of North America." Golden Press, New York.

ANALYSIS OF COMMUNITY CENSUS DATA:
A MULTIVARIATE APPROACH

Leon J. Folse, Jr.

Department of Wildlife & Fisheries Sciences
Texas A&M University
College Station, Texas

A method of comparing sample census data between biotic communities is described and illustrated with data from grassland
bird communities. The method is based on multivariate contrast
analysis and takes into account variation within species (over
replications or subsamples, depending on the experimental design)
and of covariation among species. Three methods of constructing
contrasts are given, Roy-Bose, Bonferroni t-interval, and
Scheffe's S methods. A sample SAS program illustrating the
Bonferroni t-interval method is supplied.

I. INTRODUCTION

Sample census data concerning species composition of communities and numbers of individuals for each species are becoming
increasingly important in analyses of natural communities. In
particular, comparisons are desirable between censuses taken in
different habitats, or those taken in similar habitats in which
experimental treatments have been used. Factors which compli-
cate satisfactory comparison of censuses include lack of adequate measures of variation of census estimates and failure to
adequately account for covariation among several species.
    Estimates of population densities and species compositions
are subject to two basic types of error, bias and lack of precision. Bias can be reduced by keeping the probability of
detection for each species constant among habitats. Variation,
however, affects the precision with which we can make estimates,
and in turn affects our ability to recognize differences between
sample censuses. There are many potential sources of variation

9

in avian census work, including variation due to sampling procedures and also variation due to changes in bird behavior with
time of day, weather, differences in detectability among species,
etc. These factors result in variation in avian density and
composition estimates, even when populations being sampled
remain unchanged. Many of these sources of variation can be
handled with a sound experimental design. However, many species'
populations are relatively less dense, clumped and/or highly
mobile; there would be a great deal of day to day variation in
population density estimates for these species on the small
areas which are usually censused. Regardless of the source of
this variation, be it sampling error alone or sampling error
compounded with "natural" variability, decisions about changes
in species composition and abundance between habitats, in
response to treatments, over seasons, etc., require that this
natural variability be taken into account.

   I claim that a satisfactory approach to this problem,
applicable with many different sampling techniques, is through
either empirical replication or subsampling with a suitable
analysis of variance design in which the variance (which is a
measure of this natural variability) may be partitioned among
design factors and residual error terms.. These residual error
terms are then used to judge significant differences among treatment combinations associated with the design. These residual
error terms incorporated our "ignorance" concerning natural
variability and thus provide a suitable measure for deciding
which differences are "significant."

   The second factor which complicates sample census analysis is
that of covariation between species. This factor requires multivariate techniques. Hence a first step is to design data collection for multivariate analysis of variance (MANOVA). One may
have a simple one-way design with "sample census" as the treatment variable, or a complex multiway design (e.g., with habitat
type, experimental treatment and season as factors through which
sample censuses are classified in a 3-way design).

                              II. EXAMPLE

   I will discuss an example from work I've done with censusing
grassland birds in the Serengeti National Park in Tanzania, in
order to illustrate a multivariate contrast solution to these
problems. I had 5 study areas of fairly open grassland, each 1
km^2 in area. During each sample census, birds were flushed from
strip transects of 20 m width. Three km of transect were run
for each subsample, and 3 subsamples (a total of 9 km of transect
with 18% of the area sampled) were used for each sample census.
The 5 study areas were sampled once a month for 11 months for a
total of 55 sample censuses. I estimated the density (number/

km^2) of each 17 species of birds for each subsample (Folse, 1978). The sampling design was thus a MANOVA (17 species) with site (A through E) and time (months 1 through 11, May 1975 through March 1976) and site by time interaction as factors.

Biological abundance and biomass data frequently need to be transformed prior to statistical analysis (cf., Sokal and Rohlf, 1969), and logarithmic transformation is often suitable. I assume that data have been properly transformed for the following analysis (mine were log-transformed biomass values).

I wished to do a series of comparisons among censuses or combinations of censuses. This was done by setting up specific contrasts, where each contrast was the mean biomass of one census combination (e.g., mean of site A over 11 months) minus the mean biomass of another census combination (e.g., mean of site B over 11 months). If there was no difference in mean biomass between the two census combinations, then the value of the contrast should be zero. However, just due to chance alone, or to sampling error, it is unlikely that this difference (i.e., the contrast) be exactly zero, even when species compositions and numbers are identical. How different from zero must it be before we are convinced that there is a real difference in species composition or biomass between census combinations, and not just random sampling error? To answer this question, we need to construct a confidence interval centered on the value of the contrast. If this confidence interval includes zero in its range, then we accept that there is no significant difference between biomasses of the two census combinations, otherwise we accept that they are significantly different. There are several methods available to construct these confidence intervals, and there are several factors we must consider. These factors are:

1) How many contrasts do we wish to do?;
2) How does the structure of each contrast affect the confidence interval width?;
3) How do we use the error structure of the MANOVA design to construct the confidence intervals?; and
4) What groups of species do we wish to consider (all species, by guild, etc.)?

In the Serengeti example, I considered 16 specific contrasts for each group of birds; the questions of interest were:

1) What were the relationships among sites averaged over the 11 months? This required 10 pairwise contrasts for the 5 sites (number of contrasts = 1/2(5x4)).

2) Were there differences between the wet season (a 7 month period) and the dry season (4 months) at each of the sites (5 contrasts) and at all sites considered together (1 contrast)?

These contrasts establish relationships among censuses. One can analyze these relationships for each species independently (univariate), all 17 species simultaneously (multivariate), and for any subgroup of species of interest (also multivariate). Correlations (either positive or negative) between species in a group require a multivariate approach to set correct confidence intervals. I considered the following groups in the example (note, they are not mutually exclusive):

1. All Species (17 species)
2. Larks (7)
3. Graminivores (3)
4. Omnivores (2)
5. Specialized Insectivores (2)
6. Generalized Insectivores (11)
7. Stationary Species (6)
8. Mobile Species (6)

The basic approach to the contrast analysis is to conduct a MANOVA on the data, produce an error sums of squares and cross products (error SSCP) matrix, construct the series of contrasts of interest, and then test these contrasts by creating suitable simultaneous confidence intervals based on the error SSCP matrix.

In establishing the MANOVA statistical design, I would like to partition variability into as many factors as possible, so that the residual variance (corresponding to our "ignorance") is as small as possible. This will maximize our ability to discriminate between censuses. Balanced designs with equal replication for each treatment combination are desirable. However, unequal replication (or subsampling) is not a serious fault, and causes very little increase in complexity of the analysis. Incomplete designs (not all treatment combinations present in a multiway design) may require analysis as a simpler design (such as a one-way design), but they can still be analyzed.

The principal function of doing a MANOVA is to create either a residual SSCP matrix or to create residual values from the design which may be used to produce a residual SSCP matrix. This residual SSCP matrix may, in some cases, be singular. Should this be the case, the formal MANOVA cannot be completed (because the matrix cannot be inverted), but it will have no effect on the contrast analysis.

I will first present results of the contrast analysis with the Serengeti bird data, and then discuss detailed procedures by which the contrasts were constructed. The Appendix contains a sample program written in SAS (Barr et al., 1976), by which these data were analyzed. These results are summarized in Table I (based on the Bonferroni method in the technique section below). Considering all species simultaneously, site B had the greatest bird biomass and site E the least. All sites were statistically distinguishable from one another. Lark biomass, however, was greatest at site C and sites D and B were not statistically separable. It is possible to get complex results such as occur with the Stationary Species group. In this case, sites D and C and sites C and A were not statistically separable, but site D was greater than site A in biomass of Stationary Species.

Results of the seasonal contrasts (Table I) are also different among the different groups of birds. Insectivores (both

TABLE I. Contrast results for Serengeti censuses (55 censuses) [a]

Group	Sites	Seasons [b]		
		Wet	Dry	No Change
All Species	B>C>A>D>E	B, C, All		A, D, E,
Larks	C>D=B>A>E	B, C, All	D	A, E,
Graminivores	E=B>C=D>A	A, C		B, D, E, All
Omnivores	C=A>B>D>E	B, C, All	D	A, E
Specialized Insectivores	A>B>C>D=E			A, B, C, D, E, All
Generalized Insectivores	D>B=C>E>A			A, B, C, D, E, All
Stationary Species	B>D=C=A>E		D	A, B, C, E, All
Mobile Species	C>B>A>D>E	B, C, All		A, D, E

[a] All statistical tests were based on α = 0.01.

[b] The wet season was 7 months (May 1975 and October 1975 through March 1976) while the dry season was 4 months (June through September 1975).

Specialized and Generalized) showed no change between wet and
dry seasons at any of the sites. Stationary Species were more
abundant at site D during the dry season, but showed no between
season variation at any of the other sites. All species considered together were more abundant during the wet season at
sites B and C with no significant change of sites A, D, and E.
However, for all sites considered together (i.e., averaged over
all sites), all species were more abundant during the wet season.

III. CONSTRUCTING CONTRASTS

This analysis is based on treatments by Miller (1966) and
Steinhorst (1977). Consider the set of censuses, I in number
(I = 5x11 = 55 for the Serengeti example), the ith of which
yields a 1xJ vector $\bar{Y}_i$ of mean values (averaged over the subsamples, 3 in this example). Note that $\bar{Y}_i$ has 1 row and J
columns. Each of the I census mean vectors has J components,
one for each of the J species (J = 17 for the Serengeti example):

$$\bar{Y}_i = (n_{i1}\ n_{i2}\ \cdots\ n_{iJ}),\qquad\qquad (1)$$

where n_{ij} is the mean value of the jth species in census i.
Since I shall be interested in several different groups of
species, I need a Jx1 vector (J rows, 1 column) of 1's and 0's,
called A, which will select species of interest from the set of
J species. If I wanted to consider all species as a group, then

$$A' = (1\ 1\ 1\ \cdots\ 1),\qquad\qquad (2)$$

with a "1" representing each of the J species. Note that the
"'" means that the vector is transposed (i.e., the rows and
columns are interchanged). If, on the other hand, I wanted to
consider a group consisting of the first 3 species, then

$$A' = (1\ 1\ 1\ 0\ 0\ \cdots\ 0),\qquad\qquad (3)$$

and the contributions of the remaining species will be subse-
quently deleted. Matrix multiplication with this vector at
appropriate stages in the analysis will select those species
represented by 1's and delete those represented by 0's (the
species are considered here in the same sequence as in the $\bar{Y}_i$
vector).
    A contrast is any linear combination of census means, called
$\hat{\Theta}$, such that,

$$\hat{\Theta} = \sum_{i=1}^{I} c_i A' x \bar{Y}_i,\qquad\qquad (4)$$

and with the restriction that,

$$\sum_{i=1}^{I} c_i = 0. \tag{5}$$

As an example, suppose that I wanted to compare the biomass of all species at site A between months 1 and 2. The matrix product $A'xY_i$ results in the sum of all species biomasses for site-time combination i. In the Serengeti example, there were 55 of these sums (I = 55; data are assumed to be sorted by site-time). I can construct this contrast with $c_1 = 1$, $c_2 = -1$, and all other $c_i = 0$. Thus

$$\hat{\Theta} = A'x\bar{Y}_1 - A'x\bar{Y}_2, \tag{6}$$

or the difference in total bird biomass (log-transformed values) between site A, time 1 and site A, time 2. If the vector A had selected only 3 particular species, then only their biomasses would be compared. Note that

$$\sum_{i=1}^{I} c_i = 1 - 1 = 0. \tag{7}$$

IV. CONSTRUCTING CONFIDENCE INTERVALS

Recall that I expect $\hat{\Theta}$ to be approximately equal to zero if there is no difference between censuses. In order to test this we need the following. Let S_i be the residual mean square and cross product (MSCP) matrix. That is for H subsamples (H = 3 in the example), h = 1, 2, ..., H, and with a completely balanced design (no missing values), I have for census i,

$$S_i = \frac{1}{H-1} \sum_{h=1}^{H} (Y_{ih} - \bar{Y}_{i.})'x(Y_{ih} - \bar{Y}_{i.}), \tag{8}$$

where Y_{ih} is the (JX1) vector of biomasses from subsample h of census i, and

$$\bar{Y}_{i.} = \frac{1}{H} \sum_{h=1}^{H} Y_{ih}. \tag{9}$$

An assumption of the MANOVA, and one that makes this technique quite useful, is that the residual variance-covariance matrices (related to MSCP matrices) are homogeneous among censuses. Hence I can pool them to get

$$S = \frac{1}{I} \sum_{i=1}^{I} S_i$$

$$= \frac{1}{I(H-1)} \sum_{i=1}^{I} \sum_{h=1}^{H} (Y_{ih} - \bar{Y}_{i.})' x (Y_{ih} - \bar{Y}_{i.}). \qquad (10)$$

In practice, I would not calculate this directly but would let a statistical program produce S (or with SAS (Barr et al., 1976), let the program produce a set of residual values, $Y_{ih} - Y_{i.}$, from which I would construct S).

The matrix product,

$$A'xSxA \qquad (11)$$

results in a single number which is the sum of variances and twice the covariances of the species group selected by the vector A. The square root of this number functions like a standard deviation in setting widths of confidence intervals. However, in this case, it also corrects for the effects of non-zero correlations among species in the group of interest. It widens the confidence intervals to correct for positive correlations and narrows the confidence intervals to correct for negative correlations. Consequently, "compensation" of one species' absence by the presence of another is fully corrected for.

There are several approaches to establishing widths of confidence intervals for the contrasts. The method chosen depends on the size of the design and the number of simultaneous confidence intervals being considered. In each of the following methods, I will consider K contrasts simultaneously with the kth contrast given by,

$$\hat{\Theta}_k = \sum_{i=1}^{I} c_{ik} A'x\bar{Y}_i. \qquad (12)$$

A. Roy-Bose Method

This method is appropriate for fairly small designs where one is interested in all possible contrasts (or more than I-1 of them). The confidence interval limits are then given by

$$\hat{\Theta}_k \pm (I\lambda_{max}^{\alpha} \sum_{i=1}^{I} c_{ik}^2)^{\frac{1}{2}} (\frac{1}{H} A'xSxA)^{\frac{1}{2}}, \qquad (13)$$

where λ_{max}^{α} is the $100(1-\alpha)^{th}$ percentile point of the distribution of the maximum root of the S determinental equation (cf., Miller, 1966). This upper percentile point is provided by Morrison, (1967, see Heck charts) and others. The probability is $1-\alpha$ that the K simultaneous intervals contain the true parametric values, θ_k, for all vectors A and C (where C is a vector of contrast coefficients). This technique actually tests for significant separation of points (centroids of the multivariate distributions) representing contrast alternatives in the multivariate space defined by the J species (or the appropriate subset as determined by the vector A). Unfortunately, Heck charts are limited to small ANOVA design sizes. For many censuses with several factors in the design, the following techniques are available.

B. Bonferroni t-interval Method

If one is considering K specific simultaneous contrasts, then this approach is appropriate. In this case, the probability is greater than or equal $1-\alpha$ that the K parameter values, θ_k, occur in their corresponding confidence intervals whose limits are given by

$$\hat{\theta}_k \pm t_{df}^{\alpha/2K} \ (\sum_{i=1}^{I} c_{ik}^2)^{\frac{1}{2}} (\frac{1}{H} A'xSxA)^{\frac{1}{2}}, \qquad (14)$$

where $t_{df}^{\alpha/2K}$ is the $100(1-\alpha/2K)^{th}$ percentile point of the t-distribution with df corresponding to the error degrees of freedom in the MANOVA design (in the Serengeti example, df = 55(3-1) = 110). These values of the t-distribution can be approximated from suitable formulas or interpolated from Bailey's (1977) tables.

C. Scheffe's S Method

This method is appropriate for all possible contrasts (or $K \geq I-1$). The confidence limits for θ_k are

$$\hat{\theta}_k \pm \{(I-1)F_{I-1,N-I}^{\alpha}\}^{\frac{1}{2}} (\sum_{i=1}^{I} c_{ik}^2)^{\frac{1}{2}} (\frac{1}{H} A'xSxA)^{\frac{1}{2}}, \qquad (15)$$

where $F_{I-1,N-I}^{\alpha}$ is the $100(1-\alpha)^{th}$ percentile point of the F-distribution with I-1 and N-I degrees of freedom (N is the total sample size).

Leon J. Folse, Jr.

ACKNOWLEDGMENT

This study was supported by the Caesar Kleberg Program in
Wildlife Ecology at Texas A&M University. I am grateful to R. K.
Steinhorst for tutoring me in the technique and N. J. Silvy for
his suggestions on a previous draft. This is contribution number
TA 14377 of the Texas Agricultural Experiment Station.

REFERENCES

Bailey, B. (1977). "Tables of the bonferroni t-statistic." J. Am.
     Stat. Assoc. 72, 469.
Barr, A., Goodnight, J., Sall, J., and Helwig J., (1976). "A
     user's guide to SAS." SAS Institute, Inc., Raleigh, N. C.
Folse, L., Jr. (1978). "Avifauna-resource relationships on the
     Serengeti plains." Ph.D. Dissertation, Texas A&M Univ.,
     College Station, TX
Johnson, N. and Kotz, S. (1970). "Distributions in statistics:
     Continuous univariate distributions." Vol. 2. Wiley, New
     York.
Miller, R., Jr. (1966). "Simultaneous statistical inference."
     McGraw-Hill, New York.
Morrison, D. (1967). "Multivariate statistical methods." McGraw-
     Hill, New York.
Sokal, R. and Rohlf, F. (1969). "Biometry." Freeman, San
     Francisco.
Steinhorst, R. (1977). "Significance tests for niche overlap."
     Paper presented at the 2nd Inter. Stat. Congr., Satellite
     Prog. in Stat. Ecol., July-August 1977, College Station, TX
     and Berkeley, Ca. (Proceedings in preparation.).

APPENDIX

A. Sample Multivariate Contrast Program

Below is a set of SAS (Barr et al., 1976) program statements
which takes a data set called BIRD containing data on SITE, TIME,
REP and log-transformed biomasses of 17 species of birds and performs 1) ANOVAs and univariate contrast analyses for each
species, 2) MANOVA and multivariate contrast analyses for all
species as a group, and 3) multivariate contrasts for each of
several subgroups of these 17 species. The data set represents
a balanced design with 5 sites, 11 times and 3 samples within
each site-time combination. Following the program is a discussion on calculating the constants required in the program.

```
* THESE STATEMENTS PRODUCE ANOVAS, MANOVA, AND A DATA SET CON-
TAINING RESIDUALS - SET BIRD MUST BE SORTED BY SITE TIME;
PROC GLM DATA=BIRD;
CLASSES SITE TIME;
MODEL N1-N17 = SITE TIME SITE*TIME / SS1 SS3;
MANOVA H = SITE TIME SITE*TIME / PRINTH PRINTE;
OUTPUT OUT = E1 RESIDUAL = R1-R17;
DATA E2;
SET E1;
KEEP R1-R17;
* THESE STATEMENTS PRINT STANDARD STATISTICS AND PRODUCE A DATA
SET CONTAINING TREATMENT MEANS;
PROC MEANS DATA = BIRD;
VARIABLES N1-N17;
BY SITE TIME;
OUTPUT OUT = T1 MEAN = M1-M17;
DATA T2;
SET T1;
KEEP M1-M17;
* THIS PROGRAM SECTION PERFORMS THE CONTRAST ANALYSES (UNIVARIATE
AND MULTIVARIATE);
PROC MATRIX;
FETCH RESID DATA = E2;
E = FUZZ(RESID'*RESID);
X = E#/110;   * NOTE, 110 IS ERROR DF = (5)(11)(3-1);
FREE E RESID;
NOTE X = RESIDUAL MEAN SQ XPX MATRIX;
PRINT X;
FETCH T DATA = T2;
TRT = FUZZ(T);
NOTE TREATMENT MEAN VALUES;
PRINT TRT;
* THIS NEXT SERIES OF STEPS SET UP THE CONTRAST DESIGN MATRIX,
CD. SINCE THIS PROBLEM HAS 5 SITE VALUES AND 11 TIME VALUES, MANY
OF THE CONTRASTS CAN BE EXPRESSED AS AN OUTER PRODUCT OF A SITE
TERM WITH A TIME TERM. THE FIRST 10 CONTRASTS ARE PAIRWISE SITE
CONTRASTS AVERAGED OVER 11 MONTHS.;
TC1 = J(1, 11, 0.090909);
C1 = (1 -1 0 0 0)@TC1;
C2 = (1 0 -1 0 0)@TC1;
C3 = (1 0 0 -1 0)@TC1;
C4 = (1 0 0 0 -1)@TC1;
C5 = (0 1 -1 0 0)@TC1;
C6 = (0 1 0 -1 0)@TC1;
C7 = (0 1 0 0 -1)@TC1;
C8 = (0 0 1 -1 0)@TC1;
C9 = (0 0 1 0 -1)@TC1;
C10 = (0 0 0 1 -1)@TC1;
```

```
* THE NEXT 5 CONTRASTS ARE WET-DRY SEASON COMPARISONS WITHIN EACH
SITE. JUN (MONTH 2) THRU SEP (MONTH 5) DESIGNATE THE DRY SEASON;
TC2 = (0.142857||J(1,4,-0.25)||J(1,6,0.142857));
C11 = (1 0 0 0 0)@TC2;
C12 = (0 1 0 0 0)@TC2;
C13 = (0 0 1 0 0)@TC2;
C14 = (0 0 0 1 0)@TC2;
C15 = (0 0 0 0 1)@TC2;
* THE LAST CONTRAST IS A WET-DRY SEASON CONTRAST OVER ALL SITES;
C16 = (0.2 0.2 0.2 0.2 0.2)@TC2;
CD = C1//C2//C3//C4//C5//C6//C7//C8//C9//C10//C11//C12//C13//C14
//C15//C16;
FREE T TC1 TC2 C1 C2 C3 C4 C5 C6 C7 C8 C9 C10 C11 C12 C13 C14 C15
C16;
NOTE CONTRAST DESIGN MATRIX;
PRINT CD;
CONTRAST = CD*TRT;
NOTE TREATMENT CONTRAST MATRIX - UNIVARIATE CONTRASTS;
NOTE EXPECTED VALUES OF TREATMENT CONTRASTS ARE ZERO UNDER THE
NULL HYPOTHESIS OF NO TREATMENT DIFFERENCES;
* THE TREATMENT CONTRAST MATRIX PROVIDES UNIVARIATE CONTRASTS FOR
EACH OF THE SPECIES;
PRINT CONTRAST;
DI = J(1,17,1)*DIAG(X);
*THE CONFIDENCE INTERVALS REQUIRED TO TEST THE CONTRASTS MUST BE
SPECIFICALLY MADE FOR EACH CONTRAST, DEPENDING ON THE NUMBER OF
TREATMENTS INVOLVED IN EACH CONTRAST AND ON THE WEIGHTING SCHEME.
THIS PROBLEM REQUIRES 3 SEPARATE CONSTANTS TO SET INTERVALS, ONE
EACH FOR CONTRAST GROUPS (1-10), (11-15), (16).  CONSEQUENTLY,
THE FOLLOWING CONSTANTS (ONE FOR ALPHA = 0.05, AND ONE FOR ALPHA
= 0.01) MUST EACH CONTAIN 3 TERMS;
CON_05 = VAL1/VAL2/VAL3;
CON_01 = VAL4/VAL5/VAL6;
MHSD_05 = SQRT(CON_05@DI);
MHSD_01 = SQRT(CON_01@DI);
NOTE SKIP=3 CONTRAST INTERVAL HALF-WIDTHS;
NOTE USE 1ST WITH CONTRASTS (1-10), 2ND WITH (11-15), 3RD WITH
(16);
NOTE CENTER INTERVAL ON TREATMENT CONTRAST VALUE;
NOTE IF INTERVAL INCLUDES ZERO, THEN ACCEPT NULL HYPOTHESIS;
NOTE MHSD_05 and MHSD_01 FOR ALPHA = 0.05 AND 0.01 RESPECTIVELY;
NOTE SKIP=3;
PRINT MHSD_05;
NOTE SKIP=3;
PRINT MHSD_01;
* THE FOLLOWING STEPS PERFORM A MULTIVARIATE CONTRAST ANALYSIS
FOR ALL SPECIES;
MACRO MVCON
CONT = CONTRAST*SPECIES';
AEA = SPECIES*X*SPECIES';
```

```
H1 = SQRT(CON_05@AEA);
H2 = SQRT(CON_01@AEA);
HSD = H1||H2;
NOTE SKIP=3 TIT1;
PRINT CONT;
NOTE SKIP=3 1ST WITH CONTRASTS (1-10), 2ND WITH (11-15), 3RD WITH
(16);
NOTE ALPHA FOR COL1 = 0.05, FOR COL2 = 0.01;
PRINT HSD;
%
MACRO TIT1 CONTRASTS FOR ALL SPECIES%
MACRO SPECIES J(1,17,1)%
MVCON;
* TO DO MULTIVARIATE CONTRASTS FOR ONLY THE 1ST 5 SPECIES, YOU
WOULD DO THE FOLLOWING;
MACRO TIT1 CONTRASTS FOR 1ST 5 SPECIES%
MACRO SPECIES (1 1 1 1 1 0 0 0 0 0 0 0 0 0 0 0 0)%
MVCON;
* FOR THE 4TH AND 9TH SPECIES;
MACRO TIT1 CONTRASTS FOR 4TH AND 9TH SPECIES%
MACRO SPECIES (0 0 0 1 0 0 0 0 1 0 0 0 0 0 0 0 0)%
MVCON;
* YOU MAY SELECT ANY GROUP YOU WISH FROM AMONG THE 17 SPECIES;
```

B. Confidence Interval Constants

I wished to construct 128(=16x8) simultaneous confidence intervals for this problem. I used the residual SS&CP matrix from
the MANOVA procedure to construct these intervals, and consequently there were no truly orthogonal contrasts as the error
terms for the various contrasts were not independent of one
another. One technique for constructing these confidence intervals in the Roy union-intersection approach (Morrison, 1967).
However, a design of this size (55 treatments, 3 subsamples of
each, with 110 df) is much too large for existing tables of the
upper α percentile of the distribution of the maximum characteristic root of the determinant equation (Heck charts, Morrison,
1967). An alternative, then, is to piece together individual
t-intervals by the Bonferroni inequality (Miller, 1966). This
approach gives a probability of α or less that any one of the
contrasts falls outside its confidence interval when the null
hypothesis is true.

For each species group, the SPECIES*X*SPECIES' operation
selects and sums the mean squares and cross products associated
with that species group, and the square root of this is used to
set the confidence interval. However, some provision has to be
made for each contrast to account for the weighting factors used
to construct the contrast (refer C1, C2, etc.). This is the

square root of the sum of squares of the weighting factors (Miller, 1966). For example, with C1: SQRT($11*(0.090909)^2$ + $11*(0.090909)^2$) = 0.426401. This same factor applies to each of the first 10 contrasts.

A second factor, associated with the t-value for 128 simultaneous contrasts, needs to be calculated. This is $t_{df}^{\alpha/2K}$, where K = number of contrasts 128 in this case), and df is the degrees of freedom for error ($5*11*(3-1)$ = 110 in this case). For α = 0.05, this becomes $t_{110}^{0.0001838}$, a value not tabulated. Since I have a large number of error df, I have used a normal approximation with $t\sqrt{(1-2\nu^{-1})} \sim N(0, 1)$ (Johnson and Kotz, 1970), where ν = 110. Thus $t_{110}^{0.0000184} \approx 3.537$. Hence the magnitude of VAL1 in CON_05 is $(3.03*0.426401)^2$, or VAL1 = 2.2743. The remaining constants are calculated in a similar fashion.

AVIAN CENSUS TECHNIQUES:
AN ANALYTICAL REVIEW

William M. Shields[1]

Department of Zoology
Ohio State University
Columbus, Ohio

Many avian census techniques involve tallying the number of individual birds, seen or heard while an observer walks a fixed path (transect) through a representative sample area. The number of contacts registered (N) during a census is a function of the number of birds present in an area (density = C), and the probability that each will be registered, given that it is present (detectability = d). N can be used alone to estimate relative abundance (index census); if d is calculated or controlled, density can be estimated (transect-density census) using the relation,

$$C = N/d.$$

In this paper I examine the methods, properties, and problems associated with the major techniques founded on this relationship. Based on observer effort and the accuracy and consistency of each technique's results, I suggest that, for most purposes, relative indices could be abandoned in favor of transect-density techniques. All contact censuses suffer from the factors that influence detectability, including; cue attenuation and emission frequency, species differences in bird behavior, observer behavior, habitat structure, season, time of day, and weather. To minimize potential biases and increase each technique's validity, accuracy, and consistency, I have recommended "standard" census procedures. These include, a constant walking speed (1 km/h in breeding season deciduous

---

[1]Present address: Department of Environmental and Forest Biology, State University of New York, College of Environmental Science and Forestry, Syracuse, New York 13210.

forests, 2 km/h in more open, or some winter habitats),
and when possible comparative and replicate censuses
should be performed by the same observer, in the same
habitat, during the same stage of the target birds'
annual cycle, in as short a period (< 1 week) as
possible. All should begin at the same time of day,
with respect to local sunrise, or if more than one
sample is run daily, starting times must be rotated
equitably and should be included as a factor in later
statistical analysis. High winds (> 20 km/h),
precipitation and extreme seasonal temperatures should
be avoided.

I. INTRODUCTION

In the title of their classic compendium,
Andrewartha and Birch (1954) distill animal ecology to
its essence, the "distibution and abundance of
animals." The importance of estimating avian numbers
had been recognized long before this (for a historical
view, see Kendeigh, 1944), but as with much in ecology
a quickened interest in bird censusing coincided with
Lack's (1937) emphasis of the importance of censusing
as a tool in ecological study.

Increasing interest in community stability and
structure, population dynamics, ecosystem energy flow,
and in environmental monitoring requires raw numbers
as grist for more sophisticated analyses. Interest in
census methods for obtaining such numbers has not
lagged behind. State of the art reviews have appeared
with regularity (Kendeigh, 1944; Enemar, 1959; Emlen,
1971; Berthold, 1976). Many investigators have
addressed the problems of single techniques (for
mark-recapture, see Lincoln, 1930; Winkler, 1930; for
territory (spot) mapping, see Williams, 1936; Enemar,
1959; Williamson, 1964; Hogstad, 1967; Robbins, 1970;
and Berthold, 1976; for variations on strip (transect)
censuses, see Breckenridge, 1935; Colquhoun, 1940;
1941; Davis, 1942; Hayne, 1949; Merikallio, 1958;
Enemar, 1967; Eberhardt, 1968;1978; Emlen, 1971; 1977;
Järvinen and Väisänen, 1975; 1976; and for relative
indices, see Linsdale, 1928; Dice, 1930; Kendeigh,
1944; Robbins and Van Velzen, 1967). Some have
directly compared census techniques, their properties
and problems, either on a theoretical basis (e.g.
Enemar, 1959; Emlen, 1971; 1977; Berthold, 1976;

Järvinen, 1976), or through corresponding censuses of
the same area (e.g. Amman and Baldwin, 1960; Snow,
1965; Kenny, 1972; Emlen, 1977; Dickson, 1978).
    I plan to examine selected census techniques which
I feel are suitable and convenient for estimating bird
abundance in forest ecosystems. I limit my discussion
to transect techniques, ignoring alternatives such as
mark-recapture and spot mapping which have recently
been reviewed elsewhere (e.g. Berthold, 1976; Emlen,
1977). For the transect techniques, I will present a
brief descriptive analysis, noting special properties
and limitations, and concentrate on (i) efficiency,
defined as the amount of observer time and effort
needed for adequate censusing, (ii) accuracy, the
closeness of a census's results to the actual number
of birds present, (iii) consistency, the comparability
of different censuses using a single method, and (iv)
biasing factors, all factors affecting the variability
of results, and therefore potential biases. For each
method I will examine potential interactions among
these factors and examine potential methods for
improving the quality of results.

II. CENSUS TECHNIQUES

    Practical forest census techniques rely on using
the number of times an observer sees or hears
(contacts) a bird species during a sample to estimate
relative or absolute abundance. The number of contacts
grows with increases in (i) the number of birds
present in an area (density), (ii) the probability
that a bird will be contacted, given that it is
present (detectability), or (iii) both (i) and (ii).
    A "typical" relation between the number of
contacts registered, actual numbers present, and
detectability can be illustrated graphically (Fig. 1).
This example assumes that birds are dispersed
regularly over the study area, so that density is
equal in each constant-width belt parallel to the
observer's transect. Mathematically this can be
represented by the simple equation,

$$n_i = c_i d_i \qquad (1)$$

where n_i is the number of contacts in belt i, with each
belt of specified width (x), at some increasing
distance (0-10 arbitrary units in this case) from, and
parallel to the transect, c_i is the number of birds

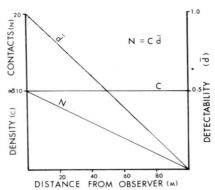

FIGURE 1. The frequency of observer-bird contacts
(N) as a function of distance from observer and actual
density (C). A linear decrease in detectability (d)
and a regular dispersion of birds is assumed.

actually present in belt i (in this case 10), and d_i
is the average detectability for each belt i (range,
1.0-0.0). The relation for an entire study area is,

$$N = C\bar{d} \qquad\qquad\qquad\qquad (2)$$

where N = total contacts for an entire census (Σn_i =
55), C = total density (Σc_i = 110) and $\bar{d}$ = average
detectability for an entire census ($\Sigma d_i/i$ = 0.5, see
Fig. 1). As detectability increases, the number of
contacts will approach the actual number of birds.
(Care must be taken to register each bird once or the
number of contacts may exceed actual density). The
actual shape of the decreasing detectability function,
and the distances involved depend on many factors
(Table I). If we assume a linear decline (as in Fig.
1), these factors will determine the scale of the
abcissa, the position of the x-intercept, and the
slope of the line. In the same habitat, a secretive
species would produce a steeper slope, a closer
x-intercept or both if compared to a more active and
conspicuous species. In contrast, a single species
should show lower detectability in dense forest than
when present in more open habitat. All the factors in
Table I can have similar substantial effects on
detectability.

TABLE I.   Factors Influencing Avian Detectabilty.

A. Extrinsic factors: (affect observer behavior).
   Observer speed
   Habitat structure
   Weather
   Observer's condition
B. Intrinsic factors: (affect bird's behavior).
   1. Interspecific behavioral differences
   2. Intraspecific behavioral differences
      Seasonal
      Diel
      Weather-related
      Competitor-related
      Predator-related

With any of the methods relying on observer-bird
contact, estimation or control of detectability is the
only way to assure reasonably accurate results
(Colquhoun, 1940; Emlen, 1971; Järvinen and Väisänen,
1975).   Yet the cost of doing either effectively can
lower efficiency enough to make ignoring detectability
attractive.   The major contact census techniques
include two distinct classes.   Those which (i) do not
estimate or control detectability (relative indices),
and (ii) those which make the attempt (transect-density
methods).

A.   Relative Indices.

Most indices are based on the number of contacts
per unit census effort.   Contacts can be tallied at
the species level (species presence/unit effort), or
at the individual level (individuals/unit effort/
species).   The latter can be used to determine the
former and is preferred by reason of its greater
information content.

Unit effort can be measured temporaly (census days,
Linsdale, 1928, or better, census half-hours, see Dice,
1930), or in distance, usually in contacts per km.

Whichever effort indicator is used, the other is controlled through observer speed during censusing. This helps assure consistency in the area size, and therefore the true population size being sampled. This increases the comparability of censuses, when such comparison is legitimate (see below).

If observer speed is held constant, then the effort indicator will determine the spatial scheme(s) appropriate for sampling. If time is used sampling may be (i) plot intensive, walking a scheduled or random route through a plot without regard to repeat samples, or (ii) via a transect, a relatively straight path, with each portion sampled once. If effort is indicated by distance the transect is more suitable.

Normally, multiple (4-12; Avg. = 8) censuses are run, one every few days on each transect. The number of daily contacts is averaged for a final index for each species. Alternatively, the maximum number of individual contacts on a daily census can be used as an estimate of minimum numbers for that species (Palmgren, 1930; cited in Emlen, 1977).

Repeat sampling will increase the cumulative species total as the probability of registering rare species increases. Under "average" conditions, most species present in a forest are likely to be tallied by the eighth visit (Robbins, 1972). The number present can be roughly predicted by plotting the cumulative total against the number of visits, as a "straight" line on semi-log paper. Total species can also be predicted mathematically by the regression,

$$y = a + b \sqrt{x} \qquad\qquad (3)$$

where y = species number, x = number of visits, and a and b are constants estimated after 4 or 5 visits (Robbins, 1972). The resulting data permit estimates of the number of censuses needed to determine the presence of most rare species, a simple diversity index, or both.

The design of an index census does not allow simultaneous calculation of the density and the detectability values which lead to a specified result (N = number of contacts). Thus contact frequency can never be translated into bird numbers. This lack of accuracy is a strength as well a weakness of indices. It permits the investigator to ignore distances and areas, since density estimates are impossible in any case, allowing relatively efficient (effortless) censusing. This benefit is often balanced by the limited analysis permitted by an index's relativity.

By ignoring species' differences in detectability
an index is rarely useful for anything but within
species comparisons. The lack of control on
detectability makes it imperative that the factors
known to influence detectability (Table I), be handled
in a way that ensures minimal biasing of each species'
results.   Thus it is best to compare censuses of a
single species begun at the same time of day, under
the same weather conditions, during the same season,
in the same or extremely similar habitat, and performed
by the same observer. Less care regarding potential
biasing factors can invalidate any conclusions
implied by the census results (see Biasing Factors,
below).
     Given these constraints, the primary role of an
index is in following the qualitative (increasing,
decreasing, or unchanged) local population changes of
a single species from year to year. Typical results
for such an index will usually illustrate some yearly
variation (Fig. 2).   In my example, both species A and
B show a general decline in abundance, especially
after 1970.   The North American Breeding Bird Survey
(Robbins and Van Velzen, 1967) and the Winter-Bird
Population Studies (Robbins, 1972) exemplify this role
on a grand scale.
     Indices do permit limited between species
comparisons. If two species are counted on the same
censuses over a number of years, their abundance
relative to one another can be compared. If, from one
year to the next, both species A and B show a decline

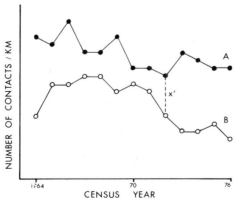

FIGURE 2.   Annual contact averages for two
hypothetical bird species from an index census.   X' =
N_A/N_B is used to compare the relative abundance of both
     species.

in contacts, but the interspecific ratio of contacts $X' = N_A/N_B$ is larger in the second year, we can tentatively assume that species B is declining at a faster rate than A (Fig. 2). If there is knowledge of a biological relation between the two species (e.g., if they are predator, prey, or competitor to one another, or if they use same nest sites and different food resources) this information can be useful.

Finally, indices have been used to examine the effects of local environmental changes on bird numbers through "before and after" censuses. Temporal comparisons are possible when an investigator knows of a "treatment" in advance, or has been censusing an area that undergoes an alteration. Treatments include technological factors such as pesticide application, intentional habitat management, including burns, cutting and planting, commercial logging, and strip mining, as well as natural factors like plant disease, insect outbreaks, fire, storm damage, and flood (e.g., see Emlen, 1970 on forest fire). Some consider such environmental monitoring a prime function of avian censusing (Svensson, 1970).

Pre- and post-treatment indices do suffer from a serious interpretational drawback. Regular changes in species' detectability are known to occur seasonally (e.g., decreasing song frequencies noted in many passerines during the breeding season, Jarvinen et al., 1976; Weber and Theberge, 1977). These behavioral differences imply that the time span including pre- and post-treatment censuses must exclude periods of normal detectability flux, if one is interested in detecting a numerical treatment response. If homogeneity of detectability can not be assured, the investigator is left with uninterpretable results. Figure 3A displays a hypothetical, but realistic, series of census results for a species during its breeding season. Given these results alone, we can only conclude that the post-treatment decline in contacts could be a numerical response to the treatment. It could also be a natural decline in detectability without a change in density. It is impossible to distinguish between these interpretations.

If it is impossible to assure homogeneity in detectability during a census (it usually is without measurment) then consecutive annual censuses, during the same stage of the life cycle, may be compared. For example, if in Fig. 2, A and B were results from two years and a single species (the abcissa would be weeks instead of years), and a treatment was

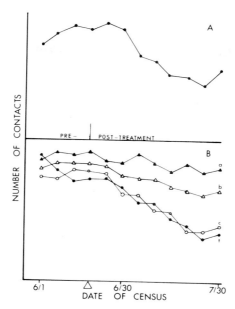

FIGURE 3. Results of a simulated pre- and
post-treatment index census for a single species
during the breeding season. A. a treated area with no
control. B. a treated area (t) compared to a series
of control results (a,b,c). See text for further
details.

administered in year B, at the point designated 1970,
the decline in B relative to A could be attributed to
treatment effects, rather than a natural change in
detectability. Alternatively, validity can be
increased by adding a separate control area. The
control should be left untreated and must be carefully
matched to the treated area in habitat structure,
species composition, and preferably local densities.
By comparing the pattern of census results between
areas, one can distinguish between changes resulting
from natural behavioral cycles and treatment effects.
    Figure 3B illustrates an example for a single
treated area and three possible controls (for
empirical examples, see Moulding, 1976 and references
therein). Our conclusion about treatment effects
depends on which of the three controls (a,b, or c)
we compare to the treated area (t). If c results, we
conclude that there was no treatment effect, just a
seasonal change in detectability, density or both. If
a results, the reverse is true. If b occurs, there is

a treatment effect, which is superimposed on a natural
seasonal change. Even with the addition of controls,
care must be taken to ensure that contact changes
result from density changes and not from treatment
effects on detectability (e.g., Doane and Schaffer,
1971). By adding controls, valid interpretation
becomes possible, but the efficiency of before and
after indices is greatly reduced, without a
corresponding increase in accuracy or consistency.

B. Transect-Density Methods

Many ecological analyses require a data base of
absolute, rather than relative, abundance. Density
calculations require estimates of numbers and areal
distribution. Valid estimates of numbers can only be
made if the relation between detectability and density
is known, or in some way controlled. This end can be
achieved by using census techniques which are
unaffected by detectability (e.g., mark-recapture,
territory mapping, see Emlen, 1971), or detectability
can be measured, and eq. 2 used to estimate density
from contact frequency,

$$C = N/\bar{d}.$$

In either case, sampling areas must be measured (and
therefore limited in size), and the investigator must
estimate distances in the field.

Of the many methods available for estimating bird
density, variations of the transect (strip) census
have recently received much attention (Merikallio,
1958; Emlen, 1971; 1977; Järvinen and Väisänen, 1975;
Järvinen, 1976). Since these seem to offer equal
accuracy and consistency and are usually more
efficient than their common alternatives (e.g., spot
mapping, see Emlen, 1971; 1977; Kenny, 1972; Berthold,
1976), they seem optimally suited for most forest
censusing. (Exceptions include small tracts, or
mixed habitat of any size, or any situation when
individual ranging behavior is important to the
investigator). When properly controlled, transect
censuses offer the additional advantage of generality,
being suitable for any homogeneous habitat, during
any season (Emlen, 1977). In the rest of this section,
I will examine the mechanics and properties of the
three variations on the transect theme. The Finnish

method (A) was originally developed by Merikallio
(1958), and later improved by Järvinen and Väisänen,
(B, 1975; 1976). A similar American method (C) was
developed by Emlen (1971) and later modified by him
(1977).

Merikallio's method (1958) is essentially an
attempt at a complete count of all the individuals
present in a narrow strip (belt) of habitat, centered
on, and extending equal lateral distances from, an
observer's transect. All cues within this main belt
are registered, those outside the belt may be tallied
but are ignored for density calculations. Total main
belt contacts are assumed to equal main belt absolute
numbers, with density calculated for the main belt
area,

$$C_A = N/2TW \qquad\qquad\qquad (4)$$

where T is the length of the transect in km, and W is
half the width of the total belt centered on the
transect. Multiple censuses can be performed and then
averaged. This will tend to balance the effects of
localized movements across the main belt boundary, if
there is no bias in direction. The calculated density
is assumed to be the true density for the entire area
sampled by the transect.

Method A's major assumptions are, (i) individual
birds are distributed (dispersed) homogeneously
throughout the target area, and (ii) detectability is
perfect ($\bar{d}$ = 1.0) through the entire main belt. To
help assure assumption one, which is also assumed in
Methods B and C, censuses are usually limited to large
(> 10 ha) patches of relatively homogeneous habitat
(e.g. a large forest). To control for assumption two,
Merikallio chose a main belt width (50 m, with W = 25
m on either side of observer), that he hoped would
yield a close approximation of perfect detectability.
When individual species are considered, few show
constant detectability out to 25 m (Emlen, 1971).
Thus the method has a built-in error which severely
limits its potential for accuracy. Subsequent study
showed that this error could be subtantial, since
detectability was unlikely to be perfect anywhere
in the main belt, and often showed large declines
within 25 m (Fig. 1; Emlen, 1971; 1977; Järvinen and
Väisänen, 1975).

   1. Cue Attenuation. There was a need for a method
controlling or measuring the magnitude of cue
attenuation with distance and determining its effects
on detectability. Järvinen and Väisänen (1975) tried
to solve the problem by assuming that declines in
detectability from cue attenuation begin near the
transect, at which it is perfect ($\bar{d} = 1$). The total
decline is then estimated from the proportion of
total contacts observed in the main belt (MB). Their
main belt is also 50 m wide (W = 25 m), and all
contacts, both within and beyond the boundary, are
registered. Using MB, and one of several possible
detectability decline functions (including linear,
exponential and normal decay curves), average
detectability and density can be estimated from any
census results. Using the linear model (for
explanation of the exponential and normal models, and
why the linear is best in most situations, see
Järvinen and Väisänen, 1975) and p = MB, the
regression coefficient k can be estimated by,

$$k = (1 - \sqrt{1 - p})/W \qquad (5),$$

where W is the linear extent of half the main belt
(25 m). Density, in birds/km^2, can then be
calculated by using k, such that,

$$C_B = 1000 \ Nky/T \qquad (6),$$

where y the only undefined variable is a correction
factor to counteract the density-dependent bias in MB,
such that MB is higher than predicted at high
densities, and lower at low densities (Järvinen and
Väisänen, 1976). To correct for this bias, y is
calculated using the fitted regression,

$$y = 0.0346x + 0.6963 \qquad (7),$$

where x is the number of main belt contacts/km of
transect in each census (for the derivation of this
relation, see Järvinen and Väisänen, 1976). The range
of probable densities can also be calculated using
the standard deviation of MB (s) to estimate three
different k's for use in eq. (6). s is obtained by,

$$s = [MB(1 - MB)/N]^{0.5} \qquad (8).$$

Then p = MB and MB $\pm$ s are all used in eq. (5) to
calculate three k's (see Järvinen and Väisänen, 1975
for details). The variance of C_B declines with

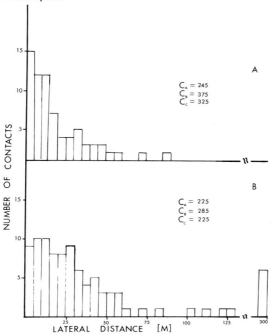

FIGURE 4. Hypothetical distribution of contacts for two bird species (A and B). Densities (C_A, C_B, and C_C) were calculated by the three transect-density methods. See text for further details.

increasing sample sizes, making the use of a "memory" a long term average MB, attractive (Järvinen, 1976). Use of this memory does assume a species' constant MB.

The original American method (C, Emlen, 1971) also assumed perfect detectability at the transect, with declines at unspecified lateral distances. Rather than using the fixed main belt of methods A and B, Emlen suggested a variable belt, with extent dictated by each species' behavior. He divided the 30 m nearest the transect into 3 m (10 foot) belts, then used wider belts from 30-60 and 60-125 m. Each contact was positioned in a belt before examining contact distribution to determine the boundary. The boundary was defined as the distance at which detectability began its <u>marked</u> decline. All boundaries were characterized by steep declines in contacts beyond, and relative constancy within their position (e.g., in Fig. 4, the boundaries are 5-15 m for species A, and 30 m for B).

A minor problem with this method is that some species show fewer contacts in the belts nearest the transect, than in those greater distances away (Kenny, 1972). This is expected if some individuals "freeze" as the observer approaches within some critical distance. Freezing stops cue emission, lowering the number of registrations nearest the transect, while the behavior of birds further away is unaffected. The only control for freezing behavior is determination of cue emission frequencies (see below), but averaging the contacts over the entire main belt will increase accuracy. A second behavior leading to fewer contacts near the transect is flushing. If individuals move away, at the approach of an observer, numbers are depleted near the transect and artificially increased further away. Valid control for the effects of flushing requires averaging contacts for the entire main belt.

Once averaged the contact density in each species' main belt is equated with absolute density for the entire area using eq. (4). To determine the main belt, the position (and lateral distance) of each contact out to 125 m is estimated, with those falling outside this outer limit ignored for main belt and detectability determinations.

Like Järvinen (1976), Emlen (1971) hoped that a "memory" would simplify subsequent censusing. He used main belt density (C_c), and the number of contacts within the fixed outer boundary (usually 125 m) to calculate a "coefficient of detectability" (CD_{125}), equivalent to my ($\bar{d}$), by the equation,

$$CD_{125} = N_{125}/C_{C125} \qquad (9).$$

He hoped that the coefficient would reflect a species-specific constant for each habitat type. Once enough were calculated, each species' grand mean could be tabled and used for density calculations from fixed main belts (W = 125 m) without the bother of distance estimations. This would increase the method's efficiency and improve the censusing of rare species, which appeared in single censuses in low numbers.

Emlen coefficients vary extensively with many non-habitat factors (Table I; and see Shields, 1977). This variation makes a theoretically ideal species' constant detectability unlikely. This is also true of average MB's (Järvinen and Väisänen, 1976). In his latest revision Emlen (1977) abandons CD's entirely. He falls back on the densities estimated by

contact frequencies in variable main belts. He
suggests that this is sufficient control for distance
cue attenuation to estimate relatively accurate
densities. This retreat to a "complete" count in a
narrow strip is reminiscent of Merikallio's (1958)
method, with the exception of the variability of the
belt.

Since Emlen (1977) found that empirically
determined main belts, in southern Wisconsin, extend
between 9 and 60 m, users of his method may profit by
limiting their attention to the smaller area bounded
at the new distance this suggests ($\approx$ 75 m versus 125).
This 40 % reduction in sampling area should lead to
fewer errors in distance estimations, because of
greater accuracy at shorter distances (Järvinen and
Väisänen, 1976). The revised method still requires
that the position of each contact be located precisely
within the new main belt.

In order to compare the properties of the three
transect-density methods, I have generated a series
of hypothetical contacts for two bird species (Fig. 4).
Species A is relatively inconspicuous (MB = .74;
CD_{125} = .203), with a steep detectability decline
beginning near the transect, and no long-distance
(> 100 m) contacts. Species B is considerably more
conspicuous (MB = .49; CD = .404), with a constant
contact frequency out to 30 m, a less precipitous
decline beyond that, and some long distance contacts.
Using these hypothetical distributions, I estimated
densities using all three methods for each species
(Fig. 4).

This analysis demonstrates that method A
underestimates the density of an inconspicuous species,
if detectability declines before reaching the fixed
boundary (25 m) of the main belt. This is not
surprising since its density estimate is based on the
contact frequency in the entire belt, regardless of
detectability changes. Since detectability declines
continuously for Species A, methods B and C yield
similar density estimates (Fig. 4).

If detectability follows a step function, with a
significant plateau before the inflection point,
rather than continuous decline, method B is less
accurate. It appears to "overestimate" the actual
density (Fig. 4; species B). This results from the
assumption of a continuous decline in the main belt.
Methods A and C are coincident in their estimates for
species B, but for different reasons. A is accurate
because of the chance inclusion of the fixed boundary

(25 m) in the initial detectability plateau, C because
it bases its estimate on observed detectability,
rather than on a convenient, but occasionally
unrealistic, theoretical distribution.

The hypothetical species' characters I used (Fig.
4), fall within the detectability limits reported from
nature (Emlen, 1971; Järvinen and Väisänen, 1975).
Thus the inconsistencies of the Finnish method are
likely to make comparisons between conspicuous and
inconspicuous species suspect. Method A's assumptions
are untenable, usually incorrectable, and perhaps the
method should be abandoned. Method B will
overestimate density for conspicuous species, but all
transect density techniques seriously underestimate
absolute abundance (see Cue Frequency, below). While
this tendency may decrease its internal consistency,
it may actually improve its accuracy. For within
species comparisons, no harm is done.

Emlen's method (C) is the only internally
consistent method, but only at the cost in decreased
efficiency necessitated by determining each contact's
position. Since careful estimation of distances
lowers an investigator's ability to register more
distant contacts (Järvinen and Väisänen, 1975) and
increases the probability of erring at any distance,
error will enter the resulting density estimates.

In direct comparison, method B is certainly easier
and more practical to use than C. The ability to
estimate a single distance, and decide whether each
conatct is within or beyond this boundary, is more
easily learned and competently performed than making
the numerous estimations of varying distances
required by C (Järvinen and Väisänen, 1975 and pers.
observ.). In addition, B yields consistently larger
sample sizes, since all contacts are used, regardless
of their distance from the transect. Any choice
between the methods (B and C) requires an analysis
of the trade-off between accuracy and consistency,
and efficiency, so characteristic of all census
methods (Emlen, 1971; 1977; Kenny, 1972), in light of
the needs and goals of each investigator.

2. Cue Frequency. The actual relationship between
the number of birds present, and the number of contacts
registered by a moving observer, depends on the
emission frequency of detectable cues, as well as on
the Cue attenuation function. As implied by eq. 2,
all methods assume that at zero distance all birds
will emit at least one detectable cue.

This was a simplifying assumption, known to be untrue to each method's originator (Merikallio, 1958; Emlen, 1971;1977; Järvinen and Väisänen, 1975). Cue production depends primarily on bird motion, for visual cues, and on vocal and mechanical sound production for auditory cues. A moving observer is likely to pass a stationary and silent bird without registering a contact, even if it is very close. This is obviously accentuated in "closed" habitat, such as forests with well-developed understories. The result is that transect-density methods yield minimum density estimates. How closely they approach true local densities will depend on how closely "basal" detectability approaches one (Emlen, 1971).

It is unlikely that detectability is ever perfect at the transect (Enemar, 1959; Järvinen and Väisänen, 1975; Emlen, 1977). Basal detectability will depend on the average probability of a bird emitting a detectable cue while the observer is in detection range. This probability is less than one and varies with the season, time of day, weather, species, sex and status (e.g. territory holder vs. vagrant) of each bird (Emlen, 1977).

Initially, both Emlen (1971) and Järvinen and Väisänen (1975) suggested using a correction factor to adjust calculated densities upwards. It was hoped that this would lead to "best" rather than minimum estimates. Based on concurrent censuses using different methods, Emlen (1971) suggested that in southern Wisconsin counts of wintering birds should be multiplied by 1.1 or 1.2, while summer counts required corrections of 1.1-2.5 (avg. = 1.5) depending on species. Using similar reasoning, but an analytic rather than empirical procedure, Järvinen and Väisänen (1975) suggested a breeding season factor between 1.3 and 1.6 in Finland, consistent with Emlen's estimates.

The small magnitude of the non-breeding season factor suggests either (i) basal detectability be ignored and minimum densities be reported, or (ii) a single correction factor (1.1?) be used for all species and censuses. The latter choice would facilitate comparison between studies and increase accuracy. The former would lessen the possibility of arbitrarily increasing species differences resulting from biasing factors (see below).

The greatly increased magnitude and interspecific variation in the correction factor during the breeding season stems from variability in male singing behavior and the relative inconspicuousness of females

in many species. Incubating or brooding females will
often remain on a nest, sitting silently, as an
observer passes nearby.
     Emlen (1971) suggested that to mitigate this
effect a tally of singing males be made, along with
the tally of all-cue contacts. This permitted
a choice between the greater of the two estimates,
twice the singing males in the main belt, or C_C from
equation 4. More recently Emlen (1977) has
developed a technique for monogamous, singing species
during the breeding season to replace the old method.
Rather than tallying all cues, only song contacts are
used.
     Since song is consitently detectable over greater
distances than combined cues, the main belt is fixed
at an arbitrary distance (Emlen uses either 200 ft.
(60 m) or 100 ft. (30 m) if the species is "soft-
singing). The inevitable decrease in contacts from
using one cue type is then balanced by a general
increase in main belt width (Emlen, 1977). This leads
to sample sizes equivalent to the all-cue method. The
number of singing males in the main belt is the
uncorrected male density. To correct for undetected
males, basal detectability is estimated by calculating
cue production frequencies. Cue frequency will depend
on the size of the transect and the speed of the
observer, which jointly determine the time a moving
observer will be within detection range of each bird.
In practice once the duration of a sample period is
determined, focal samples of the cue emission of
representative males are made and the proportion of
the total sample periods during which they emit a cue
is determined. This cue frequency (f; ranging from 0
to 1) is used to correct male densities by multiplying
the raw density by $1/f$. Emlen found (1977:464)
frequencies ranging from .06-.80, averaging 0.48 for
16 species in the last week of June, in southern
Wisconsin. This corresponds to correction factors of
1.2-16.0 (mean = 2) for song contacts alone, versus
1.1-2.5 for all cues (Emlen, 1971). Final density is
calculated by assuming an equal sex ratio and doubling
to account for females. This density includes both
the stationary (breeding) and floating (but singing)
populations and approaches the maximal accuracy likely
for a transect technique during the breeding season.
One important attribute of cue emission frequencies is
that they can be used to correct Finnish density
estimates. All that is necessary is restricting
MB estimates to song cues.. Then the density estimated

through equation (6), can be corrected by the simultaneously generated f's according to Emlen's technique.

Emlen (1977:468) expressed a hope that cue frequency might well be a species-specific constant for a given set of conditions (habitat, time of day, weather). If true this would permit tabled average frequencies to be used for density calculations, in place of concurrent determinations of frequency and song contact registrations. Only further field tests will permit us to determine whether this will be possible, or if the variation in emission frequencies will lead us to another species-specific dead end.

III. BIASING FACTORS

There are numerous factors which influence patterns of cue attenuation and emission frequencies (Table I). Given the relation between detectability and density implied by equation (2), it is obvious that changes in detectability might be misinterpreted as changes in density. It is important to control, as precisely as possible, any of the factors capable of influencing detectability in order to ensure relatively accurate and unbiased density estimates (Colquhoun, 1940; 1941; Enemar, 1959; Robbins and Van Velzen, 1967; Emlen, 1971; Berthold, 1976).

Control of interspecific factors (Table I) is achieved in two ways. Transect density techniques attempt to reduce problems by estimating detectability as a species-specific character. Indices are controlled by limiting analysis to intraspecific comparisons, unless elaborate, but not foolproof, precautions are added. In either case control is achieved through post-census analytic adjustments. Intraspecific and observer related factors (Table I) are usually ignored in analysis. For them, pre-census design constraints are the rule.

Such prescriptive constraints vary in the rigor and consistency of their application from census to census and investigator to investigator. Many of the common constraints are vague, permitting, and at times promoting unnecessary fluctuations in potential biasing factors. Possible effects of these fluctuations include, an inability to compare the results of different investigators, even if they census the same area (Robbins and Van Velzen, 1967) and subtantial increases in the variability of

replicate density estimates (e.g., Moulding, 1976;
Shields, 1977). If these effects are sufficiently
biased, then density estimates will be biased as well,
rendering valid interpretation and comparison
impossible.

I believe that more precisely designed and
"constant" control of potential biasing factors would
facilitate inter-individual comparison and increase
valid interpretation. It would also decrease the
variability in density estimates, given a constant
density. To pursue this goal, I will examine each
factor's effects on each census technique. When
appropriate I will offer tentative design controls
and analytical treatments to minimize their effects.
If my suggestions hold up in the field, we are part
way to "standard" censusing (cf., Berthold, 1976).

A.   Observer Behavior

1. Census Speed. On fixed length transects,
observers who vary their speed of advance, vary the
time they are within cue-detection range of each bird.
Increasing speed will decrease the total number of
contacts tallied, thereby decreasing the probability
of registering rare or inconspicuous species.
Decreasing walking speed has an opposite effect, but
too slow a speed increases the possibility of density
inflating double registrations from single birds.

Habitat structure also influences the choice of an
optimal census speed. Dense closed habitats (e.g.,
mature forest with a high, thick shrub layer in late
spring) obstruct a census-taker's view, while
partially absorbing sound cues. Such habitats
require more careful censusing at slower speeds than
more open habitats (e.g., the same forest in winter).

Based on these considerations, my own experience,
and the recommendations of others (Colquhoun, 1940;
Hogstad, 1967; Robbins and Van Velzen, 1967; Emlen,
1971; 1977; Järvinen and Väisänen, 1975 and Robbins,
pers. comm.), I recommend a standard speed of 1 km/h
in the breeding season and 2 km/h in winter or open
non-forest habitat.

In practice it is difficult to control walking
speed. Average speed is controlled by limiting the
duration of the census over a fixed transect. An
observer should spend 1 (1 km/h) or $\frac{1}{2}$ (2 km/h) an hour
per km of transect. Different portions of each
transect are likely to be travelled at different
speeds, but the average should be maintained within
20 % (between .8-1.2 km/h = 75-50 min/km) of optimum.

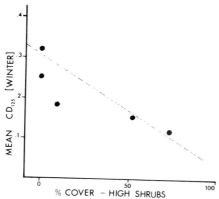

FIGURE 5.   Inverse relationship between habitat
density and detectability (CD) observed by Emlen
(1971) on Grand Bahama in winter. The line (---) is
the least squares regression.

    2. Habitat Structure.   Besides influencing the
choice of census speed, habitat structure has a direct
effect on census results. A habitat's vegetational
profile determines the degree of interference with an
observer's cue detection. Increasing vegetation
densities decreases detectability for most bird
species. Emlen (1971) found a 50 % reduction in
detectability going from open to closed habitat, on
Grand Bahama Island in winter (Fig. 5). This
occurred, even though the species assembleges were
similar in all the habitats sampled.
    Thus transect-density censuses should be limited
to relatively homogeneous habitat. Since indices
ignore detectability, they are suitable (along with
territory mapping and mark-recapture) for censusing
mixed habitat. They are limited to temporal
comparisons of the same area (Weber and Theberge, 1977)
and are less useful for inter-area or interspecific
comparisons in mixed habitat. The only exception
would be comparison of two areas in which the
proportion of habitat types is equal. Indeed, this is
one of the criteria for choosing index control areas.

    3. Observer Condition.   Interobserver differences
in visual and auditory acuity, and in tendencies to
over- or underestimate distances lead to wide
variability in census results (Enemar, 1962; Hogstad,
1967; Robbins and Van Velzen, 1970), and affect all
the techniques discussed. It is always safest to
control this factor by limiting comparisons to

censuses performed by the same observer with standard
techniques. Even with a single observer, levels of
alertness and fatigue will influence daily results and
should be controlled.

    4. Weather. As a factor influencing the
observer, weather is most important when it is extreme.
High winds or heavy rains will limit an observer's
likelihood of registering contacts, regardless of
the birds' behavior. Bad weather's effects are both
physical and psychological, and the best defense is
to limit censusing to dry days with gentle winds
(< 20 km/h).

        B.   Bird Behavior- Intraspecific Factors

    1. "Season". Seasonal differences in detectability
are commonplace for most bird species (Emlen, 1971;
1977; Weber and Theberge, 1977; but cf. Järvinen et
al., 1976). These differences result from changes in
weather and habitat structure, as well as changes in
bird behavior. The latter include differences in the
kinds of cue emitted, in emission frequency and
attenuation (Emlen, 1971; Robbins and Van Velzen,
1970). Differences between human scale seasons (e.g.,
winter vs. breeding season), are unlikely to cause
serious problems for any of the techniques. Indices
should be restricted to single season comparisons, and
calculation of detectability functions should be
limited to data from single seasons. Calculations of
seasonal detectability values will then permit
comparison of seasonal density differences.
    More difficult to control are the within-season
fluctuations in detectability characteristic of
breeding birds. On a bird time scale, the breeding
season is not homogeneous. It is divided into stages
devoted to different parts of the reproductive cycle
(e.g. territory establishment, courtship, incubation,
feeding nestlings or fledglings, and renesting). This
regular succession of "roles" leads to changes in
behavior which are reflected in detectability. A
general decline in detectability, as the breeding
season progresses, appears to be the rule for most
forest species (Fig. 3 and, see Moulding, 1976; Weber
and Theberge, 1977).
    These rapid changes appear to be absent in winter,
but some control is necessary during the breeding
season. The primary control is to census during a
single reproductive stage, performing replicates within

a short period of time. The problem with this is that
different species, and less often different
individuals within a species, will be in different
stages at the same calendar time. In order to
effectively census all species, the census period must
be extended. This leads to inevitable error in the
intraseason estimates of detectability for some
species. Intrinsic variability in detectability
translates into variation in its estimates (CD, MB, f),
as well as in the number of contacts registered.
Thus, the possibility of species-specific
detectability constants becomes an optimistic goal
(see Time of Day and Weather, below). Some control
can be achieved by limiting replicate census periods
(< 1 week, Emlen, 1977), while repeating series of
replicates through the season to control for species
differences in cycle time.

     2. Time of Day and Weather. Both time of day and
weather (air temperature, wind velocity, humidity,
and solar radiation) are correlated with significant
changes in avian detectability (Colquhoun, 1940; 1941;
Davis, 1965; Hogstad, 1967; Robbins and Van Velzen,
1967; Berthold, 1976; Weber and Theberge, 1977;
Shields, 1977). The joint correlation leaves
interpretation of the actual cause of detectability
changes ambiguous. Weather, especially air

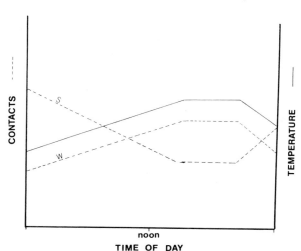

FIGURE 6. Idealized relationship between air
temperature, time of day, and detectability during
winter (w) and summer breeding seasons (s).

temperature is closely correlated with time of day, because of temporal changes in levels of solar radiation (Fig. 6). A dawn census will usually be performed at lower temperatures than one begun later in the day. The degree of difference will depend on local conditions (e.g., maritime or continental climate), and the temporal separation between censuses.

During the breeding season, detectability is inversely correlated with air temperature. Maximum detectability is observed around dawn and decreases at species' variable rates to a diurnal minimum around midday (Fig. 6; Robbins and Van Velzen, 1970; Weber and Theberge, 1977; Shields, 1977). The magnitude of the decline varies with the species and the delay between censuses. When quantified, it has ranged from 67 % reported within half an hour for Rufous-sided Towhees (Pipilo erythrophthalmus) during a roadside index (Davis, 1965), to 66 % for Mourning Doves (Zenaida macroura), but over 4 h (Weber and Theberge, 1977). In more extensive studies, Robbins and Van Velzen (1970) and Shields (1977), observed an average decline of about 20 % for 18 species (not the same species in each study) over 1.5 hours. Shields also presented results for individual species with declines ranging from 3-67 %.

The evidence that this variation is a response to weather and not some other correlate of time of day (e.g., light intensity), is circumstantial but convincing. The increase in detectability in late afternoon and early evening (Fig. 6) occurs at decreasing light intensities, opposite to the morning effect. Temperature does decrease, at this time, as would be expected if it is controlling detectability. Further support for the temperature (weather) hypothesis are the numerous anecdotal reports that the decline in detectability is of greater magnitude and faster on "hotter" days (e.g., see Robbins and Van Velzen, 1970). During the breeding season, detectability declines appreciably and regularly after dawn, probably in response to increasing air temperatures (and solar radiation?).

In cold temperate regions (above 40° N Lat.), the relation between time of day and detectability reverses (Fig. 6). The magnitude of this effect is also a function of species and delay between samples. Quantified, 8 species showed an average increase in contacts of 81 % (range = 11-250 %), between 0800 and 1000 e.s.t. (Shields, 1977). In winter, in this region detectability increases with increasing temperature.

The interaction between temperature and detectability is more complex than implied by the data from either season alone. It appears that warmer temperatures in summer, and colder in winter, can both have a depressive effect on detectability. This suggests a modified explanatory hypothesis, that increasingly "stressful" weather conditions affect detectability negatively. This hypothesis is consistent with the patterns of detectability seen in nature and with the reported effects of stressful weather on vocal emissions (Berger, 1961; Dorst, 1974), and locomotor behavior (Grubb, 1975; 1977; 1978; Austin, 1976).

Although time may not be the ultimate cause of variation in detectability, its close correlation with weather requires our control of time of censusing. The traditional prescription for census times, "early morning, when birds are most active" (Pettingill, 1970; see also Moulding, 1976; Emlen, 1977) is not adequate. In fact, in colder regions, late morning censuses may be most efficient in winter (Shields, 1977). In either season, it is not enough to haphazardly limit censusing to a random portion of an extended (e.g., 4 h) census period.

Replicate censuses, or comparative censuses should be run at the same time of day, with respect to local sunrise. If more than one transect is to be sampled each census day, starting times should be rotated to prevent bias. It is even preferable to start a transect on opposite ends on alternate replicates (Weber and Theberge, 1977).

Once a census has been designed to assure that all transects are affected equally by time-correlated weather changes, potential bias from this factor is removed. If time of day is then included as a factor for statistical analysis, the extraneous variability caused by regular changes in census times can be removed analytically. This will increase the statistical sensitivity of subsequent comparisons, permitting the detection of smaller density differences than would otherwise be possible (Shields, 1977).

Besides varying daily, weather shows larger temporal variability. It is not sufficient to control time of day. General weather conditions should also be distributed equally among replicates and between comparison areas to reduce bias. An example of a weather effect, even after time of day is controlled, is shown by a study of bird "distribution" on north and south slopes during a winter in New Jersey (Shields

and Grubb, 1974). A transect-density census was used
to demonstrate that the colder north slope had
fewer birds resident during the coldest weather.
Since coefficients of detectability were calculated
from data combined from both slopes, the fewer
contacts registered on the north slope were taken as
evidence that density was also lower. An alternative,
overlooked during the original study, is that the
difference in contacts could have resulted from
lower north slope detectabilities, because of its
harsher weather. The magnitude of the contact
difference between slopes makes it likely that there
is a true density difference, but at least some of it
should be attributed to weather related differences
in detectability between the slopes. The simplest
procedure for controlling macro-climate differences
is to keep weather conditions homogeneous, avoiding
unusual weather conditions.

    3. Predator and Competitor. Speculatively, I
offer the following as possibilities for future
research. We know that competitors and predators
affect avian behavior (e.g., mobbing predators;
the instantaneous niche shifts of subordinate
competitors in mixed flocks reported by Morse, 1974).
Whenever the behavioral response is translated into
changes in detectability, the presence or absence of
particular predators or competitors will influence
census results. Currently, we have no handle on the
importance of this effect, but it is likely to be
especially important in studies of numerical
responses to specific predators or competitors.

                    IV. CONCLUSIONS

    Given the wide variation expected in avian
detectability, much of the current use of index
censuses appears to be counterproductive. With the
exception of large scale monitoring of single
species' relative abundance (especially in mixed
habitat, Weber and Theberge, 1977), exemplified by
continental surveys (e.g. Robbins and Van Velzen,
1967), the slightly decreased effort does not appear
to balance the large potential for inaccuracy and
inconsistency. This is especially true for before
and after censuses, where ecologically and
economically important decisions may hang on
interpretation of census results. As currently

practiced, transect-density techniques offer higher
consitency and accuracy at a relatively slight cost
in efficiency.

In order to maximize consistency and accuracy, with
any of the transect methods, the major factors
affecting detectability must be controlled. I have
suggested that standardizing census procedures, with
respect to these factors, is desirable, offering
specific suggestions as to how this might be
achieved. To develop a truly standard technique, we
should choose between Emlen's and Jarvinen and
Vaisanen's techniques. At present we do not have
sufficient empirical evidence to make the choice. We
need simultaneous censuses comparing both techniques
with an independent, and if possible, more accurate,
control.

To facilitate such comparison, we need only change
the distances used in Emlen's method. Specifically,
by using 5 m belts within 25 m (vs. 3 m belts within
30 m), and 25 m belts out to 75 m, our contact data
can be analyzed using either method. Based on this
comparison, we may be able to choose one of the
techniques, which when performed in a standard way,
will offer a sought after efficient, consistent and
relatively accurate census technique.

ACKNOWLEDGMENTS

I am grateful to the symposium's entire editorial
board, and especially to J. Dickson and J. Jackson,
for affording me the opportunity to present my ideas
and for improving their presentation. I thank C. B.
Brownsmith and J. Crook for their gifts of time and
criticism. I am especially grateful to T.C. Grubb, Jr.
and J. T. Emlen for arousing my interest in the
problems of censusing and for commenting on earlier
versions of this manuscript.

REFERENCES

Amman, G., and Baldwin, P. (1960). Ecology 41, 699.
Andrewartha, H., and Birch, L. (1954). "The Distribu-
     tion and Abundance of Animals." University of
     Chicago Press, Chicago.
Austin, G. (1976). Auk 93, 245.
Berger, A. (1961). "Bird Study." John Wiley, New York.

Berthold, P. (1976). J. für Ornith. 117, 1.
Breckenridge, W. (1935). Wilson Bull. 47, 195.
Colquhoun, M. (1940). J. Anim. Ecol. 9, 53.
Colquhoun, M. (1941). Proc. Zool. Soc. London, Sec.
     A, 110, 129.
Davis, D. (1942). Ecology 23, 370.
Davis, J. (1965). Condor 67, 86.
Dice, L. (1930). Auk 47, 22.
Dickson, J. (1978). Amer. Birds 32, 10.
Doane, C., and Schaffer, P. (1971) Conn. Agr. Exp.
     Stat. Bull. 724, 5.
Dorst, J. (1971). "The Life of Birds." Columbia
     University Press, New York.
Eberhardt, L. (1968). J. Wildl. Manage. 32, 82.
Eberhardt, L. (1978). J. Wildl. Manage. 42, 1.
Emlen, J. (1970). Ecology 51, 343.
Emlen, J. (1971). Auk 88, 323.
Emlen, J. (1977). Auk 94, 455.
Enemar, A. (1959). Vår Fågelvärld, Suppl. 2, 1.
Enemar, A. (1962). Vår Fågelvärld 21, 109.
Grubb, T. (1975). Condor 77, 175.
Grubb, T. (1977). Condor 79, 271.
Grubb, T. (1978). Auk 95, 370.
Hayne, D. (1949). J. Wildl. Manage. 13, 145.
Hogstad, O. (1967). Nytt. Mag. Zool. 14, 125.
Järvinen, O. (1976). Ornis Scand. 7, 43.
Järvinen, O. and Väisänen, R. (1975). Oikos 26, 316.
Järvinen, O. and Väisänen, R. (1976). Ornis Fenn.
     53, 87.
Järvinen, O., Väisänen, R. and Haila, Y. (1976).
     Ornis Fenn. 53, 40.
Kendeigh, S. (1944). Ecol. Monogr. 14, 67.
Kenny, J. (1972). "A Comparison of Bird Censusing
     Techniques." Unpublished master's thesis, Rutgers
     University, New Jersey.
Lack, D. (1937). Ibis 1, 369.
Lincoln, F. (1930). U.S.D.A., Circ. 118, 1.
Linsdale, J. (1928). Condor 30, 180.
Merikallio, E. (1958). Fauna Fennica 5, 1.
Morse, D. (1974). Amer. Natur. 108, 818.
Moulding, J. (1976). Auk 93, 692.
Pettingill, O. (1970). "Ornithology in Laboratory
     and Field." Burgess Publ., Minneapolis.
Robbins, C. (1972). Amer. Birds 26, 1.
Robbins, C. and Van Velzen, W. (1967). U.S.F.and W.
     Spec. Sci. Rep., Wildl. No. 102.
Robbins, C. and Van Velzen, W. (1970). Bull. Ecol.
     Res. Comm., No. 9, Lund.
Shields, W. (1977). Auk 94, 380.
Shields, W. and Grubb, T. (1974). Wilson Bull. 86,125.

Snow, D. (1965). Bird Study 12, 287.
Svensson, S. (1970). Bull. Ecol. Res. Comm., No. 9.
Weber, W., and Theberge, J. (1977). Wilson Bull.
    89, 543.
Williams, A. (1936). Ecol. Monogr. 6, 317.
Williamson, K. (1964). Bird Study 11, 1.
Winkler, W. (1932). Verhandl. Zool. Bob. Ges. Wein
    80, 53.
Yapp, W. (1956). Bird Study 3, 93.

SAMPLING CONSIDERATIONS FOR EVALUATING THE
EFFECTS OF MORTALITY AGENTS ON BARK BEETLES[1]

Robert Coulson

Department of Entomology
Texas A&M University
College Station, Texas

Paul Pulley

Data Processing Center
Texas A&M University
College Station, Texas

Lewis Edson

Department of Entomology
Texas A&M University
College Station, Texas

Basic requirements for estimating effects of mortality
agents on bark beetle populations are examined. Emphasis
is placed on sampling within-tree populations of both host
insect and mortality agent at single points in time. The
estimation procedures described should be suitable for
obtaining information on the effects of mortality agents
on the distribution and abundance of the host insect
through space and time. When multiple samples are to be
collected the topological estimation procedure described
is superior to the other options presented. The topological procedure is suitable for obtaining estimates of

---

[1]Texas Agric. Exp. Stn. No. 14430

surface area or volume of the habitat, density of the host
insect, and density of the mortality agent. An example of
interspecific competition between Dendroctonus frontalis
and Monochamus titillator is examined. This example is
comparable to woodpecker predation of D. frontalis from
both logistical and analytical sampling standpoints.

I. INTRODUCTION

   Many insectivorous bird species have received attention because of their role as mortality agents of forest insect pests.
Woodpecker (Picidae) predation on bark beetles (Scolytidae) has
been observed and studied for many years and there is little
question of the significance of the association in the population
dynamics of both animal groups. However, avian predation
represents only one components of a complex array of mortality
agents (i.e., predators, parasitoids, competitors and pathogens)
associated with the bark beetle community.
   Few studies have been conducted which quantitatively demonstrate and describe effects of mortality agents on bark beetle
populations. Theoretical foundations for sampling and estimation
exist in the statistical literature. Likewise, the ecological
literature provides a theoretical basis for interpreting
predation, parasitism and competition. Therefore, the central
issue for evaluating consequences of mortality agents acting on
bark beetle populations, or insects in general, becomes a problem
of logistics; i.e., how, when and where to collect data that
accurately represent the process of interest. Furthermore, these
data must be collected in a format which is amenable to statistical analyses.
   The objective of our paper was to scrutinize and define basic
requirements for evaluating effects of mortality agents acting on
bark beetle populations. In pursuing this objective, we will use
the association of the southern pine beetle, Dendroctonus
frontalis Zimm. (Coleoptera: Scolytidae) and the southern pine
sawyer, Monochamus titillator Fab. (Coleoptera: Cerambycidae) as
a model. This association which is an example of interspecific
competition, results in mortality to D. frontalis as a consequence of foraging by M. titillator. Foraging by woodpeckers
results in patterns of mortality to D. frontalis that are very
similar to those produced by M. titillator. Likewise, sampling
and estimation considerations are also similar.
   It is important to recognize that in appraising interrelationships between insectivorous birds (or other mortality
agents) and the host insect, several different approaches are

possible: one can focus on (1) the role of the mortality agent
in the population dynamics of the host insect, (2) the role of the
host insect in the population dynamics of the mortality agent or
(3) the focus can be directed to the mutual roles of both the
host insect and mortality agent acting upon the population
dynamics of one another. Entomologists usually follow the first
tack, ornithologists the second, and the third option is rarely
attempted. The focus of the following discussion will be oriented
to the first option.

II.   LIFE HISTORIES AND COMPETITIVE INTERACTION
OF DENDROCTONUS FRONTALIS AND
MONOCHAMUS TITILLATOR

Since the association of D. frontalis and M. titillator will
be used as a model for scrutinizing requirements for evaluating
effects of mortality agents on bark beetles, it is necessary to
review briefly the life histories and nature of the competitive
interactions of the two insects.
The natural history of D. frontalis was originally described
by MacAndrews (1926) and has been modified and expanded by
Thatcher (1960), Dixon and Osgood (1961), Coulson et al. (1972)
and Coulson (1978). The general life history of the insect as
summarized from this literature begins with the selection of
suitable host trees by adults. While this process is not well
understood, initially attacking beetles are probably directed by
primary (host-produced) attractants. Following initial attacks,
actual host colonization is regulated by a blend of insect-
produced pheromones and host-produced attractants. Females
initiate construction of "egg galleries" by boring into the inner
bark region where they are joined by males. Mating takes place
within the galleries and eggs are oviposited in niches in the
phloem at intervals along the lateral walls. Both the males and
females reemerge and are capable of attacking and colonizing the
same or new host trees. Eggs hatch shortly after oviposition and
the ensuing larvae begin to excavate "larval galleries" at right
angles away from the egg galleries. There are four larval
instars. The first two remain in the phloem region and the last
two migrate into the outer corky bark where pupation and adult
emergence occur. Development can be completed in as few as 35
days, with six to eight generations per year occurring in some
regions of the South. The host tree must be killed before successful colonization occurs. The process of tree death is
initiated by the successful introduction of the blue staining
fungi (Ceratocystis minor Hedge and Hunt) and by girdling brought

about by attacking adults. Numerous parasites, predators and
associates have been identified to occur with D. frontalis. Infestation spots often enlarge dramatically in a short period of
time, are characterized by an active front or head, and are comprised of multiple asynchronous generations of the insect
developing concurrently. Both reemerging and parent adults
are involved in colonization of trees. Various site and stand
conditions have been associated with the occurrence of D.
frontalis. Prominent correlates include stand density and age,
stand disturbances, species composition, radial growth, and soil
type.
     Although M. titillator was originally described almost two
centuries ago, only one comprehensive descriptive study has been
conducted (Webb, 1909). The life history of the insect was observed in "storm felled timber" and is similar to the development
observed in pines which are first injured by D. frontalis. M.
titillator females respond to pines which are being attacked by
D. frontalis and oviposition by the two species occurs concurrently. An elliptical-shaped pit is excavated in the bark by the
female and six to nine eggs are deposited in the phloem. The
entire bole of the tree is utilized. Larvae forage the inner bark
region and, after several instars, bore into the xylem where a
"U" shaped, pupation tunnel is constructed. However, even after
larvae construct the gallery into the xylem, foraging normally
continues in the inner bark. Adults emerge from the pupal chamber
by excavating a characteristic circular hole to the outside surface of the tree. Foraging by M. titillator larvae takes place
throughout the life cycle of D. frontalis and in the same area
where development is occurring. The area foraged by M.
titillator is clearly detectable and often extensive.

III. DEFINING THE SCOPE AND BOUNDARIES
OF THE ESTIMATION PROBLEM

     In an investigation of the association of a mortality agent
and a host insect one goal is to evaluate and describe effects of
the agent on the distribution and abundance of the host insect,
i.e. to define the role of the mortality agent in the population
dynamics of the host. To achieve this goal it is necessary to
move through a hierarchy of organizational complexity beginning
with the identification of a potential association, proceeding to
quantification of the relationship, and culminating in description
of the temporal-spatial relationships of the association.

The first step in the hierarchy, identification of the relationship, is usually straightforward and based on observation. In the case of the D. frontalis/M. titillator association, foresters and entomologists have observed foraging activities of M. titillator in trees colonized by D. frontalis for many years. That mortality was occurring to D. frontalis as a result of the mutual occurrence of the two insects seemed a certainty.

At the second step in the hierarchy, quantification of the relationship, it becomes necessary to apply some of the crafts of science. The nature of the relationship must be defined in rather precise terms, which involves detailed information of the life histories of both the host insect and mortality agent. With this information it is possible to define where mortality likely occurs and to design an experiment, employing quantitative estimation procedures, to measure the association.

The final step, measuring the dynamic properties of the association, is unquestionably the most difficult. The sampling procedures developed for use in quantification of the relationship likely will be too costly and will require too much data to be efficiently applied over long periods of time. Often times probable dynamic properties of the association are simulated through the use of mathematical models. In effect the properties identified in the quantification of the association are extrapolated to the next higher level and we believe this practice to be less advisable than actual measurement taken at the level of abstraction.

Regardless of the level of entry into the organizational hierarchy discussed above, there are three estimation problems with which one must deal. These problems include estimation of the habitat of the host insect and mortality agent, estimation of the density of the host insect, and estimation of the density of the mortality agent or the result of the agent's activity.

In the present D. frontalis/M. titillator example, measurement of the surface area and/or volume of the habitat is rather straightforward. These insects are cryptic for most of their life cycles and occur within the bark and along the open bole of various species of pines (Pinus spp.). This discrete habitat is amenable to precise measurement. Estimating the density of the host insect and associated mortality agent (or the result of it's activity) is somewhat more involved. The first consideration is selection of a sample unit size which is sufficiently large to capture the true distribution of the insect and yet is small enough to permit practical application.

Once an adequate sample unit has been established and the sampling universe defined then it is possible to proceed with the definition of the distribution of the insect. Once this

definition is made, a specific sampling plan can be developed
using the distribution to guide the intensity of data collection
required to estimate within a prescribed level of precision.
Knowledge of the distribution will also be required in order to
appraise the appropriateness of statistical procedures used in
the subsequent analysis and interpretation of the data.

The host insect and mortality agent will undoubtedly have
different distributions. In complex bark beetle-induced
communities, mortality agents (e.g. parasites and predators)
generally have clumped distributions with sparce densities
relative to the host insect. Therefore, estimation plans for the
mortality agent generally requires larger sample units and more
intensive sampling than the host insect.

IV.   INTERSPECIFIC COMPETITION BETWEEN DENDROCTONUS FRONTALIS AND MONOCHAMUS TITILLATOR

In the remaining sections of this paper we will explore the
approach taken in estimating mortality resulting to within-tree
populations of D. frontalis as a consequence of foraging by M.
titillator. We will consider (1) procedures for estimating the
habitat utilized by these insects, (2) procedures for estimating
within-tree populations of D. frontalis, (3) procedures for estimating foraging by M. titillator, (4) measurement of mortality
occurring to D. frontalis as a result of foraging, and (5) the
nature of the association between the two insects. This example
is comparable in approach to the estimation of woodpecker
predation of bark beetles.

A.   Estimating the Habitat Utilized by D. frontalis and M. titillator

The habitat of D. frontalis and M. titillator consists of the
surface area or volume of bark on the infested bole of a pine
tree. There are a number of different ways this habitat can be
measured and several procedures have been described by Foltz
et al. (1976) and Pulley et al. (1977 a&b, 1978).

Because measuring the association between D. frontalis and M.
titillator requires data collected systematically along the
infested bole of the tree, multiple measurements of diameter and
bark thickness can be made at the time of sampling at little
added expense. Pulley et al. (1977a) have described four separate
procedures for estimating surface area based on systematically

collected data.   These procedures were designated the stacked
cone, stacked cylinder, longer cylinder and the topological estimation techniques, respectively.   The stacked cone and topological
estimation procedures produce exactly the same results and are the
most precise procedures.   Each procedure requires diameter
measurements at various intervals along the infested bole, and
the closer the intervals the better the estimates.   The stacked
cylinder and long cylinder techniques are less precise because
they ignore tree taper.
    An alternative means of estimating tree area or bark volume
is to utilize the tree geometry model developed by Foltz et al.
(1976).   This model, which is similar in principal to "tree
taper" models employed in forestry, utilizes as input parameters
the tree diameter and bark thickness at 2.0 m and the infested
bole height.   Integrating over the infested bole of the tree
yields the surface area of the infestation.   Tables of surface
area, based on this model, have been provided for loblolly pine
(P. taeda L.) of 25-40 cm diameter outside bark (DOB) for three
bark thickness classes (Coulson et al., 1976a).   This model is
particularly useful with insect sub-sampling procedures where
many trees are to be examined.
    Each procedure provides a reasonable approximation of the
habitat surface area.   The tree geometry model requires the least
data and can be used to estimate both bark surface area and
volume.   The topological mapping procedure is the most precise
procedure for estimation of both bark surface area and volume, but
requires several measurements of bark thickness and diameter.

                B.   Estimating Within-Tree Populations
              of D. frontalis and Other Bark Beetles

    Considerable research has been conducted on methods and pro-
cedures for estimating within-tree populations of bark beetles.
This research has been reviewed by Pulley et al. (1978).
Essentially there are two aspects to estimating within-tree
populations: logistical methodologies for collecting field
sample data and mathematical methodologies for summarizing and
interpreting the field data.
    The logistics of collecting field data on bark beetles has
received substantial attention and reviews on the subject have
been provided by Stark and Dahlsten (1970) and Coulson et al.
(1975).   There are many different options available which vary in
extent of data collected and hence in information obtained.

Sample unit size has been examined by a number of workers
(Safranyik, 1968; DeMars, 1970; Mayyasi et al., 1976). A 100 cm^2
unit has been found to be adequate for most bark beetle species
that infest the "open bole" of the tree.

Distribution of sampling effort, i.e. the number and location
of sample units, has been investigated by Pulley et al. (1978).
Again the selection of a particular format for data collection
will affect the accuracy, precision, and informational content of
the estimate. A common format used for D. frontalis is to collect
four 100 cm^2 bark samples at the NE, NW, SE and SW aspects each
1.5 m beginning at 2.0 m and continuing to the top of the infested
bole. This intensity of data collection is more than adequate
for estimating within-tree populations of D. frontalis. The major
disadvantage of the format is that it requires the expenditure of
considerable time and energy.

Methodologies for collecting field data on bark beetles have
become more efficient over the last ten years. If infested trees
are to be sampled more than one time, it is necessary to climb the
tree. Tree climbing spikes, tree climbing bicycles and ladders
all permit access to the infested portion of the tree. Electric
drills powered by portable power generators permit the use of hole
cutting saws for extracting bark samples. Various types of traps
have been developed to capture reemerging and emerging adults.
These traps are normally designed to sample a 100 cm^2 area and
are thus compatible with the disk sampling procedures.
McClelland et al. (1978) have reviewed the various types of traps
used for bark beetles and have suggested a design to standardize
the procedure in the future.

Remote sensing techniques using X-rays have been developed
and refined specifically for identifying insect inclusions in
bark disks. The X-ray procedure is suitable for detecting adults,
egg galleries, large larvae, and pupae. It may also be possible
to detect small larvae by trimming excess bark scales from the
disk samples and by the use of high contrast films and intensi-
fying screens. In addition to bark beetle inclusions in the
sample, activity of other insects, such as foraging by M.
titillator, is detectable. Likewise, foraging by woodpeckers is
observable (Fig. 1).

The mathematical methodologies for summarizing and interpreting sample data for D. frontalis within-tree populations have
been extensively studied and a number of different estimation
procedures have been described that utilize systematically collected data. Details relating to the specific options available
and a discussion of appropriate use(s) of the procedures have been
reviewed by Pulley et al. (1978).

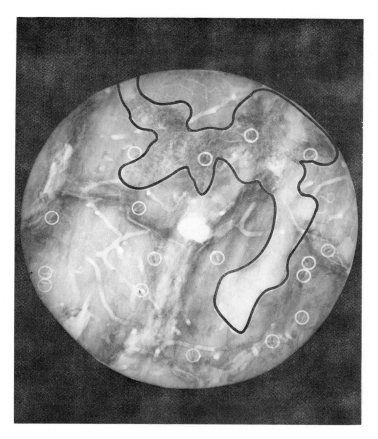

FIGURE 1. Radiograph of 100 cm^2 bark sample illustrating
Dendroctonus frontalis tenoral adults and pupae (circles) and
foraging activity by Monochamus titillator. SPB_1 = number of
D. frontalis occurring in the foraged area, A_1 = area foraged by
M. titillator.

There were several steps involved in development of the estimation procedures for D. frontalis. First, the distribution of
beetles within the 100 cm^2 disks was investigated. Analysis of
the distribution of various life stages within the sample disks
provided no reason to reject the hypothesis that beetles were
uniformly distributed within the average disk. Furthermore,
there was no consistent directional bias associated with aspect.

Next, the distribution of life stages along the infested bole
was described for each life stage using non-linear mathematical
models. The general pattern was for beetle density to be greatest
toward the central portion of the infested bole and to diminish
towards the extremes (Pulley et al., 1978). It is possible to
obtain an estimate of the total number of beetles occurring on the
tree by integrating the curve of beetles/100 cm^2 vs. tree height.

With this knowledge of the distribution of beetles within
sample disks and along the infested bole, Pulley et al. (1976)
developed an alternative methodology for computing the total
number of beetles on a tree. This methodology, termed the
"topological estimation procedure," minimizes errors resulting
from smoothing and extrapolation, utilizes systematically collected data, and was consistent with the previously defined
information on beetle distribution.

Using the estimates obtained from the topological procedure as
target values, several subsampling procedures were developed that
required fewer sample units and had defined accuracy and precision.

The topological estimation procedure is particularly well
suited for investigation of the association of D. frontalis and
mortality agents because the sparce distribution of the various
agents, relative to the host, require extensive sampling. The
sub-sampling procedures are more suited for extensive sampling of
many infested trees when a premium is placed on collecting as few
samples as possible.

C. Estimating Foraging by M. titillator

As with D. frontalis population, estimating foraging activity
of M. titillator entails consideration of logistical and mathematical procedures. The general procedures outlined for D.
frontalis are also suitable for estimating foraging activity by
M. titillator.

The logistical protocol described above for estimating
within-tree populations of D. frontalis, i.e. four 100 cm^2 bark
disks collected at the NE, NW, SE and SW aspects each 1.5 m
beginning at 2.0 m, is also adequate for sampling the foraging
activity of M. titillator (Coulson et al., 1976a). These bark
disks are radiographed and the foraging activity of M. titillator
is easily identified. The area foraged can be measured by using
an optical planimeter (Fig. 1).

Mathematical models used to describe the distribution of
D. frontalis life stages are also suitable for describing the
area foraged/100 cm^2 in relation to normalized infested bole
height. The total area foraged on the tree can be obtained by

integrating the curve. Alternatively the topological estimation procedure can be used to estimate the area foraged. This procedure is more precise than the model fitting routine, as errors of undefined magnitude resulting from smooth, extrapolation, and regression are avoided.

D. Mortality Occurring to D. Frontalis as a Result of M. titillator Foraging

The association of D. frontalis and M. titillator is an excellent example of interspecific competition. We are defining interspecific competition as a measurable disadvantageous influence exerted on survival of D. frontalis by M. titillator through the active demands of M. titillator for food and space. This definition and usage of the term is similar to that proposed by Clements and Shelford (1939). Inconsistencies in the literature regarding the meaning and significance of competition have been reviewed by Crombie (1947), Solomon (1959), Birch (1957), Brian (1956) and Miller (1967).

The definition of interspecific competition to be conclusive must demonstrate (1) alteration of the spatial distribution of one species by another and (2) reduction in survival of the species because of the association. The entomological literature contains few examples where both of these criteria have been met and quantitatively defined (Coulson et al., 1976a).

In demonstrating alteration of the distribution of D. frontalis by M. titillator, Coulson et al. (1976) compared the distributions of D. frontalis in both the presence and absence of M. titillator foraging. The following reasoning was utilized in demonstrating interspecific competition: if M. titillator is not an interspecific competitor, the population of progeny adult D. frontalis in the foraged area of the disk samples should have a satisfactorily estimatable value based on the density in the unforaged area of the disk. A χ^2 test was performed on numbers of D. frontalis occurring in the foraged area [$(\overline{SPB_1})$ in $\overline{A_1}$] vs. expected numbers if no foraging were present $E(\overline{SPB_1})$. The null hypothesis that observed and expected values came from the same distribution was rejected, thus demonstraing an altered spatial distribution resulting from the association of the two insects.

Reduction in the survivorship of D. frontalis as a result of M. titillator foraging was demonstrated by further examining the relationship between $\overline{SPB_1}$ vs. $E(\overline{SPB_1})$ and infested bole height (Fig. 2). The difference between $\overline{SPB_1}$ and $E(\overline{SPB_1})$ = ΔSPB_1 was next plotted in relation to area foraged by M. titillator. This

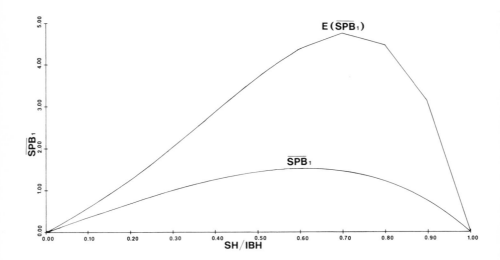

FIGURE 2.   The relationship between the number of Dendroctonus frontalis in the area foraged by Monochamus titillator (SPB$_1$ in A$_1$) and the expected number if no foraging were present [E($\overline{\text{SPB}}_1$)].   SH/IBH = sample height/infested bole height.

relationship (Fig. 3) provides an expression of mortality occurring to D. frontalis as a result of foraging.   Nonlinear mathematical models were used in each of these analyses to quantitatively describe the relationships.

E.   The Nature of the Association
Between D. frontalis and
M. titillator

In examining the life histories of M. titillator and D. frontalis, it appears that the niche requirements of the two species overlap during early and intermediate stage development. The eggs of both species are simultaneously deposited in the region of the inner-bark between the xylem and corky bark. Larval development occurs in the inner bark region and both species utilize this area as a food source and habitat.   Late

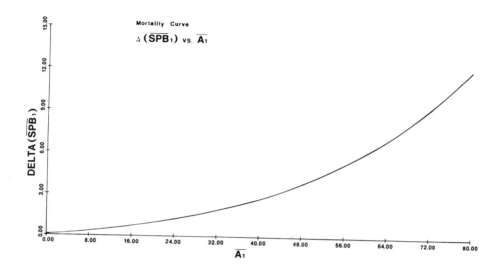

FIGURE 3. Mortality occurring to Dendroctonus frontalis as a
result of foraging by Monochamus titillator.
$\Delta(\overline{SPB_1})$ = the difference between $(\overline{SPB_1})$ and $E(SPB_1)$, A_1 =
area foraged by M. titillator.

instar D. frontalis move to the outer corky bark where pupation
occurs and adult emergence takes place. M. titillator late instar
larvae move inward into the xylem to pupate and become adults.
Competition takes place when both species occur together in the
inner bark region in what can be considered the intersection of
two fundamental niches (Hutchinson, 1957). The object of competition is the limited area of inner-bark necessary for development
of both species.
    The form of competition contains elements of both interference
and exploitation as defined by Brian (1956). The presence of
M. titillator directly and measurably limits access by D.
frontalis to the inner-bark region of the tree which constitutes
a requisite resource for both species. In this sense, the
process described herein can be considered interference or

"contest" competition (Nicholson, 1955). M. titillator, being a
much larger and more mobile insect than D. frontalis has a greater
ability to utilize the inner bark region of the tree, and in this
regard, the form of competition can be considered of the exploitation or "scramble" type (Nicholson, 1955), i.e. M. titillator
has a differential ability to obtain and utilize a greater portion of the common resource.

REFERENCES

Birch, L.   (1957).   Amer. Nat. 91, 5.
Brian, M.   (1956).   J. Anim. Ecol. 25, 339.
Clements, F., and Shelford, V.   (1939).   "Bio-ecology".   John
    Wiley and Sons, New York.
Coulson, R.   (1978).   Population dynamics of bark beetles.   Ann.
    Rev. Ent. 24, (in press).
Coulson, R., Payne, T., Coster, J., and Houseweart, M.   (1972).
    "The southern pine beetle, Dendroctonus frontalis Zimm.
    (Coleoptera: Scolytidae) 1961-1971."   Texas For. Serv. Publ.
    108.
Coulson, R., Hain, F., Foltz, J., and Mayyasi, A.   (1975).
    "Techniques for sampling the dynamics of southern pine beetle
    populations."   Texas Agric. Exp. Stn. Misc. Pub. 1185.
Coulson, R., Pulley, P., Foltz, J., and Martin, W.   (1976a).
    "Procedural guide for quantitatively sampling within-tree
    populations of Dendroctonus frontalis."   Texas Agric. Exp.
    Stn. Misc. Publ. 1267.
Coulson, R., Mayyasi, A., Foltz, J., and Hain, F.   (1976b).
    Environ. Ent. 5, 235.
Crombie, A.   (1947).   J. Anim. Ecol. 16, 44.
DeMars, C.   (1970).   In "Studies on the population dynamics of
    the western pine beetle, Dendroctonus brevicomis LeConte
    (Coleoptera: Scolytidae)."   (R. Stark and D. Dahlsten, ed.),
    Part II, Sec. 6, pp. 37-41.   U. Ca. Press.
Dixon, J., and Osgood, E.   (1961).   "Southern pine beetle.   A
    review of current knowledge."   USDA For. Serv. S.E. For. Exp.
    Stn. Pap. No. 128
Foltz, J., Mayyasi, A., Pulley, P., Coulson, R. and Martin, W.
    (1976).   Environ. Ent. 5, 14.
Hutchinson, G.   (1957).   "Concluding remarks."   Cold Spring
    Harbor Symposium on Quant. Biol. 22, 415.
MacAndrews, A.   (1926).   "The biology of the southern pine
    beetle".   M.S. Thesis, Syracuse Univ.

Mayyasi, A., Pulley, P., Coulson, R., DeMichele, D., and Foltz, J. (1976). Res. Popul. Ecol. 18, 135.

McClelland, W., Hain, F., DeMars, C., Fargo, W., Coulson, R., and Nebeker, T. (1978). Sampling bark beetle emergence: A review of methodologies, a proposal for standardization, and a new trap design. Bull. Ent. Soc. Amer. 6, 137.

Miller, R. (1967). "Pattern and process in competition." Adv. Ecol. Res. 4, 1. Academic Press, New York.

Nicholson, A. (1955). Austral. J. Zool. 2, 9.

Pulley, P., Mayyasi, A., Foltz, J., Coulson, R., and Martin, W. (1976). Environ. Ent. 5, 640.

Pulley, P., Foltz, J., Coulson, R., Mayyasi, A., and Martin, W. (1977a). Can. Ent. 109, 39.

Pulley, P., Coulson, R., Foltz, J., Martin, W., and Kelley, C. (1977b). Environ. Ent. 6, 607.

Pulley, P., Coulson, R., and Foltz, J. (1978). Sampling bark beetle populations for abundance. In "Sampling Biological Populations" (R. Cormack, G. Patil, and D. Robson, ed.). Satellite Program in Statistical Ecology. International Co-operative Publishing House, Fairland, Maryland (in press).

Safranyik, L. (1968). "Development of a technique for sampling mountain pine beetle populations in lodgepole pine." Ph.D. Thesis, Univ. British Columbia, Vancouver.

Solomon, M. (1959). J. Anim. Ecol. 18, 1.

Stark, R., and Dahlsten, D., ed. (1970). "Studies on the population dynamics of the western pine beetle, Dendroctonus brevicomis LeConte (Coleoptera: Scolytidae)." University of California Division of Agricultural Science, Berkeley.

Thatcher, R. (1960). "Bark beetles affecting southern pines: a review of current knowledge." So. For. Exp. Sta. U. S. For. Serv. Occl. Pap. 180.

Webb, J. (1909). USDA Bur. Entomol. Bull. No. 58, pt. 4, 41.

TREE SURFACES AS FORAGING SUBSTRATES FOR INSECTIVOROUS BIRDS

Jerome A. Jackson

Department of Biological Sciences
Mississippi State University
Mississippi State, Mississippi

Patterns of variability in tree surfaces are discussed and these patterns are related to the suitability of the surfaces as habitats for arthropods and foraging sites for insectivorous birds. Tree species and age diversity and the extent of forest environments are seen as important factors influencing the population dynamics of forest arthropods and insectivorous forest birds. Problems with forest insect pests often stem from human perturbations of the forest ecosystem that result in decreased forest diversity. Maintenance of this diversity contributes to forest health by providing more varied habitats and food resources for insectivorous forest birds.

I. INTRODUCTION

Insects, spiders, mites, and their relatives pre-dated the angiosperms, yet most of the genera of trees in existence today were already present by the end of the Cretaceous 100,000,000 years ago (Smith, 1962) as the earliest passerines were making their debut (Brodkorb, 1971). The earliest woodpeckers known are from the Pliocene (Brodkorb, 1970). Thus, more than 100 million years of evolution have contributed to the intricacies of co-existence in our forest ecosystems. We recognize much of the taxonomic diversity in our forests and have detailed predator-prey relationships for many species. Studies such as those of Elton (1966) have documented complex energy relationships within forest communities. Other studies have detailed partitioning of foraging niches among bird species and between the sexes of single species in forest ecosystems (MacArthur, 1958; Morse, 1968; Jackson, 1970). At the community level, the foliage height diversity within forests has been related to avian species diversity (McArthur and McArthur, 1961). These studies have dealt with niche partitioning involving such parameters as foraging height, perch diameter, and crown vs trunk vs branch vs foliage as foraging sites. Seasonal variation in foraging sites and techniques are well known for many species and have often been related to shifts in diet that are correlated with seasonal availability of prey (e.g., Jackson, 1970; Dickson and Noble, 1978).

The purpose of this paper is to look at the foraging ecology of insectivorous forest birds from a different perspective. How do the intricacies of tree surface variation influence the availability of arthropods to birds and the foraging strategies used by birds?

In seeking at least partial answer to this question, I will consider several tree surfaces: the trunk, limbs, foliage, flowers, and fruit.

II. TREE TRUNK SURFACES

Tree trunks are characteristically vertical surfaces and as such some bird species are better able to use them as foraging substrates than are others. Woodpeckers, nuthatches, Brown Creepers (Certhia familiaris), and Black-and-white Warblers (Mniotilta varia) have behavioral and morphological adaptations for tree-trunk foraging (Grinnell, 1924; Burt, 1930; Richardson, 1942; Spring, 1965; Feduccia, 1972). Others such as chickadees and titmice, other warblers, flycatchers, and hummingbirds occasionally forage on trunks by hovering briefly at them. The availability of branches which can be used as perches allows other species to exploit this surface. Variability in bark roughness within and among tree species likely influences the abundance of surface arthropods, their detectability by foraging birds, and the suitability of these tree surfaces as sites for egg-laying or overwintering by arthropods. Travis (1977) associated variability in bark roughness with seasonal differences in the foraging behavior of Downy Woodpeckers (Picoides pubescens). Differential use of tree trunk surfaces as foraging substrates for insectivorous birds has now been quantitatively described for many bird and tree species and several geographic areas (e.g., Morse, 1967a, 1970; Jackson, 1970; Willson, 1970; Austin and Smith, 1972). Some of the variation in use of these substrates can be attributed to competition within and among bird species (e.g., Jackson, 1970; Willson, 1970), as well as to variation in the availability of arthropods and ultimately to the physical variation in trunk surfaces.

A. Variation in Tree Trunk Surfaces

Many tree species are easily recognizable by the specific appearance of their bark. The plates of many pines (<u>Pinus</u> sp.), the warts on sugarberry (<u>Celtis</u> <u>laevigata</u>), and the smooth, thin scales of sycamore (<u>Platanus</u> <u>occidentalis</u>) readily distinguish these trees. Illustrations of bark patterns in many species can be found in Neelands (1974), Stephens (1969), and Symonds (1958). Chang (1954) presents an excellent discussion of the physical and chemical variation in the bark of 20 tree species. He notes that the effects of physiological and environmental changes on bark structure are very conspicuous, but quite consistent within a given species or genus.

In order to measure interspecific variability in bark roughness, I used a carpenter's contour gauge (Figure 1) to graphically reproduce the outlines of surfaces of several tree species.

FIGURE 1. Use of a contour gauge to copy the contours of bark surfaces. The gauge is pressed against the tree and the outline traced onto paper.

Jerome A. Jackson

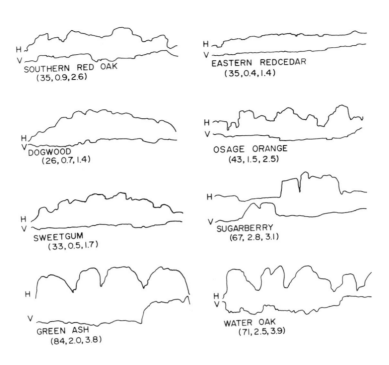

FIGURE 2.   Horizontal (H) and vertical (V) contours of eight
tree species.   Contours were traced from a carpenter's contour
gauge pressed against the tree trunk at 1.5 m.   Numbers in parentheses are diameter at breast height (cm), and bark thickness in
crevices and at ridges (cm).   The latter two measurements were
made with a bark thickness gauge.

A sample of the variability is shown in Figure 2.   At each tree
I measured diameter at 1.5 m and bark thickness at ridges and in
crevices.   The contour gauge was then pressed against the tree
vertically and horizontally at breast height and the vertical and
horizontal impression traced onto a card.   The segment of bark
surface sampled in each example reproduced in Figure 2 is 15 cm
long.   For some trees I took consecutive impressions around the
entire circumference of the tree and compared these to see how
representative a single impression might be.   In comparing
impressions, I measured the depth of fissures, the width of
fissures at the tree surface, and the width of bark plates at the
bottom of fissures.   Within tree variation in these parameters

FIGURE 3. Bark surfaces of three cottonwood trees illustrating age-size related textural differences.

among impressions was minimal for some species, but was pronounced for others (such as sugarberry). Thus, depending on tree
species, the bark environment of arthropods may be predictable or
unpredictable in "quality," and thus also might be the quality
of the foraging substrate for an insectivorous bird.

As is obvious from Figure 2, trees are generally more rugose
in a horizontal plane than they are vertically. The bark
fissures run up and down the tree. The three cottonwoods (Populus
deltoides) in Figure 3 readily illustrate this pattern. Such a
pattern to the roughness is likely a result of basic laws of
physics regarding the expansion of a cylinder. The end result is
a series of bark "valleys" that offers some protection and direct
pathways for arthropods crawling up or down the tree. The
ubiquity of vertical ridges on tree trunks may also have had an
influence on the evolution of bird locomotory patterns on tree
trunks. Horizontal ridges would be more ladder-like and might
require a modification of climbing behavior. Vertical ridges
likely also hasten the flow of rainwater from tree surfaces and
hence decrease the suitability of these surfaces for colonizing
fungi and epiphytes which, if present, might enhance the trunk as
an environment for arthropods and tree-surface foraging birds.

Certain tree surface features may decrease the suitability of
trunks as foraging sites. In a study of the foraging ecology of
Downy Woodpeckers in Kansas (Jackson, 1970), I discovered that
this species did not forage on the trunk of honeylocusts
(Gleditsia triacanthos) or older shagbark hickories (Carya ovata),
though both of these tree species were common. A critical glance
at the trunk surfaces of these species suggests that these trees
may simply be physically unsuitable as avian foraging sites -
regardless of what potential prey might exist on them (Figure 4).
The thorns of honeylocust could easily be lethal to a bird just
as the leaves of yucca have occasionally been shown to be. The
shags of shagbark hickory are very smooth and hard, perhaps making
them difficult to grasp, and they extend from the tree to such an
extent that they become an obstacle for a climbing bird.

In addition to looking at interspecific differences in bark
surface texture, I also examined age differences within species.
Three examples are illustrated in Figure 5. In each case,
rugosity and bark thickness increased with size of the tree. Such
variation must influence the species composition and abundance of
arthropods and the foraging tactics of insectivorous birds using
these trees. Considering the great forests and giant trees that
once occurred in North America, it is easy to see how the foraging
niche of a species like the Ivory-billed Woodpecker (Campephilus
principalis) could have been tied to a tree surface environment
that is essentially gone today. While vast areas of forest may
have been required by the Ivory-bill, it seems likely that changes
in the character of its foraging niche could have equally contributed to the species demise. Figure 3 illustrates the magnitude of difference indicated by the graphic representation in
Figure 5.

FIGURE 4. Rough shags on a shagbark hickory (A) and branch-
ing thorns of a honeylocust (B) present a barrier to tree surface
foraging birds.

Mechanical injuries, whether a fresh lightning strike that
provides access to the living tissues of the cambium, woodpecker
excavations, healed scars, or rotted branch stubs that collect
rainwater, all enhance the environment for arthropods and ultimately their predators. In their foraging activities, a woodpecker seldom fails to pay special attention to such irregular-
ities that it encounters on the bark surface. Similarly, vines,
lichens, fungi, and other organisms on tree surfaces increase the
surface variability and provide additional hiding places for
arthropods and foraging sites for birds. Many fungi are somewhat
host specific and hence contribute to the uniqueness of some tree
species as foraging sites for insectivorous birds. Park (1931),
Weiss (1920), and Smith (1960) discuss associations between

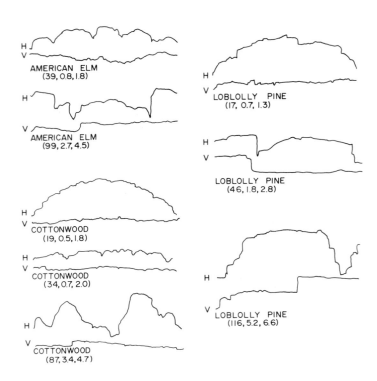

FIGURE 5. Differences in tree surface texture with tree
size in three tree species (abbreviations and numbers are as in
Figure 2).

arthropods and fungi that fruit on tree surfaces. Smith (1960)
notes that beetles (Coleoptera) and flies (Diptera) are the most
common invertebrates attacking fungi. He mentions that 346
species of beetles and 182 species of flies have been identified
from the fruiting bodies of macrofungi.

Epiphytes such as mosses also vary in their ability to
colonize surfaces of trees of different species and of the same
species at different ages and in different microenvironments
(Phillips, 1951; Billings and Drew, 1938). I examined 10 trees
of each of four species (southern red oak, Quercus falcata;
sweetgum, Liquidamber styraciflua; pignut hickory, Carya glabra;
and loblolly pine, Pinus taeda) and found all forty trees to
have lichens on the bark. The 30 hardwoods also each had mosses
growing on their bark, whereas none of the 10 loblolly pines had
epiphytic mosses. Perhaps the relative looseness or pH of the

bark or lesser flow of rainwater down the trunk of pines kept mosses from becoming established. (I have since found mosses growing on two loblolly pines, though these were exceptional.) Phillips (1951:315) notes the significance of bryophytes to other tree surface communities: "The bark under bryophyte communities is much more spongy, soft and flaky and enables new communities to establish themselves more easily."

Thus far we have looked at the natural physical structure of tree trunk surfaces and modifications of that structure as a result of fungi, epiphytes, or injury. Another modification of the bark surface that I would like to consider is that caused by foraging birds. Two extreme examples come to mind. Downy, Hairy (Picoides villosus), and Red-bellied (Melanerpes carolinus) woodpeckers forage extensively on the trunks of American elms (Ulmus americana) that are infected with Dutch elm disease and infested with bark beetles. The foraging technique used by these birds is called scaling: the woodpeckers pry the bark from the tree with their beak in order to expose the beetles. The action of the birds on the elms exposes the red inner bark, making the trees obvious and likely influencing the suitability of the trunk surface as an arthropod habitat and foraging substrate for other birds.

Similarly, the bark color of pines is normally gray, but the Red-cockaded Woodpecker (Picoides borealis), in its normal foraging activities, scales the outer bark and exposes the red inner bark. Excessive scaling on cavity trees provides an indication of activity at the tree (Jackson, 1977). In a study in South Carolina, I found that this red inner bark remains in contrast to the normal outer bark for as long as eight months (Jackson, 1978). McCambridge and Knight (1972) reported that dessication of beetle larvae and their food was enhanced by woodpecker feeding activity and contributed significantly to the decline of a spruce beetle (Dendroctonus rufipennis) outbreak.

Color differences that are due to (1) the activities of foraging birds, (2) genetic nature of the tree species, (3) the presence of epiphytes, (4) wetness of bark following a rain, or (5) other environmental factors, including pollution, may all influence the suitability of tree surfaces as habitats for arthropods and as foraging sites for birds. Behavior of arthropods is known to vary with substrate color. Sargent (1966) demonstrated that Geometrid and Noctuid moths tended to rest on backgrounds that matched the color of their forewings. Thus, variation in bark color among tree species might be expected to result in differential availability of these moths and differential use of tree species as foraging substrates by birds which prey on the moths.

Color differences coupled with exposure of the trunk surface to the sun may also influence the thermal suitability of the surface for arthropods and birds. In order to demonstrate the type of thermal variability there might be on tree trunk surfaces,

I attached telethermometer probes to similar-sized, adjacent trees of three species (sugarberry; water oak, Quercus nigra; and persimmon, Diospyros virginiana). Each tree was exposed to full morning sun. On each I placed one probe in direct sun against the bark and another probe in the shade against the bark on the opposite side of the tree. On the water oak I placed an additional probe against a lichen in the sun. The two probes in the sun on the water oak were 3 cm from one another. Temperatures were then recorded simultaneously from all probes. The 8 to 11° C difference on opposite sides of the trees (Table 1) suggests that thermal advantages and disadvantages to being in the sun at different times of day could influence arthropod distribution and avian foraging site selection. The differences among tree species and between bark and lichen also suggest the potential for related differential niche use. Grubb (1975) found temperature and wind related differences in foraging sites used by forest birds in Ohio. Roling and Kearby (1977) found that Scolytid beetles attacking oaks were more numerous on larger diameter trees and on the north side of tree trunks.

B. Tree Trunk Surface Arthropod Communities

With all the variability in tree trunk surfaces that I have indicated, it should be obvious that sampling of arthropod populations from trunk surfaces for the purpose of comparing tree species is fraught with difficulty. Perhaps this explains why I found few published studies on bark surface arthropod communities.

TABLE I. Tree Trunk Surface Temperature (°C) 0800, 10 July[a]

Tree Species	Munsell Color Notation[b]	Direct Sun	Shade
Sugarberry	7.5 N	36	27
Water Oak			
On Bark	10 yr 5/2	38	27
On Lichen	10 yr 8/2	35	27
Sassafras	7.5 yr 5/2	35	27

[a] Ambient temperature in shade at 1.5 M = 28°C.

[b] Basic substrate color against which telethermometer probe was placed. Notations are for hue, value, and chroma as given in Munsell (1942).

Most studies that have been done have been concerned with particular pest species or with spiders. One study (Smith-Davidson, 1930) suggested that the abundance of spiders on trunk surfaces decreases with height in the tree. In order to get a feeling for the composition of bark surface communities, I collected arthropods from the surface of 10 trees of each of four species (loblolly pine, pignut hickory, southern red oak, and sweetgum). The trees were all within the same mowed picnic area at Noxubee National Wildlife Refuge, Mississippi, and all were of similar size.

Arthropods were collected from the surface with small paint brushes dipped in alcohol. Larger arthropods were taken with a collecting jar. Three persons simultaneously collected arthropods from the same tree in a one meter high band around the tree beginning at one meter above the ground. Trees to be sampled were selected the night before sampling and were chosen at a distance on the basis of similarity of size and position relative to other trees and by their lack of unusual epiphytes or low branches. Tree diameter was measured prior to beginning collection and the time spent looking for arthropods on a tree was made proportional to the diameter. Five minutes were allotted for each square meter of tree surface. With successive trees sampled, we alternated tree species to avoid variability among species due to time of day. I will not detail the results of this sampling here, but will summarize some of the findings. In Table 2, I have included arthropod Orders which occurred as 10% or more of the dry weight of the combined samples for one or more tree species. With the exception of one bug, all of the Hemipterans indicated here were Largus succinctus. Half or more of the trees of each species had these bugs on them. For this species these trees seemed merely a gathering place and tree species seemed irrelevant in influencing where they occurred. Indeed, the majority of arthropods encountered seemed to be chance transients.

TABLE II. Percent (Dry Weight) Composition of Arthropod Samples Collected From the Bark Surface of Four Tree Species[a]

Order[b]	Loblolly Pine	Southern Red Oak	Sweetgum	Pignut Hickory
Hemiptera	55.0	37.7	65.2	55.0
Hymenoptera (Formicidae)	3.3	7.5	6.7	13.0
Lepidoptera	-----	-----	-----	15.6
Araneida	16.7	11.3	6.7	10.0

[a]Samples were collected from 10 trees of each species.
[b]Only those orders which included 10% or more of the arthropod sample from one or more tree species are listed.

Jerome A. Jackson

Samples taken at other times of year have revealed no such
abundance of L. succinctus, suggesting a very seasonal importance in the community. Ants were by far the most numerous
group encountered, but because of their small size, accounted
for generally small percentages of the samples. Spiders were
among the larger arthropods encountered on all tree species.

Summarizing the samples another way, ants were collected
from 38 of the 40 trees samples (Table 3). In spite of their
small size, their ubiquity may explain why the diets of woodpeckers characteristically include large quantities of ants
(Beal, 1911). The spiders also seemed to be a consistently
available resource. In similar samples collected from loblolly
pines in August, September, October, and November in South
Carolina and Mississippi, ants and spiders were similarly
ubiquitous.

I will end my discussion of tree trunk surfaces by briefly
characterizing the arthropod community that seems to occur there.
This characterization is based on my own studies and my review
of the literature. The tree trunk surface arthropod community:
(1) is dominated in terms of numbers of individuals and conspicuousness by ants that are transient on the surface but that
often nest beneath the bark; (2) is dominated in terms of resident biomass and trophic level by the spiders; (3) includes
Psocoptera, mites, and Neuropteran larvae as additional resident
taxa; but (4) seems to include a substantial component of transients to other tree surfaces or chance volant insects that
happen to land on the bark.

Such a relationship between trees and arthropods was predicted in Janzen's (1968) consideration of plants as ecological
islands. Bark crevices serve as sites for egg-laying and overwintering for numerous species, but like the proverbial suburbs,
the bark surface is just a bedroom community. The active portion
of the lives of these arthropods is carried on elsewhere.

TABLE III. Frequency of Occurrence of Arthropod Taxa on the
Bark Surface of Four Tree Species[a]

Order	Loblolly Pine	Southern Red Oak	Sweetgum	Pignut Hickory
Hymenoptera (Formicidae)	9	9	10	10
Hemiptera	6	3	3	4
Coleoptera	1	2	2	4
Diptera	2	5	3	8
Lepidoptera	1	0	0	0
Araneida	5	6	4	3

a Number of trees/tree species on which each taxon was encountered. Ten trees of each species were examined.

III. BRANCHES

Most of the preceding comments concerning the bark surface of tree trunks apply equally well to the surfaces of limbs higher in the tree. The primary function of branches is support of the crown. The nature of the crown varies among tree species and under different environmental conditions. Consequently, the arrangement of tree branches is a result of evolutionary and environmental pressures to optimize crown support (Esser, 1946). There are some major differences between the trunk and branches that influence the suitability of these structures as foraging substrates for insectivorous birds. These differences can be summarized as follows: (1) limbs often have more horizontal aspects, (2) bark rugosity decreases with limb size, (3) limbs are higher and thus subject to a different microclimate (e.g., Christy, 1952; Bergen 1971, 1974), and (4) limbs are closer to the sites of primary productivity. I will briefly detail some of the variability in branches that occurs among trees and relate this variability to the suitability of these tree surfaces as substrates for foraging insectivorous birds.

A. Branch Angle

The horizontal nature of many branches makes these surfaces available to a large number of insectivorous forest birds that do not have morphological adaptations that permit their use of vertical substrates. The frequency of horizontal branches varies with (1) tree species, (2) tree age, (3) height within a tree, (4) branch size, and (5) habitat and environmental conditions (Wilson 1970; Zimmerman and Brown 1971). Some epiphytes may attain greater development on the more horizontal surfaces (e.g., resurrection fern, Polypodium polypodioides). The physical environment for arthropods on the trunk and branches also varies with branch angle as a result of a differential flow of rainwater over these surfaces. Voth (1939) demonstrated that trees with vertical limbs funnel up to 6 or 7 times more water down trunk surfaces than actually falls per unit area around the tree. Trees with more horizontal branches and those with drooping branches, such as the northern conifers, distribute rainwater more to the periphery of the tree canopy and considerably less flows down the trunk. Horizontal limbs also provide undersurfaces that are likely more difficult for birds to reach than are vertical surfaces, and that are less influenced by temperature from direct sunlight, and probably less susceptible to colonization by epiphytes. For the most part, these surfaces are gleaned only by the vertical surface foragers.

B. Branch Diameter

A number of factors related to branch size likely influence
the use of limbs by foraging birds. Larger branches are similar
in physical characteristics to the trunk; with decreased size
limb surfaces become similar to the trunks of younger and younger
trees. Whereas the bark surfaces of older trunks and limbs are
composed of rugose, dead thick bark, smaller branches have
thinner, less rugose bark. Some have bark thin enough to be
penetrated by sap feeding insects such as larger Homopterans and
Hemipterans. Branches less than about 2 cm in diameter, whether
vertical or horizontal, become suitable foraging substrates for
many small passerines. In contrast, the smallest branches become
too small to support the weight of larger birds. Grubb (pers.
comm.) has recently experimentally demonstrated perch diameter
preferences in parids. Spanish moss (<u>Tillandsia</u> <u>usneiodes</u>) occurs
on both horizontal and vertical branches, but is found less often
on the trunk or small branches (Garth, 1964). Rainwater (1941)
found 163 species of insects and numerous spiders associated with
this southern epiphyte.

C. Number of Branches

The number of branches on a tree varies with tree size, tree
species, tree age, habitat, and perhaps other factors (Wilson,
1970; Whittaker and Woodwell, 1967). Distribution of branches on
trees is strongly associated with the openness of the habitat.
Trees in dense, shaded environments tend to lose more lower
branches through natural pruning and to grow taller than trees
growing in full sun. Each year each branch sends out new shoots
and, as Wilson (1970) suggests, proliferation of branches must
take place in some sort of exponential relationship. Turrell
(1961) used the allometric growth equation of Huxley (Huxley and
Teissier, 1936) to estimate leaf production by trees. Wilson
(1970) feels that this equation might also be applicable to the
problem of determining the increase in number of branches on a
tree with increasing age. Thus, Wilson predicts a relationship
which can be expressed by the equation $\underline{N} = c\underline{a}^{\underline{n}}$, where $\underline{N}$ is the total
number of branches, $\underline{c}$ is a constant for the particular tree, $\underline{a}$
is tree age, and $\underline{n}$ is the exponential factor describing the
number of branches added to each parent branch per year. This
annual exponential increase in number of branches as the tree
grows also represents increased surface area for arthropods and
tree surface foraging birds. While such exponential growth in
number of branches may hold approximately true for many years,
this equation does not allow for branch mortality which is an
increasingly important factor in older trees. The presence of
dead branches, however, provides yet another tree surface environment for arthropods and birds. A high frequency of dead

branches in older trees may provide a significant foraging substrate for insectivorous birds. The lesser frequency of dead
branches in younger trees may limit bird populations or community
structure in a young forest. This feature of old trees has been
overlooked in the arguments against short-rotation forestry.

IV. FOLIAGE

The trunk and branches are merely supports for what is usually
the most obvious part of a tree - its foliage. Tree foliage is
the site of most photosynthesis and the substrate for many grazing
arthropod taxa. Numerous variations in foliage characteristics
within and among tree species influence the quality and temporal
suitability of the foliage as a foraging substrate for arthropods
and for insectivorous birds. I will examine here only a few of
the major patterns as they might relate to foraging birds.

A. Deciduous Versus Evergreen

In the north temperate zone there is an apparent basic
dichotomy of deciduous versus evergreen foliage. This dichotomy
is not so evident elsewhere in the world. Axelrod (1966) points
out a complete series of intermediates to deciduous and evergreen
trees and discusses the evolutionary origin and adaptive significance of "deciduousness." Evergreens, with their relatively
continuous primary productivity, provide a more constant structural environment, resulting in continual cover and an amelio-
rated microclimate for both arthropods and foliage gleaning birds.
The leaves of evergreens are able to persist through extreme cold
and xeric conditions because they develop "tough" cuticles and
thick-walled cells which protect them from dessication and freezing. Tough cuticles can be a limiting factor for many foliage
gleaning arthropods (Southwood, 1973) and might thus secondarily
become a limiting factor for insectivorous birds.
The loss of leaves of deciduous trees necessitates a drastic
shift in foraging strategy by foliage gleaning birds or migration
to a more suitable environment. A few tree species (such as some
of the oaks) have leaves which die at the end of the growing
season but which are retained until the emergence of new leaves
the following spring. These trees provide a lesser change in the
"niche gestalt" (James, 1971) of foliage gleaning birds than do
the deciduous trees. Though lacking primary productivity, such
marcescent trees still provide hiding places for overwintering
arthropods.
Arthropod populations and their effects on trees vary between
deciduous and evergreen species. Evergreens typically suffer
more severely from defoliating insects than do deciduous trees

(Kulman, 1971). Deciduous trees tend to have a greater energy
reserve capacity and can put out new foliage in the middle of the
growing season. In contrast, many conifers are killed by defoliation. The different effects of defoliating insects on deciduous
and evergreen trees may in turn elicit diverse responses from
populations of insectivorous forest birds. ‘For example, foliage
gleaning birds of deciduous forests may have frequent, local,
cyclic, population fluctuations that are in synchrony with extreme populations of defoliating insects. In coniferous forests,
on the other hand, because of the fatal nature of extreme defoliation, avian population increases related to superabundance of
defoliating arthropods would be expected to be short-lived and
non-cyclic (or very long-cycled) in local areas. Population
fluctuations of the spruce budworm (Choristoneura fumiferana) and
its avian predators fit the prediction well for coniferous
forests (e.g., Morris et al., 1958). The forest tent caterpillar
(Malacosoma disstria) and the larch tortrix moth (Zeiraphera
diniana) exemplify cyclic population fluctuations that result in
total defoliation of deciduous trees every eight to twelve years
(Varley et al., 1974). The larch sawfly (Pristiphora erichsonii)
may defoliate larches (Larix spp.) in a local area for six to
nine years in succession before trees die (Baker, 1972:458).
Buckner and Turnock (1965) have demonstrated dramatic numerical
responses of several bird species to such outbreaks.

B. Foliage Density

Cotter and Monk (1967) quantify some of the interspecific
variation in number of leaves per branch of several tree species.
Pearson (1975) and Franzreb (1978) document the magnitude of
variation in foliage volume that can occur among species. Such
variation also occurs within species under different environmental conditions (Waring et al., 1978). In addition to number
of leaves, the arrangement of leaves on a branch may be of significance in determining the facility with which some bird species
can use foliage as a foraging substrate. For example, tight
whorls of pine needles provide hiding places for arthropods and a
likely effective barrier to some foraging birds. The needle-like
beaks of many North American wood warblers (Parulidae) and old
world warblers (Sylviidae) may be adaptations (or preadaptations?)
which have contributed to their success in coniferous forests by
allowing easier extraction of arthropods from among needles.
Pearson (1975) provides an excellent discussion of the relationship between foliage density and the physical and behavioral
characteristics of birds capable of efficiently using that foliage
as a foraging substrate. One major correlate of increased foliage
density is decreased size among foliage gleaning birds.
     Karr and Roth (1971) suggest that foliage density may play a
significant role in determining the bird species' diversity of an

area. To a point increasing foliage density seems positively
correlated with bird species diversity. In very dense forests,
however, - such as in young pine plantations - foliage density
may restrict bird mobility and reduce avian diversity. Similarly,
dense foliage may restrict dispersal of some arthropods, resulting
in aggregations or "outbreaks" in local areas. Stand density has
been associated with outbreaks of several forest insect pests -
usually with the suggestion that reduced tree vigor resulting
from overcrowding has been the primary cause of the infestation.
Perhaps aggregation of the insects as a result of the physical
barrier provided by dense foliage also plays a role in such
outbreaks.

C. Leaf Structure

The structure of leaves varies with a multitude of parameters
including such things as species, availability of water, wind,
disease, plant age, light, season, and mineral availability (e.g.,
Ashby, 1948; Stover, 1944). Leaf structure not only varies among
species and among trees within a species, but has been shown to
vary predictably within individual trees (e.g., Talbert and Holch,
1957).

Most intra-tree differences in leaf structure are environ-
mentally induced as the leaf is developing (Esau, 1953). Because
of seasonal changes in the environment, leaves produced in the
summer sometimes differ from those produced in early spring. In
sweetgum (Liquidambar styraciflua), for example, early leaves are
smaller, less-deeply lobed, and have shorter petioles than the
leaves of mid-summer. By late summer new leaves are smaller, with
shorter petioles again, but are still deeply lobed (Zimmerman and
Brown, 1971:47-49). Because of their longer petioles and larger
size, mid-summer leaves may not be suitable foraging substrates
for small insectivorous birds. Not only might they be "out of
reach" of some birds, but they are likely to be in continual
motion as a result of wind action. Trees with compound leaves
might be particularly unsuitable as foraging sites for foliage
gleaning birds by virtue of the relative inaccessibility of the
leaflets. Franzreb (1978) suggested large leaf size and leaf
movement as possible explanations for the disproportionate lack
of use of aspens (Populus sp.) by Ruby-crowned Kinglets (Regulus
calendula) in Arizona.

Within many trees there are striking differences in leaf
structure that are related to position within the tree and
exposure to the sun. In deciduous trees the "sun leaves" that
are at the top of the tree are thicker and smaller, have thicker
cuticles, and are likely to be more deeply lobed than are "shade
leaves" that are on lower and more internal branches (Wylie, 1951;
Jackson, 1967). Sun-exposure related differences have also been
documented for pines, but the largest needles are the "sun needles"

at the top of the tree (McLaughlin and Madgwick, 1968). The
"sun leaf-shade leaf" syndrome is well-documented as an adapta-
tion for reducing water loss while maximizing leaf productivity.
At the same time, these leaf differences elicit a response of
differential use by foliage grazing arthropods. For example,
Nielsen and Ejlersen (1977) considered a community of arthropods
that fed on beech (Fagus sylvatica) leaves and found that most
foliage grazers concentrated their feeding in the low canopy.
They also demonstrated age-specific differences in foraging sites
of some arthropods. Numerous authors have demonstrated differential bird use of foliage that could easily have been the ultimate
result of intra-tree variation in foliage characteristics (e.g.,
MacArthur, 1958; Morse, 1967b; Balda, 1969; Williamson, 1971).

A final structural feature of leaves which I wish to discuss
is leaf texture. There is considerable evidence (e.g., Painter,
1958; Thorsteinson, 1960; Gibson, 1976 a,b) that glandular hairs,
pubescence, very smooth surfaces, and other textural characteristics can deter some insect pests of crops. Such attributes
might also protect some trees (e.g., the pubescent leaves of
Blackjack Oak, Quercus marilandica) from foliage-grazing arthropods. On the other hand, there is also evidence (Thorsteinson,
1960) that surface characteristics that might protect a plant
from one pest might make the plant more susceptible to another.
Leaf texture must be considered as a factor influencing foliage
suitability as a substrate for arthropods and their avian
predators, but the nature of the influence must be considered
on a case-by-case basis.

D. Plant Chemistry

Much attention has been given in the past few decades to
plant chemicals which seem to render plants resistant to some
insect pests (see reviews in Beck, 1965; Dethier, 1970; Whittaker,
1970). Phenols, tannins, volatile oils, and other plant chemicals
are known which inhibit, kill, or slow the development of phyto-
phagous insects. Some of these chemicals are present only at
certain stages of plant development. For example, the tannin
content of oak leaves increases as the leaves mature. Winter
moths (Operophtera brumata) have adapted to seasonal changes in
oak leaf tannins by altering their reproductive cycle such that
the larvae feed primarily on the young leaves (Feeny, 1970). In
this and similar cases, a food resource is available to foliage
gleaning birds early in the season, but is gone by the time the
leaves mature. Bird reproductive efforts seem to be closely
attuned to such insect population dynamics. The nesting period
of titmice (Parus spp.) coincides closely with timing of maximum
availability of winter moth larvae (Varley, 1970). Lack (1966)
found that second or late broods of Great Tits (Parus major)
often die of starvation or fledge underweight.

Recently McKey et al. (1978) demonstrated that vegetation on low-nutrient soils contains relatively high concentrations of chemicals that deter herbivores. Some insects have evolved mechanisms of tolerating plant chemicals that are detrimental to other arthropods and we can reasonably expect these species to be more prevalent in areas where these "resistant" plants thrive. In many cases insects that can tolerate the plant chemicals have themselves become unpalatable (e.g., Brower and Brower, 1964). Thus, regional differences in nutrient quality of soil are mirrored in the chemical defense mechanisms of plants and foliage grazing arthropods and come to bear on populations of insectivorous birds.

V.   FLOWERS AND FRUITS

The flowers and fruits of trees are specially attractive to arthropods as foraging sites because of the concentrations of nutrients found in them. Indeed, many trees depend at least in part on insects for pollination. The availability of these arthropods to insectivorous birds is another matter, since flowers and fruits are often positioned such that they are easily accessible to only the most agile birds. During times of stress the importance of these insect "attractants" is probably greatest. In Figure 6 I have attempted to summarize available information on flowering and fruiting of the major genera of North American trees. Variability in flowering and fruiting phenology, in addition to differential attractiveness of tree species to insects, suggests that some flowers and fruits will be more important as foraging sites for insectivorous birds than will others. Birds can be put under stress by adverse weather at any time of the year, but among the more stressful periods are early spring as migrants begin arriving from their wintering areas, late fall as late migrants are leaving their breeding areas, and the nesting season. It would seem that the earliest flowering and latest fruiting trees might be most critical as attractants for insects which can be fed on by migrant birds. However, many of these early flowering trees are themselves adapted to the stressful weather by being wind-pollinated (e.g. pines and willows) - and their flowers are not so attractive to insects. In many cases the earliest arriving and latest leaving birds are not strict insectivores and in times of stress the flowers and fruit may become the fare for normally insectivorous birds (e.g., Tramer and Tramer, 1977).

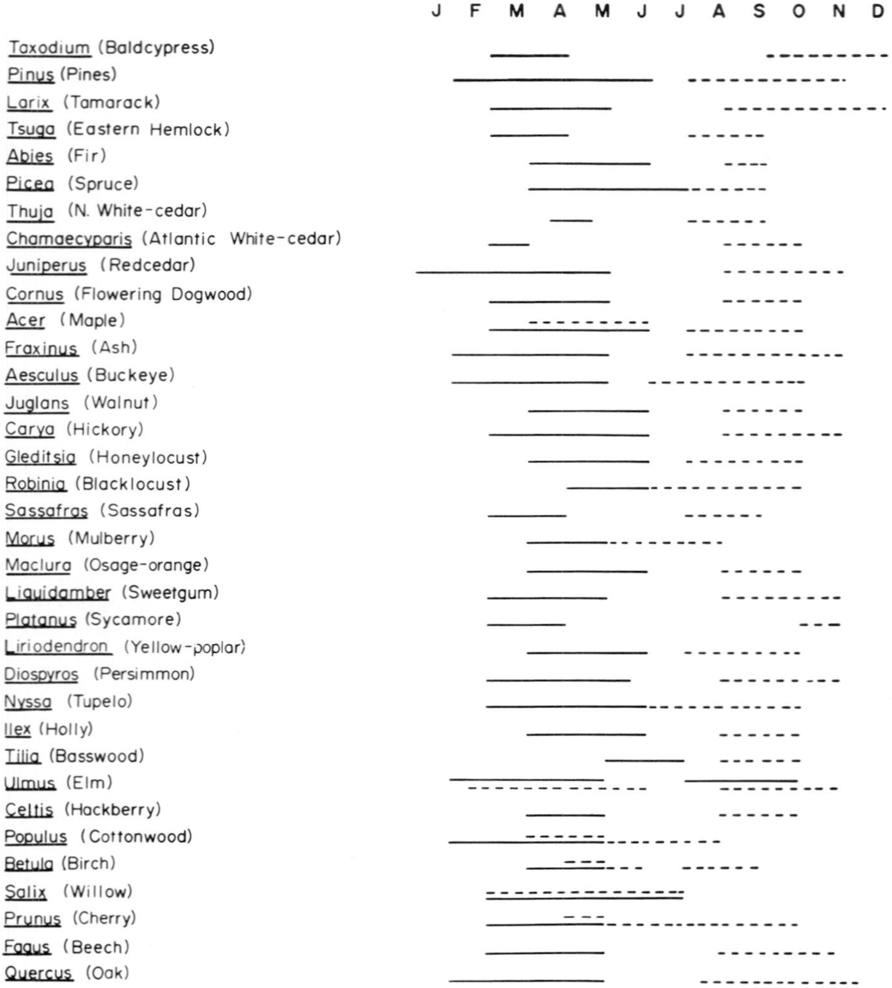

J   F   M   A   M   J   J   A   S   O   N   D

Taxodium (Baldcypress)
Pinus (Pines)
Larix (Tamarack)
Tsuga (Eastern Hemlock)
Abies (Fir)
Picea (Spruce)
Thuja (N. White-cedar)
Chamaecyparis (Atlantic White-cedar)
Juniperus (Redcedar)
Cornus (Flowering Dogwood)
Acer (Maple)
Fraxinus (Ash)
Aesculus (Buckeye)
Juglans (Walnut)
Carya (Hickory)
Gleditsia (Honeylocust)
Robinia (Blacklocust)
Sassafras (Sassafras)
Morus (Mulberry)
Maclura (Osage-orange)
Liquidamber (Sweetgum)
Platanus (Sycamore)
Liriodendron (Yellow-poplar)
Diospyros (Persimmon)
Nyssa (Tupelo)
Ilex (Holly)
Tilia (Basswood)
Ulmus (Elm)
Celtis (Hackberry)
Populus (Cottonwood)
Betula (Birch)
Salix (Willow)
Prunus (Cherry)
Fagus (Beech)
Quercus (Oak)

FIGURE 6. Approximate ranges of flowering (solid lines) and fruiting (dashed lines) dates for 35 genera of North American trees. The genera illustrated are those included in Neelands (1968). Phenological data are summarized from Schopmeyer (1974).

VI.   DISCUSSION

Throughout this paper I have explored tree surface variability that likely influences both arthropod and insectivorous bird populations in forest ecosystems. Other factors obviously

also influence these populations. Climate is perhaps the single
most important variable that influences all ecosystems and it
must be kept in mind throughout the following discussion. However, I have chosen to dwell primarily on variability in additional biotic features of forest ecosystems since these are the
components over which man has the most control and which seem
to be undergoing the greatest disturbance today.

Most forest insects are more or less restricted to single
hosts or to a few closely related species (Baker, 1972), and each
tree species has its own associated arthropod fauna. Southwood
(1961) has compared the number of insect species known from a
number of tree genera in Britain and has summarized similar data
for other areas. His analyses of these data suggest that the
diversity of insects associated with a given tree species will
to a large extent be related to the relative abundance of the
tree species in an area and the past history of the species. For
example, newly introduced tree species tend to have a depauperate
insect fauna, and uncommon tree species tend to have a less
diverse arthropod fauna than common trees. Southwood (op. cit.)
also notes that some taxa of trees seem to have relatively more
species of insects associated with them, regardless of the
status of the tree (e.g., members of the Rosaceae and Salicaceae),
while others seem to have relatively fewer insect species (e.g.,
hazel, Corylus; ash, Fraxinus; holly, Ilex; and yew, Taxus).
Perhaps these differences are related to some of the structural
variability in tree surfaces discussed in this paper.

Populations of forest insects fluctuate dramatically from
generation to generation. Ten-fold increases in numbers followed
by similar declines are not uncommon (Varley et al.,1974).
Single species stands in plantations are more susceptible to
devastating insect outbreaks than are mixed-species in natural
stands (Baker, 1972:12-13). The reason for this is in part
because of the concentration of food resources for the insect
population. However, as Anderson (1960:39) and Root (1975:86)
discuss, mixed-species stands seem to have an "associational
resistance" to some herbivores. This associational resistance
is something beyond the individual defense mechanisms of the
trees. As Root (op. cit.) phrases it, it is an added resistance
due to the "texture" of the ecosystem.

Lack of associational resistance may only partially explain
the susceptibility of forest monocultures to insect outbreaks.
Inability of single-age monocultures to support an adequate
diversity of arthropods to maintain a stable food supply and to
provide the niche gestalt for each species of a diverse population of insectivorous birds may be contributing factors. Because
of the specific host relationships of insects, increasing tree
species diversity will increase arthropod diversity. Allowing
age diversity in a forest provides more dimensions for further
increases in arthropod and avian diversity. With stratification

of tree surface types, the increased "dead branch" component in older trees, and the associated stratification of epiphytes (e.g., Hale, 1952) and arthropods (e.g., Fichter, 1939; Adams, 1941; Dowdy, 1951; Reid, 1957; Enders, 1974; Albert, 1976), it is not surprising to find the relationship between foliage height diversity and bird species diversity that numerous workers have now documented (e.g., MacArthur and MacArthur, 1961; Pearson, 1971; Karr and Roth, 1971; Willson, 1974; Moss, 1978). Not only is the food resource for birds and arthropods spread out more in space above the ground (thus facilitating vertical niche separation), but as the vegetation increases in height, the foraging surfaces increase exponentially and become more diverse. The horizontal and vertical taxonomic and structural diversity of the mixed-age, mixed-species forest ecosystem provides the opportunity for maximum avian diversity and system stability.

But the opportunity for maximum avian diversity will not be realized if we only consider taxonomic and age diversity of the forest. Forest size is a final factor which must be considered. The once expansive forests of North America are now splintered into minute parcels. This fractioning of the forest ecosystem also has a negative effect on avian populations (Forman et al., 1976).

Birds, arthropods, and trees are only part of an ecosystem. There is more. If we want to preserve any part, we must manage for the whole. I would like to conclude this paper by quoting from Graham and Knight's (1965:68) text, Principles of Forest Entomology:
"... the ecological picture gradually changes as the trees pass from infancy to old age. It has already been pointed out that trees of different ages are attacked by different insects. Likewise different plants and animals are associated with each age-class of trees. Thus, in considering any tree or small group of trees, we are dealing with only a segment of the complete environmental complex. Only on a large forest area, with a normal distribution of age-classes and their associated flora and fauna, is the environmental complex complete. Then each element of the complex is operative but in different spots at different times. As one factor ceases to operate in one spot, it is replaced by another. Thus the balance is maintained."

ACKNOWLEDGMENTS

B.J. Schardien, C.D. Cooley, and P. Ramey, helped in many ways with the preparation of this manuscript. I am also indebted to the U.S. Department of Energy and the U.S. Forest Service for financial support of my studies. Personnel at the Savannah River Plant, and at Noxubee National Wildlife Refuge have been generous and helpful in their logistical support of my research. Lyda Eubank expertly typed the early and final drafts of the manuscript.

REFERENCES

Adams, R. (1941). _Ecol. Monogr._ 11, 191.
Albert, R. (1976). _Faun-Oekol. Mitt._ 5, 65.
Ashby, E. (1948). _New Phytol._ 47, 153.
Austin, G., and Smith, E. (1972). _Condor_ 74, 17.
Axelrod, D. (1966). _Evolution_ 20, 1.
Baker, W. (1972). "Eastern Forest Insects." U.S.D.A. Forest
    Service, Misc. Publ. No. 1175.
Balda, R. (1969). _Condor_ 71, 399.
Beal, F. (1911). "Food of the Woodpeckers of the United States."
    U.S.D.A., Biol. Surv. Bull. No. 37.
Beck, S. (1965). _Annu. Rev. Entomol._ 10, 207.
Bergen, J. (1971). _Forest Sci._ 17, 314.
Bergen, J. (1974). _Forest Sci._ 20, 64.
Billings, W., and Drew, W. (1938). _Am. Midl. Nat._ 20, 302.
Brodkorb, P. (1970). _Fla. Acad. Sci._ 33, 132.
Brodkorb, P. (1971). _In_ "Avian Biology" (D. Farner and J. King,
    eds.), p. 19. Academic Press, New York.
Brower, L., and Brower, J. (1964). _Zoologica_ 49, 137.
Buckner, C., and Turnock, W. (1965). _Ecology_ 46, 223.
Burt, W. (1930). _Univ. Calif. Publ. Zool._ 32, 455.
Chang, Y. (1954). _TAPPI Tech. Assoc. Pulp Pap. Ind._, 14.
Christy, H. (1952). _Ohio J. Sci._ 52, 199.
Cotter, D., and Monk, C. (1967). Savannah River Ecology
    Laboratory, Annual Report 1967, 18.
Dethier, V. (1970). _In_ "Chemical Ecology" (E. Sondheimer and
    J. Simeone, eds.), p. 83. Academic Press, New York.
Dickson, J., and Noble, R. (1978). _Wilson Bull._ 90, 19.
Dowdy, W. (1951). _Ecology_ 32, 37.
Elton, C. (1966). "The Pattern of Animal Communities," John
    Wiley & Sons, New York.
Enders, F. (1974). _Ecology_ 55, 317.
Esau, K. (1953). "Plant Anatomy," John Wiley & Sons, New York.
Esser, M. (1946). _Bull. Math. Biophys._ 8, 65.
Feduccia, A. (1972). _Wilson Bull._ 84, 315.
Feeny, P. (1970). _Ecology_ 51, 565.
Fichter, E. (1939). _Ecol. Monogr._ 9, 184.
Forman, R., Galli, A., and Leck, C. (1976). _Oecologia_ 26, 1.
Franzreb, K. (1978). _Wilson Bull._ 90, 221.
Garth, R. (1964). _Ecology_ 45, 470.
Gibson, R. (1976a). _Ann. Appl. Biol._ 82, 143.
Gibson, R. (1976b). _Ann. Appl. Biol._ 82, 147.
Graham, S., and Knight, F. (1965). "Principles of Forest
    Entomology," McGraw-Hill Book Co., New York.
Grinnell, J. (1924). _Condor_ 26, 32.
Grubb, T., Jr. (1975). _Condor_ 77, 175.
Hale, M., Jr. (1952). _Ecology_ 33, 398.

Horn, H. (1971). "The Adaptive Geometry of Trees," Princeton Univ. Press, Princeton, N.J.

Huxley, J., and Teissier, G. (1936). Nature 137, 780.

Jackson, J. (1970). Ecology 51, 318.

Jackson, J. (1977). J. Wildl. Manage. 41, 448.

Jackson, J. (1978). Wildl. Soc. Bull. 6, 171.

Jackson, L. (1967). Ecology 48, 498.

James, F. (1971). Wilson Bull. 83, 215.

Janzen, D. (1968). Am. Nat., 102, 592.

Karr, J. and Roth, R. (1971). Am. Nat. 105, 423.

Kulman, H. (1971). Annu. Rev. Entomol. 16, 289.

Lack, D. (1966). "Population Studies of Birds," Clarendon Press, Oxford.

MacArthur, R. (1958). Ecology 39, 599.

MacArthur, R., and MacArthur, J. (1961). Ecology 42, 594.

McCambridge, W., and Knight, F. (1972). Ecology 53, 830.

McKey, D., Waterman, P., Mbi, C., Gartlan, J., and Struhsaker, T. (1978). Science 202, 61.

McLaughlin, S. and Madgwick, H. (1968). Am. Midl. Nat. 80, 547.

Morris, R., Cheshire, W., Miller, C., and Mott, D. (1958). Ecology 39, 487.

Morse, D. (1967a). Ecology 48, 94.

Morse, D. (1967b). Auk 84, 490.

Morse, D. (1968). Ecology 49, 779.

Morse, D. (1970). Ecol. Monogr. 40, 119.

Moss, D. (1978). J. Anim. Ecol. 47, 521.

Neelands, R. (1974). "Important Trees of Eastern Forests," U.S.D.A. Forest Service, Southern Region, Atlanta, Georgia.

Nielsen, B., and Ejlersen, A. (1977). Ecol. Entomol. 2, 293.

Painter, R. (1958). Annu. Rev. Entomol. 3, 267.

Park, O. (1931). Ecology 12, 188.

Pearson, D. (1971). Condor 73, 46.

Pearson, D. (1975). Condor 77, 453.

Phillips, E. (1951). Ecol. Monogr. 21, 301.

Rainwater, D. (1941). Insects and Spiders Found in Spanish Moss, Gin Trash, and Wood Trash, and on Wild Cotton. U.S. Dep. Agr. Bur. Entomol. Plant Quarantine, Series E-528, 20 p.

Reid, R. (1957). Can. Entomol. 89, 437.

Richardson, F. (1942). U. Calif. Publ. Zool. 46, 317.

Roling, M., and Kearby, W. (1977). Can. Entomol. 109, 1235.

Sargent, T. (1966). Science 154, 1674.

Schopmeyer, C. (1974). "Seeds of Woody Plants in the United States." U.S.D.A. Forest Service, Agricultural Handbook No. 450.

Smith, D. (1962). In "Regional Silviculture of the United States" (J. Barrett, ed.), p. 3. John Wiley & Sons, New York.

Smith, K. (1960). Oikos 11:43.

Smith, L. (1971). Winter Ecology of Woodpeckers and Nuthatches in Southeastern South Dakota. Ph.D. Thesis, Univ. of South Dakota, Vermillion.

Smith–Davidson, V. (1930). Ecology 11, 601.
Southwood, T. (1961). J. Anim. Ecol. 30, 1.
Southwood, T. (1973). In "Insect/Plant Relationships" (H. van
    Emden, ed.), p. 3 Symp. Royal Entomol. Soc. Lond., no. 6.
Spring, L. (1965). Condor 67, 457.
Stephens, H. (1969). "Trees, Shrubs, and Woody Vines in Kansas."
    Univ. of Kansas Press, Lawrence.
Stover, E. (1944). Bot. Gaz. 106, 12.
Symonds, G. (1958). "The Tree Identification Book." William
    Morrow & Company, New York.
Talbert, C., and Holch, A. (1957). Ecology 38, 655.
Thorsteinson, A. (1960). Annu. Rev. Entomol. 5, 193.
Tramer, E., and Tramer, F. (1977). Wilson Bull. 89, 166.
Travis, J. (1977). Condor 79, 371.
Turrell, F. (1961). Bot. Gaz. 122, 284.
Varley, G. (1970). In "Animal Populations in Relation to Their
    Food Resources" (A. Watson, ed.), p. 389. Br. Ecol. Soc.,
    Blackwell Scientific Publ., Oxford.
Varley, G., Gradwell, G., and Hassell, M. (1974). "Insect
    Population Ecology an Analytical Approach." Univ. of
    California Press, Berkeley, California.
Voth, P. (1939). Bot. Gaz. 101, 328.
Waring, R., Emmingham, W., Gholz, H., and Grier, C. (1978).
    For. Sci 24, 131.
Weiss, H. (1920). Am. Nat. 54, 443.
Whittaker, R. (1970). In "Chemical Ecology" (E. Sondheimer and
    J. Simeone, eds.), p. 43. Academic Press, New York.
Whittaker, R., and Woodwell, G. (1967). Am. J. Bot. 54, 931.
Williamson, P. (1971). Ecol. Monogr. 41, 129.
Willson, M. (1970). Condor 72, 169.
Willson, M. (1974). Ecology 55, 1017.
Wilson, B. (1970). "The Growing Tree." The University of
    Massachusetts Press, Amherst.
Wylie, R. (1951). Am. J. Bot. 38, 355.
Zimmerman, M., and Brown, C. (1971). "Trees Structure and
    Function." Springer-Verlag, New York.

SEASONAL CHANGES IN WOODPECKER FORAGING METHODS:
STRATEGIES FOR WINTER SURVIVAL

Richard N. Conner[1]

Department of Biology
Virginia Polytechnic Institute and State University
Blacksburg, Virginia

Seasonal changes in the foraging methods of five species of
woodpeckers were studied over a four year period in southwestern
Virginia. Woodpeckers using "typical" woodpecker foraging methods such as the Downy (Picoides pubescens), Hairy (Picoides
villosus), and Pileated (Dryocopus pileatus) woodpeckers used
methods that penetrated trees to greater depths during winter
than during the breeding and post-breeding seasons. In woodpeckers with "non-traditional" foraging behavior, selection may
have favored development of migratory behavior to avoid winter
food shortages.

I. INTRODUCTION

Seasonal changes in availability of woodpecker food items
might be expected to cause concurrent changes in foraging methods. During mild seasons, insects are more readily available to
woodpeckers than in colder seasons. Past studies have demonstrated that Downy Woodpeckers (Picoides pubescens) use surface
gleaning and superficial foraging methods more during mild seasons than in winter. During colder seasons foraging methods that

---

[1]Present address: Southern Forest Experiment Station,
U.S.D.A. Forest Service, Nacogdoches, Texas 75962

penetrate deeper into foraging substrates are used more often
(Jackson, 1970; Willson, 1970). Presumably, Downies are responding to the positional availability of insect prey on and in
foraging substrates.

Seasonal changes in woodpecker foraging methods have been
implied in several other studies. Hairy Woodpeckers (Picoides
villosus) shifted from a diet of arthropods during the breeding
season to one of pine seeds during the non-breeding season in
Colorado (Stallcup, 1968). Pileated Woodpeckers (Dryocopus
pileatus) typically fed on arthropods on the bark's surface or
just below it in milder seasons, but excavated trees extensively
for carpenter ants (Camponotus spp.) during winter (Munro, 1923;
Hoyt, 1941). Common Flickers (Colaptes auratus) shifted from a
mostly insectivorous diet in the spring to mainly a frugivorous
one in the fall (Burns, 1900).

Most studies of foraging methods used by eastern North
American woodpeckers are descriptive with data collected during a
single season: Downy Woodpeckers (Kilham, 1970; Kisiel, 1972;
Williams, 1975), Hairy Woodpeckers (Kilham, 1965; Williams,
1975), Pileated Woodpeckers (Hoyt, 1950, Hoyt, 1957; Kilham,
1976), and Red-headed Woodpeckers (Melanerpes erythrocephalus)
(Goodpasture, 1909; Dorsey, 1926; Judd, 1956; Reller, 1972;
Jackson, 1976). Hence, little quantitative data exists to examine seasonal shifts in woodpecker foraging methods.

The current study compares seasonal changes in foraging methods of typical woodpeckers (Downy, Hairy, and Pileated woodpeckers) with Common Flickers and Red-Headed Woodpeckers whose
ancestors in adapting to habitats with fewer trees had to deviate
from typical woodpecker foraging methods (Short, 1972; Jackson,
1976). The objective of this study was to determine if the two
groups of woodpeckers exhibited two different strategies for
overwinter survival.

                            II. METHODS

The majority of the study area consisted of 20 km^2 on the
upper Craig and Poverty Creek drainages of the Jefferson National
Forest in southwestern Virginia. Oaks (Quercus spp.) and hickories (Carya spp.) covered 60 percent of the area, and stands consisting primarily of oaks and pines (Pinus spp.) covered 20 percent. Stands of yellow-poplar (Liriodendron tupilifera), white
oak (Q. alba)and northern red oak (Q. rubra), and stands
of Virginia pine(P. virginiana), white pine (P. strobus), and
pitch pine (P. rigida), each occupied approximately 10 percent
of the area. A wide range of cover types and successional stages,
resulting from clearcutting, were present.

The second part of the study area was centered around the town of Blacksburg and the Virginia Polytechnic Institute and State University campus. This non-forest area was primarily in pasture, but it included six mature woodlots (250-350 yrs. old) of oaks and hickories that varied between 0.5 and 20 ha in size. Most of the woodlots had no understory; grass was the dominant ground cover. Red-headed Woodpeckers nested regularly in these areas (Conner, 1976).

Foraging methods of Downy, Hairy, Pileated, and Red-headed woodpeckers, and Common Flickers were observed from September 1972 through July 1976. Field observation time was distributed evenly over all daylight hours as well as habitat types and conditions in order to eliminate temporal and interhabitat biases. During winter, however, open areas were searched more intensively than other seasons because of the scarcity of Common Flickers and Red-headed Woodpeckers. I recorded the foraging methods used by the woodpeckers (revised from Kilham, 1965) and their duration until the birds disappeared from sight (Table I). Data were collected during the breeding season (15 April through 15 June), post-breeding season (July through October), and winter (December through February).

TABLE I. Foraging Methods of Woodpeckers

Code	Description of foraging methods
PP	Peer-and-poke, a surface gleaning technique without any disturbance to the substrate foraged on
PECK	Pecking on the foraging substrates without any subcambial penetration
SCAL	Scaling the bark off a tree in search of food items
EXCA	Subcambial excavation in search of food items
HAWK	Aerial forays to capture insects on the wing
VEGF	Consumption of any vegetable material
GRDF	Foraging on the ground for animal food items

III. RESULTS

Downy and Hairy woodpeckers typically spent less time at
each foraging site than did the other three species. Downies
and Hairies were encountered more frequently, but less cumulative
foraging time was observed because of relatively rapid horizontal
movement through the forest. Pileated and Red-headed woodpeckers
and flickers tended to remain at foraging sites or areas for
extended periods of time.

Foraging behavior of woodpeckers varied seasonally. During
the breeding and non-breeding seasons Downy Woodpeckers foraged
primarily by the peer-and-poke method and secondarily by pecking
(Fig. 1). By winter this relative order of preference was

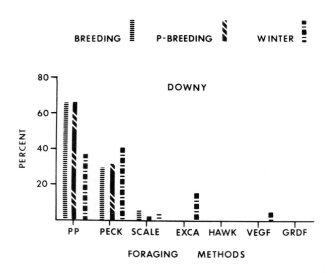

FIGURE 1. Percentages of foraging methods used by Downy
Woodpeckers during the breeding (80 min of observations),
post-breeding (98 min), and winter (93 min) seasons. (See
Table I for mnemonic variable code.

reversed, and at the same time Downies increased time spent foraging by sub-cambial excavation. Thus, during winter, Downy Woodpeckers used methods that penetrated trees to greater depths more frequently than during milder seasons as observed by Jackson (1970) and Willson (1970). During winter, Downy Woodpeckers also consumed more vegetable material than in the milder seasons (Fig. 1).

In all seasons Hairy Woodpeckers used the pecking method more frequently than the peer-and-poke method (Fig. 2). During winter, however, Hairies decreased time spent using the peer-and-poke and pecking methods, but increased time spent scaling bark and excavating. Thus, during winter, Hairy Woodpeckers also used methods that penetrated trees deeper and disturbed the substrate more than during milder seasons.

Pileated Woodpeckers varied foraging behavior slightly between the breeding and post-breeding seasons by decreasing time spent excavating and increasing scaling and consumption of

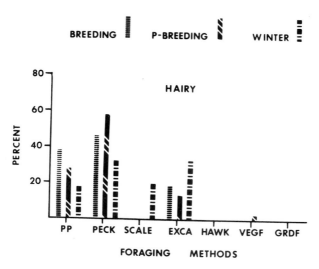

FIGURE 2. Percentages of foraging methods used by Hairy Woodpeckers during the breeding (75 min of observations), post-breeding (91 min), and winter (94 min) seasons.

Richard N. Conner

plant material (Fig. 3). Winter foraging behavior of Pileated
Woodpeckers varied greatly from behavior exhibited during the
other two seasons. Pileateds excavated more than 70 percent
of the observation time during winter demonstrating an increased
use of food items deep within trees.

Seasonal changes in foraging behavior of Common Flickers
were quite different from that observed for the Downy, Hairy,
and Pileated woodpeckers. Common Flickers foraged primarily
on the ground for ants and other invertebrates during the

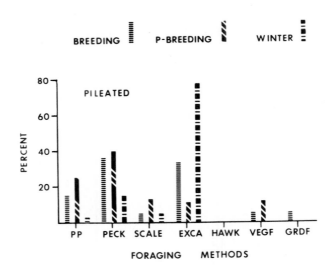

FIGURE 3. Percentages of foraging methods used by Pileated
Woodpeckers during the breeding (292 min of observations),
post-breeding (287 min), and winter (403 min) seasons.

breeding season and winter (Fig. 4). During the post-breeding season flickers foraged primarily on dogwood fruits (Cornus florida) and other fruits and berries that were available (Conner and Crawford, 1974), but still foraged extensively on the ground.

Red-headed Woodpeckers also varied foraging behavior seasonally. A comparison of the breeding and post-breeding seasons reveals a decrease in use of peer-and-poke, hawking, and ground feeding methods but an increase in foraging on

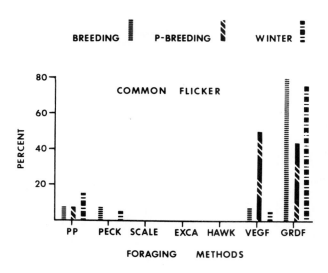

FIGURE 4. Percentages of foraging methods used by Common Flickers during the breeding (257 min of observations), post-breeding (279 min), and winter (135 min) seasons.

fruits and berries (Fig. 5). During winter Red-headed Woodpeckers further decreased their use of the peer-and-poke,
hawking and ground feeding methods, but markedly increased
feeding on plant material. Red-headed Woodpeckers were highly
dependent on acorns during late fall and winter that they
collected and ate or stored in small holes they had excavated
in dead portions of trees. Food-storing by Red-headed Woodpeckers is common (Kilham, 1958; MacRoberts, 1975).

Two of the five woodpecker species migrated out of the
study area seasonally. Common Flickers were much less abundant
during winter than during the other two seasons (See caption
Fig. 4). The abundance of Red-headed Woodpeckers in southwestern Virginia also declined during winter (See caption
Fig. 5), but not to the extent as that of Common Flickers.
Seasonal migration of flickers and Red-headed Woodpeckers is
well known (Bent, 1939; Robbins et al., 1966). Downy, Hairy,
and Pileated woodpeckers are non-migratory and I detected no
sudden decrease in abundance prior to winter.

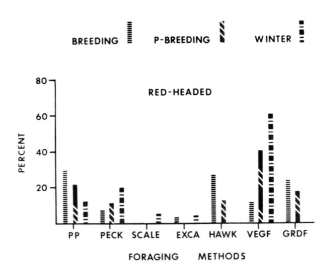

FIGURE 5. Percentages of foraging methods used by Red-
headed Woodpeckers during the breeding (153 min of observation),
post-breeding (171 min), and winter (120 min) seasons.

IV. DISCUSSION

Several factors have probably affected the evolution of resident versus migratory behavior in woodpeckers. Haartman (1968) suggested that resident birds of a species are able to occupy breeding habitats earlier than migrants of the same species, thus migrants may be forced to breed in lower quality habitat, and on the average, produce fewer young than residents.

Cavity nesting birds exhibit resident behavior at a higher frequency than do non-cavity nesting species (Haartman, 1968). If nest cavities or potential sites for cavities were severely limiting, a decided selective advantage would exist for resident individuals who would claim territories with cavities or potential cavity sites before migrants arrived. Cavities also provide excellent insulation from cold nights for birds using them as winter roost sites (Kendeigh, 1961). Thus, resident cavity nesters would spend less energy keeping warm on winter nights than similarly sized non-cavity nesting birds.

In addition to the above factors, I suggest that foraging behavior has also influenced the evolution of migratory versus resident behavior in woodpeckers. The five species of woodpeckers in this study appear to have different behavior patterns that aid in over-winter survival. The Downy, Hairy, and Pileated woodpeckers, species that use "typical" woodpecker foraging methods, demonstrated similar seasonal changes in foraging behavior. Their change to foraging methods that penetrate deeper into trees during winter appears to be a response to seasonal availability and location of insect prey. During the mild season when insects are readily available on the surface of trees, peer-and-poke, and pecking methods were used most often. In the winter when ants and insect pupae are under the bark or inside the cambium, and other insects are not abundant on the surface of trees, woodpeckers must excavate in search of prey. By changing their foraging methods in this manner I suggest that these three species have "avoided" the "need" to be migratory. Natural selection may have favored individuals that shifted foraging methods seasonally and were morphologically superior at excavation. The three species are apparently able to use a relatively unused food resource during winter by altering foraging methods and excavating deeper into trees.

When ancestral forms of Common Flickers and Red-headed Woodpeckers departed from traditional woodpecker foraging modes, natural selection may have favored individuals exhibiting seasonal movement patterns. Thus, rather than favoring an alteration of foraging methods to gain access to food deep within trees as availability of surface insects and fruits decreased, selection on Common Flickers may have favored migration to areas with warm climates where these food items were more readily available.

Red-headed Woodpecker strategy for overwinter survival has an additional component not observed in Common Flicker foraging behavior; they store food. Acorns stored during the post-breeding season when they are abundant provide Red-headed Woodpeckers a winter food source and may decrease the "necessity" to migrate to milder wintering areas. In southwestern Virginia, Red-headed Woodpeckers appeared to be less migratory than were Common Flickers. An interesting test of this hypothesis would be a comparison of fall mast crop with winter density of Red-headed Woodpeckers over a several year period.

Seasonal shifts in woodpecker foraging methods might be currently regulated by factors other than changes in food availability. Berthold (1976) suggested that seasonal food preferences in some birds is controlled by endogenous physiological requirements. Thus, physiological control as well as "success" with a particular foraging method and foraging site may be the mechanisms currently regulating seasonal changes in woodpecker foraging methods.

In summary, I suggest that the ability of Downy, Hairy and Pileated woodpeckers to shift to foraging methods that penetrate trees to greater depths during winter has enabled them to avoid the "necessity" to migrate. Since Common Flickers and Red-headed Woodpeckers have deviated from what is considered "typical" woodpecker foraging behavior, I suggest that selection concurrently favored migratory behavior as a solution to winter food shortages.

ACKNOWLEDGMENTS

I thank J. G. Dickson, J. A. Jackson, L. K. Halls, and J. C. Kroll for reviewing the manuscript and making many helpful comments, and C. S. Adkisson for advice on preliminary aspects of the study.

REFERENCES

Bent, A. (1939). U. S. Natl. Mus. Bull. 174.
Berthold, P. (1976). J. für Ornith. 117, 145.
Burns, F. (1900). "A monograph of the flicker." Wilson Bull.
    7, 1.
Conner, R. (1976). Bird-Banding 47, 40.
Conner, R., and Crawford, H. (1974). J. Forestry 72, 564.
Dorsey, G. (1926). Bird-Lore 28, 333.
Goodpasture, A. (1909). Bird-Lore 11, 196.
Haartman, L. (1968). Ornis Fenn. 45, 1.
Hoyt, J. (1941). Audubon Mag. 43, 525.
Hoyt, J. (1950). Bull. Mass. Aud. Soc. 34, 99.
Hoyt, S. (1957). Ecology 38, 246.
Jackson, J. (1970). Ecology 51, 318.
Jackson, J. (1976). Condor 78, 67.
Judd, W. (1956). Auk 73, 285.
Kendeigh, S. (1961). Wilson Bull. 73, 140.
Kilham, L. (1958). Wilson Bull. 70, 107.
Kilham, L. (1965). Wilson Bull. 77, 134.
Kilham, L. (1970). Auk 87, 544.
Kilham, L. (1976). Auk 93, 15.
Kisiel, D. (1972). Condor 74, 393.
MacRoberts, M. (1975). Auk 92, 382.
Munro, J. (1923). Can. Field Nat. 37, 85.
Reller, A. (1972). Am. Midl. Nat. 88, 270.
Robbins, C., Bruun, B., Zim, H., and Singer, A. (1966).
    "A guide to field identification: Birds of North America."
    Golden Press, New York.
Short, L. (1972). Bull. Am. Mus. Nat. Hist. 149.
Stallcup, P. (1968). Ecology 49, 831.
Williams, J. (1975). Am. Midl. Nat. 93, 354.
Willson, M. (1970). Condor 72, 169.

IMPLICATIONS OF OPTIMAL FORAGING THEORY FOR INSECTIVOROUS FOREST BIRDS

Edward O. Garton

Wildlife Resources
University of Idaho
Moscow, Idaho

Optimal foraging theory is useful in understanding foraging
behavior of insectivorous forest birds. The theory
correctly predicts qualitative changes in dietary specialization and allocation of foraging time observed in experimental and field studies. It suggests that forest birds
will exert the strongest stabilizing influence on highly
profitable insects occurring in patchy environments. The
theory currently fails in many quantitative predictions but
is useful as a unifying framework to guide research. Further development of the theory will require a better understanding of the cognitive abilities of these birds, improved
methods for quantifying foraging behavior, and more realistic theoretical approaches incorporating the birds'
imperfect and changing perceptions of insect abundances and
profitabilities.

I. INTRODUCTION

An insectivorous forest bird must make many decisions when
feeding. It must decide where to feed, what insects to pursue
and consume, how long to forage in a particular area and how to
search. What is the basis for these decisions? Schoener (1971)
pointed out that an individual will increase its fitness by
increasing its net energy and nutrient intake through: (1) increasing weight or nutritional reserves for reproduction, (2)
increasing clutch size, (3) increasing size of eggs, young and/or
their growth rate, and (4) increasing number of broods. In order
to obtain additional food the bird must devote time to foraging,
time which is therefore unavailable for other activities.

107

Fitness of a bird will also increase if it spends less time
foraging, allowing it to spend more time in activities such as
predator avoidance (surveillance and avoidance of exposed locations), social behavior (territorial defense) and energy con-
serving behavior (decreased activity and exposure to inclement
weather). Thus foraging decisions which lead to the highest
fitness will result in a bird obtaining the maximum amount of
food possible (measured in energy or nutrients) in the least
amount of time. Many investigators have examined the best or
optimal foraging decisions under various theoretical conditions
(Emlen, 1966; MacArthur and Pianka, 1966; Royama, 1970; Schoener,
1971; Rapport, 1971; Marten, 1973; Pulliam, 1974; Charnov, 1976;
Estabrook and Dunham, 1976; Oaten, 1977; Rapport and Turner,
1977). Pyke *et al.* (1977) provided an excellent review of this
literature and empirical evidence available in support or contra-
diction of the theory. My comments will be directed specifically
to the implications of optimal foraging theory for insectivorous
forest birds. The necessary decisions can be grouped into three
categories: (1) What to eat (optimal diet); (2) Where to search
(optimal allocation of foraging time); and (3) How to search
(optimal search behavior).

II.   OPTIMAL DIET

     Within a homogeneous patch of the environment, an
insectivorous bird must decide which prey to consume. The
optimal decision can be determined explicitly if we make the
following simplifying assumptions:
     1.   The rate at which the bird encounters insects while
          searching does not change during feeding.
     2.   The net value of a potential prey species can be
          measured in a single currency such as its weight, energy
          content or digestible protein content.
     3.   The bird has complete and perfect knowledge of the net
          value, handling time (see below) and abundance of each
          edible prey type occurring in the patch.
     The decision as to which insects will form the optimal diet
is based upon a ranking of profitability of insect types and upon
abundance of each type. A bird must spend time, termed handling
time (Holling, 1959), pursuing, capturing, preparing and ingest-
ing an individual of a particular type of insect. I will define
the profitability of an insect as its net value per individual
divided by its handling time per individual because this represents the food intake per unit time capturing and consuming that
insect (Royama, 1970). If not all pursuits of a particular type
of insect are successful, then its profitability must be based
upon the total time spent in attempted captures of that insect
and the total net value of insects successfully captured in those

attempts. Profitability of an insect is an excellent criterion
for deciding whether or not to pursue that insect. If an insect
is so difficult to capture, so small, or so indigestible that net
food intake per unit time handling that insect is less than the
active metabolic rate of the bird, then that type of insect
should be ignored by the bird. However, the optimal diet will
rarely include all of the insect types above this minimum
profitability as will be shown below.

Time spent foraging is not spent solely handling insects;
much of it is spent searching for insects. Density of an insect,
the bird's rate of search and the insect's ability to avoid
detection all influence how much time the bird must spend searching to find an insect. Taking these things into account,
abundance of an insect is measured as the number of insects
encountered per unit of time spent searching. If we assume that
time handling and time searching for insects are mutually exclusive then we can calculate the net rate of food intake for any
diet from the following multispecies functional response model
(Marten, 1973):

$$\frac{F_r}{t} = \frac{\sum\limits_{i=1}^{r} X_i w_i}{1 + \sum\limits_{i=1}^{r} X_i h_i} , \tag{1}$$

where F_r = amount of insects in diet (measured in grams,
            biomass, calories or other appropriate units)
       t = time feeding (including both searching and handing
            time
       X_i = abundance of insect i measured in captures per unit
            time searching
       h_i = handling time for insect i
       r = number of types of insects in the diet
       w_i = net value of an individual of insect i

Optimal diet can be determined by first ranking all types of
insects by their profitability. Beginning with the insect with
the highest profitability, insects are added to the diet until
the total net rate of intake begins to decline. This occurs when
profitability of an insect that is added to the diet is less than
the net rate of intake of the diet excluding that insect (Emlen,
1966; Schoener, 1971; MacArthur, 1972; Pulliam, 1974). Optimal
diet includes all insects of higher profitability than this last
one. This diet will yield the maximum rate of food intake in the
least amount of feeding time. This is shown for a hypothetical
example based on experimental studies of white crowned sparrows
(*Zonotrichia leucophrys*)(Table I). As insects are added to the

TABLE I.   Hypothetical Diets Predicted by Multispecies
Handling Time Model[a]

Insect i	Abundance X_i (no./sec)	Weight w_i (mg)	Handling time, h_i (sec)	Profitability w_i/h_i ·(mg/sec)	Intake[b] F_r/t (mg/sec)
A	0.05	50	10	5	1.67
B	0.1	20	5	4	2.25
C	0.4	15	5	3	2.63
D	0.6	6	3	2	2.43
E	0.9	3	3	1	1.98

[a]Parameter values are similar to actual values measured in
experimental studies of white crowned sparrows (Garton, 1976).

[b]Food intake of diet calculated from equation (1) including
insect i and all items of higher profitability.

diet beginning with A and proceeding through C, the food intake
calculated from equation (1) increases. The profitability of
insects D and E are less than the net rate of intake of the diet
excluding them so that the food intake declines when D and E are
added to the diet.
        If the diet of insectivorous forest birds is optimal accord-
ing to this theory, then certain qualitative predictions concerning their diet result. Changes in profitability of an insect
either through its net value, its handling time or both may lead
to including it in or excluding it from the diet. For example,
for insects without involved escape mechanisms (e.g., pupae,
larvae of coleoptera, scale insects) handling time probably
increases in proportion to length while weight increases in proportion to their volume (Royama, 1966: 1970). Profitability of
such insects is likely to increase as they mature. Thus insect
larvae which are not present in the diet of a bird in the
insect's early instars may be expected to occur in the diet in
later instars. This phenomenon has been reported (Royama, 1970;
Tinbergen, 1960).
        For foods brought to nestlings, handling time per item is
much longer than for the same food consumed by an adult. Extra
time may be spent in preparing the insect (dewinging, mastica-
tion), in carrying the food to the nest and in feeding it to
nestlings. Because handling time is longer, profitability of
foods brought to nestlings is lower. The effect of this is shown
in the hypothetical example when handling time for each type of
insect is increased by 10 seconds (Table II). In this case, the
optimal diet includes only the most profitable insects, A and B.
Thus optimally foraging birds would be expected to bring only the
most profitable items in the adult birds' diet to nestlings.

TABLE II.   Hypothetical Diets Reflecting Increased Handling
Time

Insect i	Abundance X_i (no./sec)	Weight w_i (mg)	Handling time, h_i (sec)	Profitability w_i/h_i (mg/sec)	Intake F_r/t (mg/sec)
A	0.05	50	20	2.50	1.25
B	0.1	20	15	1.33	1.29
C	0.4	15	15	1.00	1.11
D	0.6	6	13	0.46	0.82
E	0.9	3	13	0.23	0.58

These would commonly be the largest items in the diet.   This
pattern was observed by Royama (1966, 1970) and Davies (1977).
The same pattern would be expected for foods brought to
fledglings and to a mate, although in such cases the diet might
not be as specialized because less preparation would be needed.
Time required to carry prey to the nest or to fledglings or to a
mate will vary with distance at which the bird is foraging.   If
it is foraging close to the nest then the optimal diet will include a wider size range of items than if the bird were foraging
far away from the nest.   Davies (1977) observed this pattern in
food brought to female spotted flycatchers (*Muscicapa striata*) by
courting males.
     Optimal foraging theory predicts that the diet of a bird will
change with abundance of food in a specific manner.   Whether or
not a particular type of insect occurs in the diet depends not
upon its own abundance, but upon the abundance of all insects of
higher profitability.   If the abundance of more profitable insects
increases, we would expect the diet to become more specialized
with the least profitable insects being dropped from the diet
(Table III).   The actual proportion of the diet made up by a

TABLE III.   Hypothetical Diets Reflecting Increased Abundance
of the Most Profitable Insect Type

Insect i	Abundance X_i (no./sec)	Weight w_i (mg)	Handling time, h_i (sec)	Profitability w_i/h_i (mg/sec)	Intake F_r/t (mg/sec)
A	0.15	50	10	5	3.0
B	0.1	20	5	4	3.17
C	0.4	15	5	3	3.10
D	0.6	6	3	2	2.43
E	0.9	3	3	1	1.98

profitable insect type will change with abundance of that insect relative to all other profitable insects, but whether or not it occurs in the diet at all will depend upon abundance of more profitable insects. This pattern has been observed experimentally (Krebs *et al.*, 1977) and in the field (Tinbergen, 1960; Davies, 1977; Goss-Custard, 1977).

Optimal foraging theory yields quantitative predictions concerning composition of the diet which could be extremely useful in the fields of pest and wildlife management. At present neither theory nor field techniques for measuring necessary parameters appear to be sufficiently operational to make this practical. This will be obvious from examining one of these quantitative predictions and the possible reasons for its error. Given the assumptions outlined above, Pulliam (1974) demonstrated that optimal foraging theory predicts that an item will either be accepted every time it is encountered or rejected every time, but not accepted occasionally and rejected occasionally. Both experimental and field studies have shown that birds and other animals do not show this behavior pattern (Royama, 1970; Garton, 1976; Krebs *et al.*, 1977; Goss-Custard, 1977; Davies, 1977). Possible explanations are that the birds are not foraging optimally, that the theory is incorrect, or both. I will discuss these possibilities by examining assumptions upon which the optimal diet is based.

The optimal diet model assumes that insect abundance remains constant. Since the bird is consuming insects it is unlikely that this will be completely correct. Optimal search strategies attempt to maintain a constant encounter rate by avoiding repeated searching in the same area but only rarely will it be possible to do this. Depletion of prey will occur to a greater or lesser degree in most cases. Competing insect predators, weather changes, and behavioral changes associated with maturation of insects will also cause insect abundance to change. Given a changing food environment, the optimal diet may be expected to include some insects of low profitability at certain times and not at other times resulting in an overall diet showing partial preferences.

The second assumption is that net value of a potential prey species can be measured in terms of a single currency. This assumes that a bird's diet optimizes only one thing. For insectivorous birds the question is whether insect species differ much in their content of net digestible calories, protein or other nutrients. If they do, then the form of the optimal diet model must be modified to optimize intake of more than one component of the food at a time, or to place constraints on the optimization of one key component. Pulliam (1975) demonstrated that such constraints lead to an optimal diet in which some insects containing higher quantities of essential nutrients are occasionally taken. There is definite evidence that female birds consume unusual foods high in calcium during the egglaying period

(MacLean, 1974; Jones, 1976; Davies, 1977). The preponderance of insects in food brought to nestlings by granivorous birds probably represents selection for protein content. These obser- vations suggest that diets of insectivorous forest birds may be optimizing intake of more than one component of the food.

The most untenable assumption of the optimal diet model is that the bird has perfect knowledge of net value, handling time and abundance of each edible insect type. Obviously a bird must sample an insect type to determine its value and handling time. Thus no bird can show the perfect optimal diet. Similarly a bird's perception of the abundance of available food types is based upon its encounters with these foods. The generally stochastic nature of such encounters will lead to a varying per- ception of abundance and therefore a less than optimal diet. Finally as a bird encounters a type of insect more frequently, its skill at finding (development of a search image; Tinbergen, 1960) and capturing the insect may increase. Thus learning may lead to changes in both the abundance and profitability of an insect.

An appropriate question to ask of such a theory is whether or not it is reasonable to expect the animal to be able to gather, remember, and analyze the necessary information in order to make proper decisions. Extensive experimental studies have led to ambiguous answers concerning abilities of animals to detect and select for nutritional components of their food (Marten, 1969). For some components, such as carbohydrates, protein and calcium, birds may select items containing high quantities of these com- ponents, while for other components this may not be possible. Studies by Willson (1971) demonstrate that granivorous birds, at least, do prefer foods which yield the highest caloric intake per unit handling time and that this varies with the size of the bird's bill and size of the seed.

The choice of which foods to include in the diet is not as complicated as it first appears. If we consider only two foods and two possible optimal diets, the above assumptions led at least three authors (Emlen, 1966; Schoener, 1971; Pulliam, 1974) to conclude that the predator should specialize in prey 1 if:

$$\frac{\text{net value of prey 1}}{\text{time to find and handle prey 1}} > \frac{\text{net value of prey 2}}{\text{time to handle prey 2}} \quad (2)$$

If we assume for the moment that net values of two prey types are equal then the optimal predator rejects prey 2 if time required to handle prey 2 is greater than time required to find and handle prey 1. This decision is extremely simple. If we now include the values of the prey types as in equation (2) the decision be- comes a little more complex. We might expect that an insectiv- orous bird would make this decision properly if there is sub- stantial difference between profitability of the two foods, but

not when the difference is small. Krebs *et al*. (1977) found
that great tits (*Parus major*) specialized on a more profitable
prey in an experimental situation when the specialized diet was
40 to 60 percent more profitable.

III. OPTIMAL ALLOCATION OF FORAGING TIME

Where should an insectivorous bird forage within the area of
the forest that it inhabits? If we assume that there are easily
distinguished and identifiable homogeneous patches available,
that these patches are unlimited in number and unchanging in
their characteristics, and that the bird knows the rate of food
intake from the optimal diet in each patch type, then the best
strategy is to feed only in the patch type yielding the highest
food intake. Experiments have shown that birds rarely do this
(Smith and Dawkins, 1971; Smith and Sweatman, 1974; Krebs *et al*.,
1974, Zach and Falls, 1976; Garton, 1976). Instead they generally
allocate most foraging time to the best patch, less to poorer
patches and least time to the poorest patch types. This behavior
has also been observed in the field (Royama, 1970; Davies, 1977;
Garton, unpubl. data). It suggests that the optimal strategy is
not possible. Two primary reasons for this are that patches are
always changing and are often indistinguishable.

A bird must sample a patch type in order to determine its
profitability and it must do this repeatedly when profitability
of different patches is changing. For these reasons the short
term strategy of foraging only in the best patch type is both
impossible and suboptimal in the long term. This is particularly
true when the predator cannot tell what it will find in a patch.
In this case the animal must search some in a new patch and
decide whether or not to stay in that patch. This is part of the
larger question of the optimal allocation of time to patches.

Where the bird is depleting prey in its patch it must decide
when to leave the patch. An optimal foraging strategy developed
for this case leads to the conclusion that the bird "should leave
a patch when its rate of food intake in the patch drops to the
average rate of the habitat" (Pyke *et al*., 1977:145). Behavior-
ally a bird can do this by leaving a patch if no food is obtained
in a certain amount of time after the last capture (the "giving-
up time," Krebs *et al*., 1974). This leads to qualitative predictions that a bird should have a constant giving-up time for all
patches within a habitat and that giving-up time should be
shorter in better habitats where average capture rate is higher.
Krebs *et al*. (1974) found both of these predictions to be true
for tits foraging on mealworms in artificial pine cones. More
quantitative predictions are possible but have yet to be tested
in field or experimental studies (Murdoch and Oaten, 1975).

IV.   OPTIMAL SEARCH BEHAVIOR

Questions concerning optimal search behavior have not proved amenable to simple mathematical methods applied to other optimal foraging questions. Characteristics of the optimal search pattern such as speed of movement, frequency of turning and angle of turning depend upon the predator's sensory skills, the distribution pattern of prey and movements of the prey. All of these characteristics are very difficult to determine and to incorporate into simple models. Changing assumptions concerning these characteristics can lead to opposite conclusions. For example, if a predator is searching for a randomly or evenly distributed prey which seldom moves, then the optimal search path is one in which there is no recrossing of the path already covered (Pulliam, 1974). On the other hand, if prey have a clumped distribution, then once a prey is encountered optimal behavior is to increase the rate of turning and search the immediate area intensively (Pyke *et al.*, 1977) resulting in extensive recrossing of the search path. More quantitative analyses require complex models (Paloheimo, 1971a; 1971b). Evaluation and application of these search models seems at present a tenuous undertaking for insectivorous forest birds, given our current abilities to quantify the necessary parameters.

V.   STABILIZING EFFECTS ON INSECT POPULATIONS

Forest birds respond to changes in insect population density through changes in the number of insects consumed per bird (functional response; Holling, 1959) and changes in bird density (numerical response). Optimal foraging theory specifies a form of functional response which leads to birds exerting a stabilizing effect on insect populations by causing positively density-dependent mortality in some cases and a destabilizing effect where negatively density-dependent mortality results.

Equation (1) can be decomposed into a series of equations specifying the functional response of a bird searching in a homogeneous area for each of a number of insect species. These equations take the following form:

$$\frac{N_i}{t} = \frac{X_i}{1 + \sum_{i=1}^{r} X_i h_i} ,$$

where N_i = number of insects of species i consumed.

In this form of functional response the consumption rate increases as abundance of the insect increases but at a de-
celerating rate until a maximum consumption rate is reached.
The percentage predation on the insect declines as its abundance
increases. This form of functional response tends to destabilize
the insect population. Other factors would have to affect the
form of functional response for the optimally foraging bird to
have a stabilizing effect in this situation. If the insect at
low density had a refuge from bird predation then the functional
response would be sigmoid (Murdoch and Oaten, 1975). A sigmoid
functional response results in positively density-dependent
mortality on the insect population up to a saturation density
above which percentage mortality begins to decline (Oaten and
Murdoch, 1975). If the birds learn to search for and capture an
insect more rapidly as it increases from low to high densities
(Tinbergen, 1960), the functional response would take a
stabilizing sigmoid form also.

In sharp contrast to the situation in a homogeneous environment, birds foraging optimally in a patchy environment will tend
to have a strongly stabilizing effect on the insect species that
they consume. When a profitable insect species increases in
abundance in one type of patch, the optimally foraging bird will
increase the percentage of its time spent foraging in that patch
type until it has depleted that patch type down to a level comparable with other patch types. This switching onto profitable
patches will oppose fluctuations in insect abundance (Royama,
1970; Murdoch and Oaten, 1975).

The total interaction between a bird predator and its insect
prey populations is determined by a number of components in
addition to the functional response. Even if the functional
response is destabilizing, the overall interaction may be
stabilized by a variety of mechanisms such as a positively
density-dependent numerical response of the bird population to
insect abundance, invulnerability of one or more age classes of
insects to bird predation and self-regulatory mechanisms operating in the insect population (refer to review by Murdoch and
Oaten, 1975). Taking all these factors into account, optimal
foraging theory suggests that birds will be most likely to play
a major role in stabilizing the most profitable insects they
encounter in patchy environments and least likely for the least
profitable insects in homogeneous environments.

VI. CONCLUSIONS

Optimal foraging theory appears useful in understanding the
foraging behavior of insectivorous forest birds. It leads to a
number of qualitative predictions with regard to dietary
specialization and allocation of foraging time which have been

verified by experimental and field studies. Use of optimal
foraging theory to make quantitative predictions about the
foraging of insectivorous birds awaits two developments: (1) a
better understanding of the cognitive abilities of these birds
and (2) development of a more realistic theory incorporating
nutritional constraints and the birds' imperfect and changing
knowledge of prey abundance and profitabilities.

ACKNOWLEDGMENTS

    I would like to thank J. Kroll, J. Jackson, D. Stauffer,
D. Kaumheimer and D. Demers for helpful reviews of the manuscript. The content of the paper benefited greatly from discussions with G. Salt, T. Foin, K. Hopper, and J. Sedinger.
All errors and omissions are, of course, my own.

REFERENCES

Charnov, E. (1976). *Theor. Pop. Biol. 9*, 129.
Davies, N. (1977). *Anim. Behav. 25*, 1016.
Emlen, J. (1966). *Am. Nat. 100*, 611.
Estabrook, G., and Dunham, A. (1976). *Am. Nat. 110* , 401.
Garton, E. (1976). "The functional response of white-crowned
    sparrows in single-species and multiple-species prey
    environments." PhD Thesis. Univ. of Calif., Davis.
Goss-Custard, J. (1977). *Anim. Behav. 25*, 10.
Holling, C. (1959). *Can. Entomol. 91*, 385.
Jones, P. (1976). *Ibis 118*, 575.
Krebs, J., Ryan, J., and Charnov. E. (1974). *Anim. Behav. 22*,
    953.
Krebs, J., Erichsen, J., Weber, M., and Charnov, E. (1977).
    Anim. Behav. 25, 30.
MacArthur, R. (1972). "Geographical Ecology." Harper and
    Row, New York.
MacArthur, R., and Pianka, E. (1966). *Am. Nat. 100*, 603.
MacLean, S. (1974). *Ibis 116*, 552.
Marten, G. (1969). "Measurement and significance of forage
    palatability. Proceedings of the National Conference on
    Forage Quality Evaluation and Utilization." Nebraska Center
    for Continuing Education, Lincoln.
Marten, G. (1973). *Ecology 54*, 92.
Murdoch, W., and Oaten, A. (1975). *Adv. Ecol. Res. 9*, 1.
Oaten, A. (1977). *Am. Nat. 111*, 1061.
Oaten, A., and Murdoch, W. (1975). *Am. Nat. 109*, 289.
Paloheimo, J. (1971a). *Biometrika 58*, 61.
Paloheimo, J. (1971b). *Math. Biosci. 12*, 105.

Pulliam, H.   (1974).   *Am. Nat. 108*, 59.
Pulliam, H.   (1975).   *Am. Nat. 109*, 765.
Pyke, G., Pulliam, H., and Charnov, E.   (1977).   *Quart. Rev. Biol. 52*, 137.
Rapport, D.   (1971).   *Am. Nat. 105*, 575.
Rapport, D., and Turner, J.   (1977).   *Science 195*, 367.
Royama, T.   (1966).   *Ibis 108*, 313.
Royama, T.   (1970).   *J. Anim. Ecol. 39*, 619.
Schoener, T.   (1971).   *Ann. Rev. Ecol. Syst. 2*, 369.
Smith, J., and Dawkins, R.   (1971).   *Anim. Behav. 19*, 695.
Smith, J., and Sweatman, H.   (1974).   *Ecology 55*, 1216.
Tinbergen, L.   (1960).   *Archs. Neerl. Zool. 13*, 265.
Willson, M.   (1971).   *Condor 73*, 415.
Zach, R., and Falls, J.   (1976).   *Can. J. Zool. 54*, 1863.

FACTORS CONTROLLING FORAGING STRATEGIES OF INSECTIVOROUS BIRDS

Thomas C. Grubb, Jr.

Department of Zoology
The Ohio State University
Columbus, Ohio

Factors are reviewed which are known or suspected to exert control over the foraging techniques of insectivorous birds. Optimal foraging strategies are postulated as describing the behavioral baseline subject to proximate modification by the internal factors of sex, age, intraspecific dominance status and hunger state, and the external factors of food distribution and abundance, habitat type, weather, interspecific social environment, proximity of predators, time of day, season of the year and age of young. Future research will involve controlled studies of the effects of individual factors on foraging. Important, also, will be the effort to refine the current first-generation models of optimal foraging.

I. INTRODUCTION

Why do birds look for food the way they do? Previous work has addressed this question both on the community level by examining ways that members of foraging guilds (Root, 1967) have evolved methods of partitioning resources to reduce competition, and on the species level by considering proximate controlling variables.

This paper will examine those factors which have been shown to modify foraging techniques at the proximate level. Also

Thomas C. Grubb, Jr.

considered will be potential control of foraging by other agents,
as yet unexplored in birds. In keeping with the theme of this
symposium, I will use insectivorous birds of forested habitats as
illustrative examples wherever possible and where these are
lacking, will wander no further taxonomically than necessary. My
intention is not so much to provide an extensive review of the
literature, but more an effort to illustrate categories of
variables known or suspected to determine how birds look for
insects. I will suggest topics and methods where research effort
could prove particularly fruitful.

II. OPTIMAL FORAGING: THE BASELINE STRATEGY

By stating above that various components, external and
internal to a bird, act as proximate modifiers of its food-
searching behavior, I imply that some behavioral baseline, subject
to such modification, must exist. The concept of optimal foraging
appears likely to explain such an ultimate, i.e. evolutionarily
determined, ground plan.

As close attention to optimal foraging strategies is beyond
the scope of this paper (cf., Schoener, 1971; Charnov, 1976), I
will only describe the theory briefly as it relates to resource
patch use and diet selection. In general, optimal foraging
models rest on the assumption that predators are geared evolutionarily to maximize their average rate of prey capture. They
do this by measuring and discriminating among alternative sources
of energy. Furthermore, predators "figure" into the equation not
only rates of energy intake, but also rates of energy expenditure
(measured, by human observers at least, in terms of search time,
pursuit time or handling time) involved in obtaining prey from
various sources. Royama (1970) presented evidence that Great Tits
(Parus major) apparently compared resource patches within English
woodlands and concentrated on those furnishing more calories of
insect prey per unit hunting effort. MacArthur and Pianka (1966)
have furnished a general model of optimal patch use resting on
relationships of energy and time. Specifically, alternative
habitat patches which a predator might search are ranked by their
energy yield over a given search time. Total number of patches
to be searched was thought to represent a compromise between
reducing travel time between patches as new ones are added to the
search route and lower energy garnered from searching new, lower-
quality patches.

Charnov (1973; 1976) brought forward a model for optimal diet
selection among alternative prey types (within a given habitat
patch). Here, a predator is presumed to select prey types

according to their rank in a hierarchy of energy yield per unit time of hunting (E/T_{hu}). An important prediction was that the decision to take a prey item of a given E/T_{hu} depended only on the summed densities of alternative prey having higher E/T_{hu} rank. If we assume a direct correlation between energy content of prey items and their body size, then above some threshold density of large prey, a predator should ignore smaller prey even if the latter are very abundant. Among insectivorous birds, this prediction has been confirmed with captive Great Tits (Krebs et al., 1977). Five birds were exposed to large and small mealworms on a passing conveyor belt. When the belt was driven slowly, worms came into range of the tits at a low rate and both worm sizes were eaten about equally. However, when the belt was speeded up and thus more worms per unit time became available, birds, with the exception of one individual, showed a distinct preference for the larger worms irrespective of their encounter rate with the smaller ones (Fig. 1). Goss-Custard (1977; 1978) reported very similar results from studies of diet selection in a shorebird, the Red-shank (Tringa totanus).

A good deal of the remainder of this paper could be construed as an attempt to present difficulties to theories of optimal foraging. My intention is to suggest that consideration of energy inputs and time expenditures alone often will be unable to predict the "realized" strategies of foraging birds. I propose that optimal foraging tendencies may furnish the rough form of behavior which is then molded by proximate modifiers into final form. For convenience, I divide such proximate control agents into two

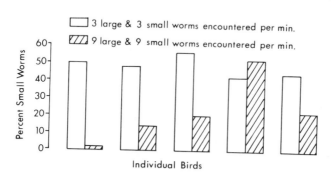

FIGURE 1. Selection of mealworms by size in captive Great Tits. The prediction from optimal foraging is supported that the decision to take a prey item of a given energy yield per hunting time (E/T_{hu}) depends only on the density of higher E/T_{hu} rank prey. (Constructed from data in Krebs et al., 1977)

groups, those internal and those external to the forager. As we
go along, I will note that this dichotomy is often blurred by
interrelationships between the two groups of factors.

III. INTERNAL FACTORS

A. Sex

Some insectivorous birds such as woodpeckers exhibit
considerable sexual dimorphism (e.g., Selander, 1966). From the
many examples of concomitant dimorphism in foraging strategies,
I will draw upon Hogstad's (1976) study of three-toed woodpeckers
(Picoides tridactylus). During the Norwegian winter, males
looked for insects by scaling bark on the main trunks of spruces
while females used several techniques (scaling, pecking, gleaning)
on a diversity of substrate heights and diameters (Fig. 2). More
recently, Hogstad (1978) extended coverage to three Dendrocopos
woodpeckers and found an inverse correlation between sexual
dimorphism and intersexual overlap of foraging niches.

We do not know how sex-specific foraging arises during
ontogeny nor do we know how it is controlled during adulthood.
Do sons learn how to forage from their fathers and daughters from
their mothers? It is intriguing that juvenile plumage among
woodpecker species is unusual in being, almost without exception,
sex-specific.

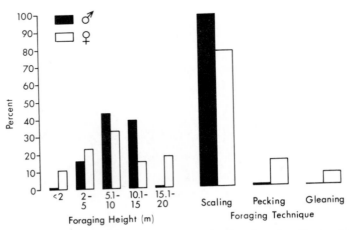

FIGURE 2. Sex-specific foraging strategies in three-toed
woodpeckers wintering in Norway. (Constructed from data in
Hogstad, 1976)

In adults, is sex-specific foraging controlled by social behavior? What would happen, for instance, if a female three-toed woodpecker had a habitat to herself rather than sharing it with a dominant male? Largely because woodpecker sexes can be identified in the field, this taxon has proven fertile ground for sex-specific foraging studies. Very little is known about this internal control factor in other avian insectivores (but see Morse, 1968).

B. Age

It seems reasonable to envision a gradual increase in foraging competence with increasing age and experience. Although documentation appears lacking for woodland insectivores, this hypothesis has been buttressed in several other birds (e.g., pelicans by Orians, 1969; herons by Recher and Recher, 1969). An interesting case with an open-country insectivore is Goldman's (1975) study of age-dependent foraging in the Eastern Bluebird (Sialia sialis). Not surprisingly, adults were significantly more successful per attempt at capturing insects than were young birds less than two weeks post-fledging. More stimulating is our glimpse of a developing strategy of optimal diet selection. While adults perched 95 seconds between chases of prey, the youngsters averaged only 71 seconds between capture attempts. Apparently, young were attempting to catch prey items outside the adults' optimal set. Prospects seem bright for developmental studies of optimal foraging in insectivorous birds.

C. Intraspecific Dominance Status

During the non-breeding season, a number of forest insectivores coalesce into rather large flocks in which stable intraspecific dominance hierarchies may occur (e.g., Dixon, 1963). Whether an individual's status in such an intraspecific hierarchy influences its foraging behavior appears unknown for wild birds. However, some evidence from laboratory studies of Great Tits is suggestive. Krebs et al. (1972) found that hand-reared tits readily formed dominance hierarchies in an aviary, and that the dominant birds visited more food baits per minute than did subordinates. A project checking the relationship between dominance status and foraging techniques in individually marked wild birds would be welcome.

D. Hunger State

Whether hunger state modifies foraging strategies of any birds in nature is unknown, partially due, I suspect, to formidable barriers of methodology. How does one find out how hungry a wild bird is? The possibility of hunger-state control needs mentioning if only because Charnov (1976) found it necessary to incorporate hunger state in predicting optimal diet selection by his captive mantids. As these predatory insects became hungrier, they stalked flies over greater distances. That is, hunger state partially controlled their optimal set of prey types. In birds, there seems promise in pursuing the general protocol of capturing individuals, feeding them to satiation, starving them for various time-spans under standard metabolic conditions, then releasing them where caught and watching them forage.

It is not obvious how such internal variables as sex, age, intraspecific dominance status and hunger state might operate, either singly or in concert, to influence foraging strategies of birds. The plot thickens still more when we overlay the complex of external factors also demonstrated or suspected to impinge on the way birds exploit their food resources.

IV. EXTERNAL FACTORS

A. Food Distribution and Abundance

The what and where of birds' food supplies have long been considered prime determinants of foraging strategies. Theories of foraging by optimization briefly described earlier as well as concepts of functional response (Holling, 1965), foraging by expectation (Gibb, 1961), search images (Tinbergen, 1960) and oddity selection (Mueller, 1977) all assume that predators respond to changes in the composition and locality of their food supply. Gibb's (1960) exhaustive study of several English titmice (Parus sp.) and the Goldcrest (Regulus regulus) will serve as an example of the kind of empirical evidence available. Over several years, locations within pine woods in which these species foraged were found to be correlated with the spatial phenology of their food supplies. To give a specific example, from July to October, Coal Tits (Parus ater) sought insect prey primarily on the dead lower portions of trees. From November on through the winter, they ranged more extensively through the upper canopy in living foliage. During March and April, this species concentrated more on taking seeds from the opening pine cones. Gibb correlated these changes in foraging with changes in the distribution and abundance of insect and plant food items.

It is worth digressing somewhat to consider nutrient content of food items as a factor controlling foraging. The major models of optimal foraging have been based on energy content of food resources. But birds may sometimes follow other than caloric criteria in constructing their diets. The predilection of Great Tits for spiders, to be detailed later in another context, appears related to some unique nutrient complex in the arachnids. Also, the herbivorous Red Grouse (<u>Lagopus</u> <u>lagopus</u>) of Scotland apparently can select heather leaves on the basis of nutrient rather than energy content (Moss, 1972). The chemical composition of heather leaves in grouse crops was compared with that in leaves sampled from the areas where the birds had been observed feeding. Heather eaten by the grouse contained significantly more phosphorous, nitrogen and calcium. Also instructive was the lack of difference in crude fat (energy) content between crop and <u>in situ</u> leaves.

To a large extent, my remaining remarks in this section speak to the influence that external factors other than those directly related to food supply exert on avian foraging behavior.

B. Habitat Type

Physiognomy and taxonomy of vegetation in a habitat appear capable of controlling avian foraging. Morse (1970) provides numerous examples of how foraging strategies of insectivorous birds changed across coniferous, mixed coniferous-deciduous and deciduous habitats. To choose one, Brown Creepers (<u>Certhia</u> <u>familiaris</u>) foraging in coniferous woods in Louisiana spent 90% of their time low (<9 m) and 8% high (>9 m) on tree trunks. In deciduous forests of the same state, this species spent only 18% of foraging low and close to 80% high. The difference here seems a case of habitat mediated foraging; creepers would have difficulty physically penetrating the maze of small branches whorling from the tops of pines. But it is conceivable that a difference in resource distribution could be partially responsible.

Somewhat afield from forest ecosystems, Brownsmith (1977) has demonstrated habitat-specific foraging rates in the Starling (<u>Sturnus</u> <u>vulgaris</u>). Birds feeding in tall grass (≥6 cm) walked more slowly, remained stationary longer at each stop and were stationary for a greater percentage of the time than those in short grass (<6 cm). Brownsmith attributed the difference in rate

of movement to tall grass physically impeding movement and/or
presenting a greater surface area to be searched for prey (Fig. 3).

C. Weather

For some years, physiologists have been concerned with
animal energetics, the dynamics of energy flow between organisms
and their environment. Four independent climatic variables,
humidity, radiation, wind velocity and temperature, surround any
terrestrial organism and control energy interchange (Porter and
Gates, 1969). Metabolic responses of animals to these climatic
variables have been studied in the laboratory. King (1974)
concluded a cogent review of animal energetics with an appeal for
studies on the importance of climatic variables to normal activities of animals. A few reports exist on how weather variation
controls foraging behavior of avian insectivores. In deserts of
the American Southwest, microhabitat selection by foraging Verdins
(_Auriparus flaviceps_) was temperature dependent (Austin, 1976).
As ambient temperature increased from 20-50° C, birds sought
insect prey in increasingly shaded microhabitats. Above 45° C,
70% of all foraging was confined to deep shade near trunks of

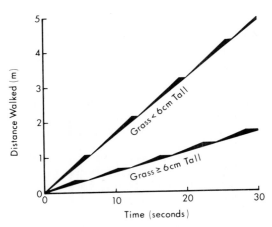

FIGURE 3. Habitat-dependent foraging rates in the starling.
Each "sawtooth" consists of a sloped component representing the
rate of an average individual movement and a flat component
denoting the time span of an average stop. The diagonal common
to all sawteeth marks the average foraging speed over 30 seconds.
(Modified from Brownsmith, 1977)

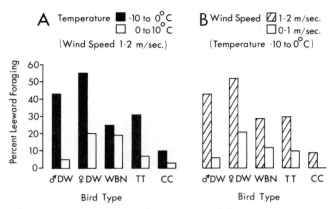

FIGURE 4. Influence of temperature (A) and wind speed (B) on extent of foraging on the leeward sides of trees by wintering male and female Downy Woodpeckers (DW), White-breasted Nuthatches (WBN), Tufted Titmice (TT) and Carolina Chickadees (CC). (Modified from Grubb, 1977)

trees. Foraging intensity was also weather dependent; number of perch changes steadily declined from about 23 per min. at $22°$ C to only 5 per min. at $43°$ C.

Grubb (1975; 1977; 1978) has recorded foraging responses to weather variation in insectivorous birds wintering in north-temperate deciduous woodlands. Downy Woodpeckers (Picoides pubescens), White-breasted Nuthatches (Sitta carolinensis), Black-capped Chickadees (Parus atricapillus), Carolina Chickadees (P. carolinensis) and Tufted Titmice (P. bicolor) were studied exclusively below their lower critical temperatures (LCT) of about $25°$ C (Helms, 1968; below LCT, metabolic rates increase autonomically to maintain body temperature). Lower ambient temperature and higher wind velocity produced significant changes in foraging height, vegetative substrates used and tree species selected. In one study in Ohio farm country (Grubb, 1977), birds were tempera-ture, wind velocity and solar radiation dependent in their horizontal distribution within woodlots. Higher winds or lower temperatures pushed them to the leeward woodlot edge and to the leeward sides of trees (Fig. 4), but higher solar radiation moderated such wind and temperature control. Carolina Chickadees were found to be excluded entirely from one mature beech-maple woodlot by temperature of -30 to $-20°$ C and wind velocities of 0.0-1.0 m/sec. They apparently took refuge from the wind in a bordering raspberry-hawthorne old-field community. Increased wind velocities and lower temperatures also significantly slowed foraging rates (Grubb, 1978). These effects were often dramatic, e.g., a 30° C increase in temperature (with wind velocity 0.0-1.0 m/sec) quadrupled foraging speed of Carolina Chickadees.

D. Interspecific Social Environment

How presence of heterospecifics might influence foraging techniques of a given bird needs more attention. To date, three main effects have been suggested: 1) foraging niche breadth is a function of social dominance, 2) foraging opportunities result from prey responses to heterospecifics, and 3) foraging behavior is imitative.

Morse (1974) reviewed effects of interspecific competition and social dominance on foraging niche width, finding that across several taxa of fishes, birds, and mammals, subordinates narrowed their niches when in the presence of dominant species. In other cases where interspecific dominance was incomplete, that is where neither of two species always dominated the other socially, both species diminished their diversity of foraging behavior. Both Pine Warblers (_Dendroica_ _pinus_) and Brown-headed Nuthatches (_Sitta_ _pusilla_) wintered in longleaf pine forests of Louisiana where they incompletely dominated each other (Morse, 1967). When foraging alone, both species spent about equal amounts of time prospecting inner and outer halves of pine branches. However, in mixed flocks the nuthatch restricted its attention much more to the distal portion and the warbler to the proximal portion of branches (Fig. 5).

Charnov _et_ _al_. (1976) touched on mixed-flock foraging strategies while considering ecological implications of resource depression. Their discussion concerned adaptive value to one predator in joining another predator of different habits. Theoretically, in the process of avoiding one of the predators, prey would become vulnerable to the other. Charnov _et al_. (1976) used Moynihan's (1962) mixed bird flocks in Panamanian rain forest to support their case. They suggested that the frequent exaggerated wing-flashing and tail-flicking of core species such as the Plain-colored Tanager (_Tangara inornata_) flushed insects, thus making them vulnerable to other species in the flock. Our interest here comes in noting that such predatory strategies would involve a change in foraging behavior from that used when foraging alone. Certainly at least, species hunting together would need to compromise on their rates of movement through the habitat.

Prompted by maneuvers of woodland insectivores on the Seychelles, Indian Ocean, Greig-Smith (1978) proposed that birds may imitate the behavior of heterospecifics resulting in a convergence of foraging niches. Seychelles Sunbirds (_Nectarinia_ _dussumieri_) foraged singly in all four tree species on the island, but joined a flock of Seychelles White-eyes (<u>Zosterops</u> <u>modesta</u>) whenever one appeared. Moreover, once in such a flock, the sunbirds became white-eye-like in their foraging. They searched only one tree species rather than four and they abandoned "hovering" and "snatching" to restrict their technique to "probing." Krebs (1973) investigated cross-species imitative foraging in captive Black-capped and Chestnut-backed (<u>Parus</u> <u>rufescens</u>) Chickadees. In an aviary, the two species obviously watched each other search for food because when a bird of one species found something to eat at a given site, members of the other species put more effort into searching near the site of the find. Rubenstein <u>et</u> <u>al</u>. (1977) concluded that wild seed-eating finches in Costa Rica were incorporating cross-species imitation into their foraging strategies.

There is a striking contrast in the examples mentioned in this section. Morse (1974) detailed numerous cases where when two species were together their foraging niches diverged, while Krebs (1973), Rubenstein <u>et</u> <u>al</u>. (1977) and Greig-Smith (1978) presented cases where species converged in foraging behavior when together. Although resolution of the disparity is uncertain, it could well turn on the abundance of the food resource being sought.

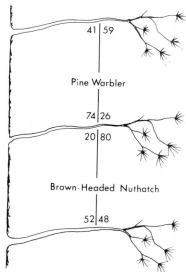

FIGURE 5. Relative percentage of foraging on the proximal and distal halves of pine branches by Pine Warblers and Brown-headed Nuthatches when alone or together with the other species. (Constructed from data in Morse, 1967 , copyright by Ecological Society of America).

E. Proximity of Predators

Essentially nothing is known of how proximity of their
predators might influence foraging behavior of birds. This is
somewhat surprising given the attention paid to predation
pressure as a possible selective factor behind flocking (e.g.,
Hamilton, 1971; Pulliamm 1973), vocalization (e.g., Marler, 1959)
and other life history phenomena. Clearly, insectivorous birds are
vulnerable to predation while foraging, particularly from
accipitrine hawks (Morse, 1973). I will wander taxonomically for
two examples suggesting that predator presence could be a significant agent controlling avian foraging. Charnov et al. (1976)
demonstrated "microhabitat depression" in mayfly nymphs in the
presence of a fish predator. The aquatic insect larvae dramatically
changed their distribution in a large aquarium when a kokanee
salmon (Oncorhynchus nerka) was introduced even though the fish
had been previously fed to satiation and made no attempt to catch
the insects. Stein and Magnuson (1976) demonstrated that proximity
of smallmouth bass (Micropterus dolomieni) influenced activity,
substrate selection, behavior and food consumption in a crayfish
(Orconectes propinquus). Predator-controlled foraging in birds is
probably an enigma because it is so difficult to keep tabs on the
predator, especially in forests. One technique warranting a
try-out would be to radio tag, say, all the Sharp-shinned
(Accipiter striatus) and Cooper's (A. cooperi) Hawks in a woodland, then monitor their whereabouts while recording the foraging
behavior of their potential avian prey.

F. Time of Day

Whether time of day controls food-seeking techniques of birds
seems very little studied. Of course, it is well known that
intensity with which prey is sought varies temporally. To cite
one example, in early April female Yellow-billed Magpies (Pica
nuttalli) spent about 90% of the hours 07:00-08:00 and 17:00-
18:00 foraging but only 30% of their behavior in that activity
during 13:00-14:00 (Verbeek, 1972).

G. Season of the Year

Travis (1977) examined substrate preferences of Downy
Woodpeckers foraging in the same woodland during winter and
summer. During winter, there was a significant shift away from
branches toward tree trunks, and toward species with more deeply
furrowed bark (Fig. 6). As no controlled experiments exist, it is

problematical whether time of day and season of year have any
effect on foraging independent of associated changes in food
distribution and abundance, weather conditons or different social
environments. Travis did not think so. He concluded that the
woodpeckers' switch to more furrowed bark substrates in winter was
a response to higher densities of insect prey found there.

H. Age of Young

One final external variable, age of young, exerts control over
what prey items parents catch and, by extension, where and how
parents forage. It is well known that adults bring progressively
larger prey to the nest as their young grow (e.g., Royama, 1970).
Unfortunately, there appears no clear documentation separating
this change in item selection from change in resource composition.
However, the remarkable and functionally mysterious preference for
spiders by parenting Great Tits is a useful exception. Across
eight nests, young tits aged three to seven days were fed over 50%
spiders by their parents; nestlings older or younger than this
received less than 20% spiders. As Royama (1970) concludes, "...so
consistent a trend....implies that it is not related to variations
in the abundance of the animals but rather that spiders have some
special nutritional value, important for the growth of nestlings
at an early stage, which is not found in other types of food."

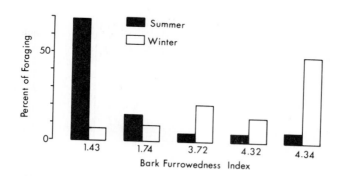

FIGURE 6. Relation between extent of bark furrowedness and
tree species (0 = perfectly smooth bark) and their use as
foraging sites by Downy Woodpeckers during summer and winter in
the same habitat. (Constructed from data in Travis, 1977)

Thomas C. Grubb, Jr.

V. DIRECTIONS OF FUTURE RESEARCH

Control of foraging is clearly a complex subject. The above categories total twelve and there are almost certainly others. Without exception to date, workers have homed in on one or two factors in searches for behavioral effects while ignoring or deemphasizing all the others.

In the forseeable future, work on foraging control will likely focus on more careful study of the influence of individual variables. The more complicated question of how independent variables interact in modifying foraging techniques should also receive attention.

Two methodologies, multivariate analysis of foraging in nature and records of foraging in captive birds under semi-natural conditions, have potential for isolating single-variable effects. The multivariate approach has not been applied as yet, but as an example, multiple regression analysis appears promising. Here, variation in predator proximity, weather, social milieu, habitat, and so on, accompanying each record of foraging could be noted. Analysis would furnish extent of correlation between each independent variable and any given foraging response of interest. Grubb (unpubl. data) found that 27.6 percent of variance in foraging rate of Carolina Chickadees in winter was accounted for by fluctuations in wind and temperature. Presumably, if other independent variables had been quantified and subjected to similar step-wise multiple regression, much of the residual, unexplained variance could have been accounted for.

A recently completed study (Grubb ms) anticipates that aviary foraging of newly-caught wild birds can give us useful leads to processes in nature. In an aviary, isolate Carolina Chickadees and Tufted Titmice selected substrates on which to forage from among 60 American elm (Ulmus americana) branches of six different diameters, but of equal lengths. Food distribution and abundance, habitat type (e.g., tree species and the height, orientation and abundance of branches), weather, proximity of predators, intra- and interspecific social environment, season of the year and time of day were controlled.

Within each species, there were significant variations among individuals in branch sizes and surfaces (top, side, bottom) used, and in the efficiencies with which new branches were incorporated into the search pattern (new branches used/total of previously unsearched branches). In each species, the mean proportions of branch sizes and surfaces foraged on differed significantly from expectations based on total lengths or total surface areas: intermediate sized branches and branch tops were used more than expected from chance (Fig. 7). Branch selection efficiency was significantly lower than expected if landings had been at random.

In the chickadee, but not in the titmouse, mean proportions of branch sizes and branch surfaces landed on differed significantly with experience (landings 1-100 against landings 101-200). For both species, branch selection efficiencies declined steadily with increasing numbers of landings. Finally, chickadees used small branches and the sides and bottoms of branches significantly more than did titmice, and were significantly more efficient at locating new branches.

These results showed that branch selection patterns differed between two species members of a foraging guild under conditions where confounding extrinsic variables were controlled. The next step will be to examine similarly nuthatch (Sitta), woodpecker (Picoides, Melanerpes) and creeper (Certhia) members of the guild. Then with this behavioral standard, we can begin analyzing foraging responses to controlled variation in the internal and external factors described above. Although quantitative identity between these laboratory findings and the behavior of birds in nature cannot be assumed, ways in which foraging strategies are modified by environmental conditions should be made clearer.

More sophisticated techniques will be required to answer the question, how do internal and external variables interact during

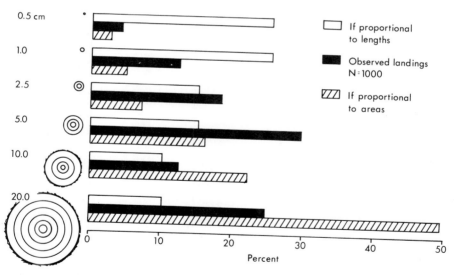

FIGURE 7. Branch sizes (drawn to scale) selected by Carolina Chickadees foraging in an aviary, and distributions expected if landings had been proportional to total lengths or total surface areas of the six branch diameters. (Constructed from data in Grubb, ms)

foraging control. For instance, fitness to be gained by feeding
nestlings at almost any cost may cause birds to ignore weather
stressful enough to control significantly their foraging behavior
outside the breeding season (M. L. Cody, pers. comm.).

In conclusion, several factors have been identified which
structure foraging techniques. Results of work with disparate
taxa indicate that others (e. g., hunger state, proximity of
predators) may also be important. We are now entering a new era
of investigation where more carefully designed studies will pinpoint
more precisely operationally important factors while controlling
others in the variable complex. Important, also, will be work
delineating how optimal foraging strategies are influenced by
external variables. Grubb's (1975l; 1977; 1978) results, for
instance, strongly suggest that optimal patch use and searching
strategies of several woodland birds are profoundly modified by
winter weather. The goal of this process, of course, is the
prediction of avian foraging strategies given any set of
environmental and internal conditions.

REFERENCES

Austin, G. (1976). Auk 93, 245.
Brownsmith, C. (1977). Condor 79, 386.
Charnov, E. (1973)."Optimal foraging theory: some theoretical
    explorations." Ph.D. Thesis, University of Washington.
Charnov, E. (1976). Am. Nat. 110, 141.
Charnov, E, Orians, G., and Hyatt, K. (1976). Am. Nat. 110, 247.
Dixon, K. (1963). Proc. 13th Intern. Ornithol. Congr. 1, 240.
Gibb, J. (1960). Ibis 102, 163.
Gibb, J. (1961). Ibis 104, 106.
Goldman, P. (1975). Auk 92, 798.
Goss-Custard, J. (1977). Anim. Behav. 25, 10.
Goss-Custard, J. (1978). Ibis 120, 230.
Greig-Smith, P. (1978). Ibis 120, 233.
Grubb, T., Jr. (1975). Condor 77, 175.
Grubb, T., Jr. (1977). Condor 79, 271.
Grubb, T., Jr. (1978). Auk 95, 370.
Hamilton, W. (1971). J. Theoret. Biol. 31, 295.
Helms, C. (1968). Am. Zool. 8, 151.
Hogstad, O. (1976). Ibis 118, 41.
Hogstad, O. (1978). Ibis 120, 198.
Holling, C. (1965). Mem. Entomol. Soc. Canada 45, 1.
King, J. (1974). In "Avian energetics."(R. Paynter, Jr. ed.),
    P. Nuttall Ornithol. Club, No. 15.
Krebs, J. (1973). Can. J. Zool. 51, 1275.

Krebs, J., MacRoberts, M., and Cullen, J. (1972). Ibis 114, 507.
Krebs, J., Erichsen, J., Webber, M., and Charnov, E. (1977).
    Anim. Behav. 25, 30.
MacArthur, R., and Pianka, E. (1966). Am. Nat. 100, 603.
Marler, P. (1959). In "Darwin's biological work." (P. Bell ed.),
    Cambridge University Press.
Morse, D. (1967). Ecology 48, 94.
Morse, D. (1968). Ecology 49, 779.
Morse, D. (1970). Ecol. Monogr. 40, 119.
Morse, D. (1973). Ibis 115, 591.
Morse, D. (1974). Am. Nat. 108, 818.
Moss, R. (1972). J. Anim. Ecol. 41, 411.
Moynihan, M. (1962). "The organization and probable evolution of
    some mixed species flocks of neotropical birds." Smithsonian
    Misc. Collections No. 143.
Mueller, H. (1977). Am. Nat. 111, 25.
Orians, G. (1969). Anim. Behav. 17, 316.
Porter, W., and Gates, D. (1969). Ecol. Monogr. 39, 227.
Pulliam, H. (1973). J. Theoret. Biol. 34, 419.
Recher, H., and Recher, J. (1969). Anim. Behav. 17, 320.
Root, R. (1967). Ecol. Monogr. 37, 317.
Royama, T. (1970). J. Anim. Ecol. 39, 619.
Rubenstein, D., Barnett, R., Ridgely, R. and Klopfer, P. (1977).
    Ibis 119, 10.
Schoener, T. (1971). Ann. Rev. Ecol. Syst. 2, 369.
Selander, R. (1966). Condor 68, 113.
Tinbergen, L. (1960). Arch. Neerl. Zool. 13, 265.
Travis, J. (1977). Condor 79, 371.
Verbeek, N. (1972). Auk 89, 567.

THE COMPARATIVE FORAGING BEHAVIOUR OF YELLOW-THROATED AND SOLITARY VIREOS: THE EFFECT OF HABITAT AND SYMPATRY

Ross D. James[1]

Department of Ornithology
Royal Ontario Museum
Toronto, Ontario.

In southern Ontario Yellow-throated Vireos usually occupy a very open deciduous forest, but with mature trees. Solitary Vireos breed in relatively mature and dense mixed forest, allopatric to Yellow-throated Vireos. In western Pennsylvania, however, these vireos may be found sympatrically occupying deciduous forest. But detailed analysis of habitats reveals important differences, although both require relatively mature stands of trees. The two species exhibit foraging behaviour which is very similar in several respects and there appears to be less divergence where the two species are sympatric. Both take food largely from branches in the interior of trees, but the Solitary Vireo secures a greater portion of its food by hovering. Yellow-throated Vireos are not restricted to crown layer foraging, but the presence of Red-eyed Vireos appears to influence the depth of their foraging activity. Male and female Yellow-throated Vireos exhibit divergence in foraging strategies. Comments on the evolution of foraging behaviour are discussed.

---

[1] 100 Queens Park, Canada M5S 2C6

I. INTRODUCTION

The northern subspecies of the Solitary Vireo (<u>Vireo</u> <u>s</u>.
<u>solitarius</u>) has generally been considered a bird of coniferous
forests (or mixed forests) and the Yellow-throated Vireo (<u>V</u>.
<u>flavifrons</u>) an occupant of deciduous forest (Bent, 1950). Yet a
segment of the Solitary Vireo population in western Pennsylvania
occupies pure deciduous woodland in sympatry with the Yellow-
throated Vireo.

The Yellow-throated Vireo has been characterized mainly as a
gleaner of arthropods from the leaves of trees (Hamilton, 1962;
Williamson, 1971). However, in southern Ontario I found them
to be gleaning largely from the branches of trees (James, 1976)
as has been observed of Solitary Vireos (Kendeigh, 1947).

Since these two species are considered to be closely related
(Hamilton, 1958; James, 1973) an examination of habitat and foraging differences is relevant to the question of competition between
them. The extent of this competition is important to our understanding of the evolution of these species.

This paper examines the habitat occupied by these two species
where they are sympatric in western Pennsylvania, and compares
that habitat with those occupied in southern Ontario where they
live in allopatry. The foraging strategies employed by Yellow-
throated and Solitary vireos are also examined within and between
these habitats.

Neither species is considered common and Yellow-throated
Vireo numbers have apparently declined significantly in this
century. Through this study, insight is gained into the role of
these species in forest ecosystems, and the habitat management
needed if their numbers are to be maintained.

II. METHODS AND MATERIALS

Observations were made in southern Ontario between 1969 and
1971, in the months of May to July, and in western Pennsylvania
in June 1974.

A. The Study Area

Solitary Vireos were studied in Ontario about 5 km southeast of the town of Dwight, Muskoka District (45°20'N 79° 01' W) in an area of mixed forest. Yellow-throated Vireos in Ontario were studied mainly 5 km north of Campbelleville, Halton County (43°29' N 79°59' W) in open deciduous woodland. In western Pennsylvania both species were observed in or near the Alleghenay National Forest in Warren Co. (41°50' N 79°10' W). In this area Solitary Vireos also occupied mixed forest, but observations presented here were from birds whose territories were wholly within deciduous forest.

B. Measurements and Analysis

The foraging parameters recorded were those set out by James (1976). In addition the sex of the bird was noted where possible.

Data were collected on the habitat surrounding 9 Solitary Vireo and 10 Yellow-throated Vireo nests in Ontario and around 3 nests of each species in Pennsylvania. Habitat data were obtained using the point—quarter method of Cottam and Curtis (1956). In addition, the volume of tree crowns was estimated for the sampled trees using the technique of Balda (1969). The canopy cover of continuous forest was recorded as 90% (Park, 1931) and for more open areas by the use of a vertical viewer (Emlen, 1967) at each of the points sampled on the habitat survey. Heights of trees were obtained by triangulation with a protractor mounted on a tripod at measured distances from the tree. All distances less than 3 m were measured with a steel tape and those over 3 m with a steel tape or a stout cord marked at one meter intervals.

A statistical test of the equality of percentages (Sokal and Rohlf, 1969:608) was used to test differences in foraging methods, sites and zones. Frequency distributions of foraging heights, tree heights and crown volumes were compared using a Kolmogorov-Smirnov test (Sokal and Rohlf, 1969:573). Mean values of various habitat statistics were compared using tests for the equality of means (Sokal and Rohlf, 1969: 222 and 374 for parametric and non-parametric data respectively).

III. RESULTS

A. Habitats Occupied

The habitat occupied by Yellow-throated Vireos in southern Ontario varies from mature deciduous forest to more open woodland with scattered trees. This latter type of habitat (Fig. 1) is more typical of that occupied in Ontario and is characteristic of the Campbelleville study area. Generally there are a number of tall mature trees, with trees of varying ages scattered about, and often patches of grassland among the groves of trees.

In Pennsylvania, however, I found Yellow-throated Vireos in mature deciduous forest. The majority of trees were tall and old, with most of the live branches concentrated near the top of the trees. There may be very little understory, but the trees may

FIGURE 1. Typical habitat occupied by Yellow-throated Vireos in the Campbelleville study area, Ontario. Tall mature trees are scattered among groves of smaller trees and open areas.

also be spaced sufficiently to permit smaller trees and even grass to grow (Fig. 2). This is similar to the type of forest where Williamson's (1971) study of this species was made in Maryland.

FIGURE 2. Tall mature deciduous forest with widely spaced trees but often with little or no understory is typical of the habitat occupied by Yellow-throated Vireos in Western Pennsylvania. Photo taken with a wide angle (28 mm) lens.

Solitary Vireos in Ontario were found in mixed forest, generally on sloping ground with conifers in low lying areas gradually
being replaced by deciduous trees higher up on the slopes (Fig.
3). The forest was always of tall trees, although the trees were

FIGURE 3. Solitary Vireo habitat near Dwight, Ontario.
Closely spaced and tall, but relatively young deciduous trees
higher on the slope grade into coniferous trees in lower areas.
Some understory, notably Balsam Fir, is present.

usually, but not always, old. There was some understory, particularly young balsam fir. Such habitat is readily distinguished
from Yellow-throated Vireo habitat by the presence of coniferous
vegetation.

In Pennsylvania, Solitary Vireos occupying pure deciduous
forest were also in tall, but younger deciduous forest with some
understory (Fig. 4). The presence of understory is important for

FIGURE 4.Dense and tall, but submature deciduous forest
with some understory is typical of the habitat occupied by a
segment of the Solitary Vireo population in the area of Warren,
Pennsylvania.

nest sites, but the arboreal vegetation is always so dense that
the forest floor is nearly devoid of greenery. In either Ontario
or Pennsylvania, Solitary Vireos were never found in areas of
very young forest or where there was dense understory.

A more detailed comparison of these habitats is presented in
Table I. The density of trees per hectare is considerably greater
on Solitary Vireo territories in both localities. This reflects
the less mature nature of Solitary Vireo habitat in Pennsylvania
(trees more crowded) and the more dense forest occupied in
Ontario. But, the mean basal area per tree, while smaller, is not
greatly different on Solitary Vireo territories, indicating that
there are relatively large trees in both habitats.

In Ontario, the crowding of trees in Solitary Vireo habitat
gives a lower average crown volume per tree, but a greater overall volume per unit area than for Yellow-throated Vireos. In
Pennsylvania, the less mature forest of Solitary Vireo habitats
gives a much greater crown volume per hectare and the greater
vertical spread of foliage gives a mean crown volume slightly
greater than on Yellow-throated Vireo territories.

The mean crown volume per tree is nearly identical for
Solitary Vireo territories in Ontario and Pennsylvania. Yellow-
throated Vireo habitat in Pennsylvania is nearly twice as dense
as in Ontario. This is also reflected in the canopy cover figures.
But the crown volume of trees on Yellow-throated Vireo territories in Ontario is much greater than in Pennsylvania, indicative
of the more open nature of the habitat.

The plant species recorded on vireo territories and a modified
importance value are given in Table II. The list of species
recorded on territories in Ontario is longer for both species,
reflecting in part the number of territories censused in Ontario.
But also the habitat in Ontario is more varied. Yellow-throated
Vireo territories are open, allowing the growth of trees which
would not be found in a forest, and Solitary Vireo territories
are on mixed rather than pure deciduous forest. However, in
Ontario, half a dozen species make up the majority of the trees
in Yellow-throated Vireo habitats, and four species dominate in
Solitary Vireo habitat.

TABLE I. Values for tree densities, areas, distances, crown volumes and canopy cover on territories of Yellow-throated and Solitary vireos in Ontario and Pennsylvania.

	Ontario		Pennsylvania	
	V. flavifrons	V. solitarius	V. flavifrons	V. solitarius
Density (trees/ha)[a]	297	742	525	873
Mean area/tree (m^2)[b]	33.64	13.47*	19.05	11.46
Mean distance between trees (m)[c]	5.80	3.67	4.36	3.39
Mean basal area/tree (cm^2)[d]	567.84	548.15	446.94	395.44
Basal area/ha (m^2)[e]	16.88	40.89	23.46	34.52
Mean crown volume/tree (m^3)[f]	140.71	95.08*	80.83	96.50
Crown volume/ha (m^3)[g]	41829	70572	42436	84245
Canopy cover (%)	62	90	90	90
Sample size	160	144	48	48

a $\dfrac{\text{number of m}^2 \text{ per hectare}}{\text{mean area per tree}}$

b (mean distance)^2

c $\dfrac{\text{sum of all distances}}{\text{number of trees sampled}}$

d $\dfrac{\text{total basal area sampled}}{\text{number of plants sampled}}$

e basal area per tree x density

f $\dfrac{\text{sum of crown volumes}}{\text{number of trees sampled}}$

g crown volume per tree x density

* (P $<$ .05)

TABLE II. Tree species recorded on vireo territories, with the importance value (IV)[a] for each species divided by three, and the relative frequency of foraging (FF) in each species.

Species	Vireo flavifrons				Vireo solitarius			
	Ontario		Pennsylvania		Ontario		Pennsylvania	
	IV	FF	IV	FF	IV	FF	IV	FF
Sugar Maple (Acer saccharum)	28.7	19.3	20.5	0.0	30.8	45.0	33.9	27.6
Red Oak (Quercus ruber)			10.4	28.7	4.0	3.3	32.6	21.5
White Oak (Q. alba)	7.2	13.8	30.8	29.9		9.9[b]	17.3	33.6
Balsam Fir (Abies balsamea)					23.7	5.0[b]		
Yellow Birch (Betula allagheniensis)		.3			14.6	18.7		
Hemlock (Tsuga canadensis)	1.9				13.4			
White Elm (Ulmus americana)	12.5	17.4	2.8	3.4	0.7	0.0		
Basswood (Tilia americana)	3.3	0.3	11.5	24.0	.6	0.0	8.8	13.8
Trembling Aspen (Populus tremuloides)	7.2	7.7			1.4	0.0		
Ash (Fraxinus sp.)	3.7	5.8	8.8	2.3	3.3	0.0	7.4	3.4
Sassafras (Sassafras albidum)								
Balsam Poplar (P. balsamifera)	7.2	11.9			3.2	5.0		
White Birch (B. papyrifera)	6.0	1.3	4.9	1.1				
White Pine (Pinus strobus)	3.5	0.0	4.7	2.3				
Black Walnut (Juglans nigra)	.7	2.1	1.8	0.0				
Ironwood (Ostrya virginiana)	3.3	1.9			1.5	3.3[b]		
White Cedar (Thuja occidentalis)	2.9	0.3	2.0	8.0				
Hickory (Carya sp.)	1.2	0.3			0.6	0.0		
Beech (Fagus grandfolia)	2.0	1.0						
Cherry (Prunus sp.)	0.6	1.0	1.8	0.0				
Hawthorn (Crataegus sp.)	4.3	8.7						
White Walnut (J. cinera)	2.9	2.1						

TABLE II. Tree species recorded on vireo territories, with the importance value (IV)[a] for each species divided by three, and the relative frequency of foraging (FF) in each species.(continued)

| | Vireo flavifrons | | | | Vireo solitarius | | | |
| | Ontario | | Pennsylvania | | Ontario | | Pennsylvania | |
	IV	FF	IV	FF	IV	FF	IV	FF
White Spruce (Picea glauca)								
Apple (Malus sp.)	0.9	2.2			2.2	5.5[b]		

a – an importance value is the sum of the relative dominance, relative frequency and relative density value of the various species in each of the four areas surveyed.

b – 37.4% of foraging done in coniferous trees.

B. Foraging Behaviour

Figures for the relative use of each tree species for foraging are also given in Table II. There is not a direct correlation
between foraging frequency and modified importance value because
foraging observations may not have been sampled in the same
proportions as the trees in each territory. In general, however,
these vireos will forage in any species of tree within their
territories and use them roughly in proportion to their relative
density. The exception is perhaps a disproportionately high use
of oak trees. Solitary Vireos in Ontario, although occupying
mixed habitats, use deciduous trees for by far the largest part
of their foraging.

The relative use of different foraging methods by the two
species are outlined in Table IIIa. In Ontario, Yellow-throated
Vireos make significantly greater use of stalking, (P<.01)
while the Solitary Vireo uses hovering almost as much as stalking. However, in Pennsylvania there are no significant differ-
ces in the foraging methods used. Although the figures indicate
that there is less difference between these species in Pennsylvania, there are no significant changes between the figures for
Ontario, and Pennsylvania for either species.

The sites within trees from which food is gleaned are summarized in Table IIIb. Both species show a significantly greater use
of branches than leaves (P<.01). An increased use of branches
in Pennsylvania over Ontario may reflect that the observations
were taken early in the year when arthropod populations may be
lower on leaves. In most cases, live branches are used more than
dead ones, but live branches are also available in greater
numbers. The high use of dead limbs by Solitary Vireos in Ontario
may reflect the high number of such branches in the lower parts
of coniferous woods. The most surprising result, however, is that
the two species foraging sites are similar in Pennsylvania, whereas they are much more distinct in Ontario, where the species are
less likely to be competing for food resources.

The use of different foraging zones by Yellow-throated and
Solitary vireos is presented in Table IIIc. Both species always
make greater use of the central areas of trees to forage. In
Pennsylvania Solitary Vireos appear to make significantly greater
use of central areas than do Yellow-throated Vireos (P<.01).
This is reflected in part from the greater use of tree trunks and
branches by Solitary Vireos in Pennsylvania over Ontario (Table
IIIb).

A comparison of the foraging height of Yellow-throated Vireos
with the distribution of the tree crown volumes on their territories is presented in Fig. 5, and a similar comparison for Solitary
Vireos in Fig. 6. In all cases the birds forage significantly

TABLE III. The percentages of various foraging methods, foraging sites, and foraging zones used by Yellow-throated and Solitary vireos. Figures for tree trunks in parentheses are included in the preceeding live and dead branch categories.

A

Locality	Foraging method	% used by V. flavifrons	% used by V. solitarius
Ontario	Stalking	69.7	50.5**
	Hovering	28.8	45.4**
	Hawking	1.2	3.6
	n	322	112
Pennsylvania	Stalking	66	55 ns
	Hovering	33	42 ns
	Hawking	1	3
	n	101	128

B

Locality	Foraging site	% use by V. flavifrons	% use by V. solitarius
Ontario	Leaves	30	25
	Live branches	51	36**
	Dead branches	19	39**
	Trunks	(15.8)	(16)
	n	316	112
Pennsylvania	Leaves	21	14
	Live branches	72	74
	Dead branches	7	12
	Trunks	(6)	(24)
	n	100	128

TABLE III. The percentage of various foraging methods,
foraging sites, and foraging zones used by Yellow-throated
and Solitary vireos. Figures for tree trunks in parentheses
are included in the preceeding live and dead branch
categories. (continued)

C

Locality	Foraging zones	% use by V. flavifrons	% use by V. solitarius
Ontario	Interior	66	68
	Periphery	34	32
	n	228	57
Pennsylvania	Interior	60	81**
	Periphery	40	19**
	n	95	119

** indicates P <.01.

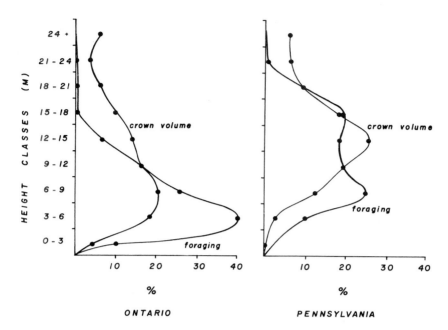

FIGURE 5. Comparison of the foraging height of Yellow-throated Vireos in Ontario (n = 320; mean = 6.7 m) and Pennsylvania (n = 101; mean = 11.7 m) with the vertical distribution of the volume of the crowns in Ontario (n = 22514 m³; mean = 10.2 m) and Pennsylvania (n = 3880 m³; mean = 14.3 m) respectively.

lower than expected on the basis of crown volume distribution (P < .01), but in keeping with their preference for branches and tree trunks.

These results do not appear to be biased by the ease of observation at lower levels. When foraging heights are expressed as percentage of tree heights (Fig. 7), there is no strong tendency to forage in any specific part of trees, and all means are above but near 50%. In Ontario, Yellow-throated Vireos foraged lower in trees than Solitary Vireos (P < .05) as might be expected from the open nature of the habitat in Ontario, but again in more nearly identical habitat in Pennsylvania there are no differences evident between the two species. In both localities however, there appears to be a bimodal distribution of foraging for both species.

The foraging heights in Ontario compared to those in Pennsylvania are given for each species in Fig. 8. Both species forage

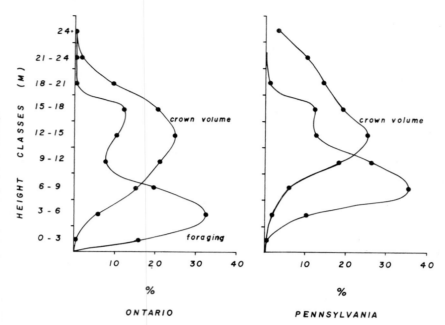

FIGURE 6 .Comparison of the foraging height of Solitary
Vireos in Ontario (n = 107; mean = 7.7 m) and Pennsylvania
(n = 128; mean = 10.0 m) with the vertical distribution of the
volume of tree crowns in Ontario (n = 13192 m^3; mean = 12.6 m)
and Pennsylvania (n = 4632 m^3; mean = 15.2 m) respectively.

significantly higher in Pennsylvania (P < .01) where habitat is
taller. The Yellow-throated Vireo forages significantly lower
than the Solitary Vireo in Ontario (P < .01) but significantly
higher in Pennsylvania (P < .01) as might be expected on the basis
of habitat differences. Again there is a tendency for a bimodal
foraging pattern by both species (except Yellow-throated Vireos
in Ontario). Such bimodal foraging patterns may be the result of
different foraging heights used by different sexes or the use of
two layers of the forest (canopy and understory) while foraging
in the same relative position in individual trees.

When foraging heights are separated by sex (Fig. 9 and Fig.
10) there are no significant differences between male and
female Solitary Vireos foraging heights in either Ontario or
Pennsylvania. However, male Yellow-throated Vireos forage significantly higher than females in both localities (P < .01). This

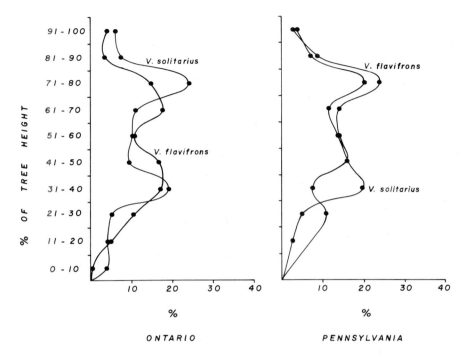

FIGURE 7. Comparison of the foraging heights of Yellow-throated (n = 291; mean = 51.2%) and Solitary (n = 99; mean = 55.7%) vireos in Ontario with Yellow-throated (n = 99; mean = 58.6%) and Solitary (n = 123; mean = 56.1%) vireos in Pennsylvania, when foraging heights are expressed as percentages of the total tree height in which the observations were made. This indicates the amount of foraging done in different parts of trees regardless of the height of trees.

seems to indicate that the bimodal foraging pattern of Yellow-throated Vireos is explained by sexual differences in foraging.

The heights of trees used for foraging compared with the heights of trees recorded on the habitat survey are illustrated in Fig. 11 and Fig. 12 for Ontario and Pennsylvania respectively. Both species in both localities apparently choose trees for foraging which are significantly shorter than would be expected on the basis of the habitat samples (P < .05).

A comparison of the heights of trees used for foraging by the two species in Ontario (from Fig. 11) shows no significant

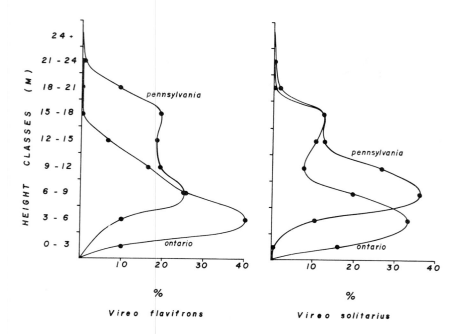

FIGURE 8 . Comparison of the foraging heights of Yellow-throated Vireos recorded in Ontario (n = 320; mean = 6.7 m) and Pennsylvania (n = 101; mean = 11.7 m) and the foraging heights of Solitary Vireos recorded in Ontario (n = 107; mean = 7.7 m) and Pennsylvania (n = 128; mean = 10.0 m).

differences, while a comparison of the trees sampled in the habitats (Fig. 11) indicates significantly taller trees in Solitary Vireo habitats (P < .01). This would indicate that the Yellow-throated Vireo tends to forage in taller trees within the occupied habitat.

Trees in Yellow-throated Vireo habitats in Pennsylvania (Fig. 12) are significantly taller than those in Solitary Vireo habitat (P < .05). A comparison of the heights of trees used for foraging by the two species shows Yellow-throated Vireos choose trees which are significantly taller than those used by Solitary Vireos (P < .01). While this is expected because of the habitats occupied, it also confirms the use of taller trees within a habitat by the Yellow-throated Vireo.

In Ontario, Solitary Vireos show a bimodal pattern of tree use which does not correspond to the habitat trees samples

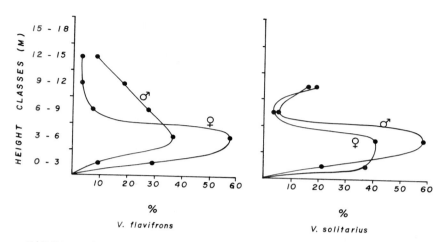

FIGURE 9. Comparison of the foraging heights of Yellow-throated Vireo males (n = 76; mean = 6.9 m) and females (n = 28; mean = 4.4 m) and Solitary Vireo males (n = 19; mean = 5.0 m) and females (n = 27; mean = 4.6 m) in Ontario.

(Fig. 11). They thus appear to be selecting canopy and understory trees, since there is no apparent difference in foraging between males and females (Fig. 9). No such trend is evident in Pennsylvania, however.

In Pennsylvania the heights of trees selected by Yellow-throated Vireos show a bimodal pattern which is not evident from the habitat sample (Fig. 12). A similar trend is apparent in Ontario, although not as pronounced (Fig. 11). Since male and female Yellow-throated Vireos tend to forage at different heights (Fig. 9 and 10) they may also use different tree heights, foraging in the same relative position whether canopy or understory trees. A comparison of the tree heights selected by male and female Yellow-throated Vireos (Fig. 13) indicates that while the female shows a bimodal pattern of use, there is singificantly greater use of lower trees by females (P < .01). This tends to confirm a selection of tall and short trees by male and female birds respectively.

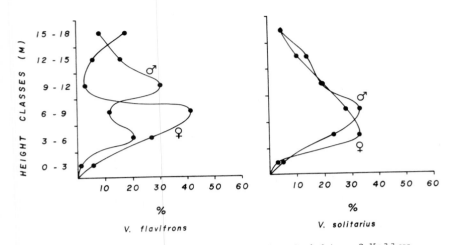

FIGURE 10.   Comparison of the foraging heights of Yellow-throated vireo males (n = 58; mean = 10.0 m) and females (n = 34; mean = 8.4 m) and Solitary Vireo males (n = 96; mean = 8.0 m) and females (n = 21; mean = 8.4 m) in Pennsylvania.

IV. DISCUSSION AND CONCLUSIONS

The closely related and morphologically similar Yellow-throated and Solitary vireos are largely allopatric in distribution except in various parts of the northeastern United States and southern Ontario. In this area the Solitary Vireo is largely restricted to higher elevations or more northerly latitudes, and occupies mainly mixed coniferous-deciduous woodlands. However, in several areas such as western Pennsylvania there is potential overlap in habitat use with both species occupying deciduous forest in close proximity to each other. But detailed analysis of the habitats occupied by these two species indicates that despite outward appearances, they occupy rather distinctly different forest configurations.

Yellow-throated Vireo habitat is always characterised by the presence of tall and old trees. These trees may be widely scattered, or closely spaced where understory is virtually lacking. Solitary Vireo habitat, while it may contain very old trees is always a dense arrangement of trees, usually tall but less mature, and with varying sizes and numbers of understory trees which are necessary to provide nest sites.

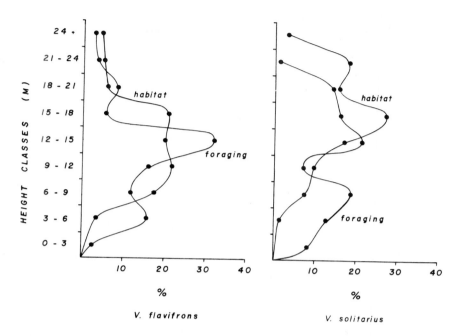

FIGURE 11. Comparison of the heights of trees in which foraging observations were recorded (V. flavifrons n = 302, mean = 11.9 m; V. solitarius n = 98, mean = 11.6 m) with tree heights sampled in the habitat survey (V. flavifrons n = 160, mean = 13.5 m; V. solitarius n = 144, mean = 16.3 m) for Yellow-throated and Solitary Vireos in Ontario.

Although Yellow-throated Vireos tend to forage in the taller trees within a habitat, either vireo species will forage in any type or size of tree within a suitable habitat configuration. It is important to note, however, that both species require relatively mature stands of trees. Since the average number of avian species in a forest tends to decline as the forest approaches a climax condition (Bond, 1957), the importance of these two species in the control of insects becomes evident. Conversely, the continuing presence of such mature stands of trees is necessary to maintain viable populations of these species.

In terms of foraging methods, sites and zones, Solitary and Yellow-throated vireos are very similar. Surprisingly, there appears to be less divergence in foraging strategies in an area where they are sympatric. This is the converse of what would be

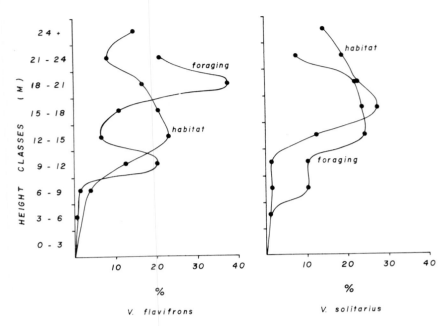

FIGURE 12. Comparison of the heights of trees in which foraging observations were recorded (V. flavifrons n = 99, mean = 17.2 m; V. solitarius n = 135, mean = 15.1 m) with tree heights sampled in the habitat survey (V. flavifrons n = 48, mean = 17.0 m; V. solitarius n = 48, mean = 18.1) for Yellow-throated and Solitary Vireos in Pennsylvania.

expected in view of competition theories, but it is not intolerable since the two occupy different habitats. One foraging difference is the increased use of the interior of trees for foraging by Solitary Vireos in Pennsylvania. But this does not seem significant as both species still forage predominately on tree branches.

A second and major point of departure between these two species is the increased use of hovering to secure food by Solitary Vireos (particularly in Ontario). Knight (in Bent, 1950) reporting on birds in Maine says that 'flycatching' (hawking) is commonly used by Solitary Vireos.While my data indicate only a slightly increased use of hawking, as compared to Yellow-throated Vireos, taking food on the wing is obviously employed more by Solitary Vireos. The two species show very similar

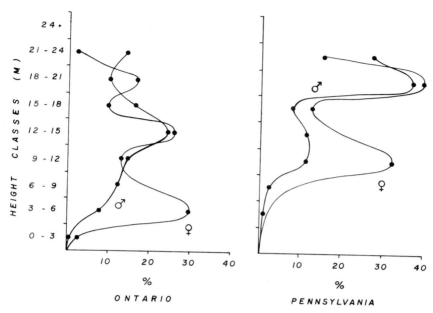

FIGURE 13. Comparison of the tree heights in which foraging observations were recorded for Yellow-throated Vireo males (n = 80; mean = 13.8) and females (n = 31; mean = 11.7 m) in Ontario, and males (n = 60; mean = 17.7 m) and females (n = 33; mean = 16.6 m) in Pennsylvania.

morphologies, including bill size and proportions, but Selander (1966) has suggested that divergences in foraging behaviour may occur without obviously correlated morphological differentiation. Indeed, Morse (1968) has shown behavioural differences in foraging between the sexes of certain warbler species which have no obviously correlated structural differences.

However, Orenstein (1970) reports comparatively weak jaw musculature in Solitary Vireos from Ontario when compared to Yellow-throated Vireos. The latter has short and heavy adductor muscles, though both species are considered heavy billed birds (among vireos). Orenstein suggested that such differences in jaw musculature cannot be correlated with habitat differences among vireos, but appear to relate to foraging preferences; that is, either the source of food or the method of obtaining it. Since no differences are evident in the diet of these two species

(Chapin, 1925), the differences in jaw musculature would appear to relate to foraging differences. Comparatively weaker muscles would be suited to faster adduction (Orenstein, 1970), increasing the accuracy and speed of securing a prey item. The more quickly a bird can return to a perch, the less energy it will have to spend in vigorous flight activity (Morse, 1968). This is of obvious advantage in reducing the expenditure of energy of a bird securing its food in flight, as does the Solitary Vireo as much as 45% of the time.

Hamilton (1958) has argued that the Yellow-throated Vireo forages strictly in the crown layer of mature deciduous forest, and as such is able to occupy, on a different stratum, the same forests as the Red-eyed Vireo (V. olivaceous). Such an impression might be gained by watching male birds in early spring. At such a time they are vocal and readily located and, as I have shown, tend to forage higher than females. This would be particularly true when they were still unmated. However, my data do not show any confinement to the upper layers of the forest, particularly by females which could easily be overlooked on casual observation. Although Williamson (1971) considered Yellow-throated Vireos to be crown layer foragers, she also reported them foraging almost to ground level in the early summer.

Both species use trees which are shorter than expected on the basis of the habitat sampled, and forage significantly lower than expected on the basis of crown volumes. This suggests further that they are not habitual crown layer foragers. Balda (1969) using similar methods of crown volume analysis and foraging height, found a disproportionately high use of the upper parts of tree crowns among foliage gleaning birds. He concluded that this 'over-use' of the upper crowns probably reflected the greater proportion of foliage (as opposed to bare branches) in the tree tops, and a truer picture of actual foliage distribution. The under use of the upper parts of tree crowns by Solitary and Yellow-throated Vireos is in keeping with a low use of foliage and a high use of branches as foraging sites.

Morse (1968) and MacArthur (1958) were able to correlate the height of foraging activity of sympatric warblers with the height of the nest above the ground level. Williamson (1971) also correlated the height at which female Red-eyed Vireos foraged with nest height. If Yellow-throated Vireos were habitual crown layer foragers, one might expect their nests to be in the crowns of trees as found for other arboreal foraging species. However, Williamson gave the range of height at which nests are situated as from 3 to 18.3 m with an average of under 9.1 m or in the range of height of the greatest amount of foraging carried out by Red-eyed Vireo males in her study area. This is inconsistent with the hypothesis that the Yellow-throated Vireo is a crown layer forager always remaining above the Red-eyed Vireo.

The presence or absence of Red-eyed Vireos overlapping Yellow-throated Vireo territories does appear to influence the height of Yellow-throated Vireo activities, however. The average nest height of 34 nests in Ontario was 11.2 m. But where Yellow-throated and Red-eyed Vireos occurred in the same forest the average was 14.4 m (n = 10) and where Yellow-throated Vireos occurred alone the average was 9.5 m (n = 24), a statistically significant difference (P $<$ .001). In the open habitat of Ontario, Yellow-throated Vireo nests were as low as 4.5 m and Sutton (1949) reported nests as low as 1.5 m. Thus the levels at which Yellow-throated Vireos forage appears to be partly dependent upon the presence or absence of Red-eyed Vireos in the same area. When Red-eyed Vireos inhabit the same woodlands, aggressive encounters between the two species, particularly in the vicinity of the nests, might keep the two more widely separated, even though the two do not compete directly for food (James, 1976). The absence of Red-eyed Vireos would constitute an 'ecological release' (Schoener, 1967) allowing Yellow-throated Vireos to increase the depth of their foraging range.

The present study and James (1976) clearly indicate that Yellow-throated Vireos do not glean predominately from foliage as suggested through casual observation by Bent (1950) and Hamilton (1962). Even in areas of tall mature forest in Pennsylvania no such restriction is seen. Williamson (1971) although indicating that Yellow-throated Vireos forage preferentially in the upper canopy, also recorded a high use of branches as foraging sites for this species.

Insectivorous birds are not dependent upon any specific food, as long as it belongs to a general type (Kendeigh, 1947:56). Therefore, if restricted somewhat to crown layers by Red-eyed Vireos where, except for singing males, they are difficult to locate, a higher proportion of gleaning from leaves might be expected. This is because of the greater availability of leaves in the upper canopy.

Numerous authors have commented on the advantages of sexual differences in foraging. There are definite advantages to having male Red-eyed Vireos forage above females as reviewed by Williamson (1971:146). However, in the case of the Red-eyed Vireo the female performs most of the nesting activities alone. For a species like the Solitary Vireo which nests close to the ground (mean height = 3.1 m; n = 24), and where both sexes share equally in nesting duties, there is no advantage to having one sex consistently flying higher to forage, and no observed differences in the foraging heights of the two sexes.

Yellow-throated Vireos on the other hand nest high in trees (mean height in Ontario = 11.2 m, n = 43; mean height in Pennsyl-

vania = 15.7 m, n = 5) and there might be considerable advantage
to have one sex forage above and the other below nest height
where both sexes share nesting duties evenly. Observed foraging
differences indicate a tendency for the sexes to forage at
different levels, but the separation is poor. The birds, therefore
do not appear to be taking full advantage of such a divergence.
However, in both study areas, population levels were low, elimin-
ating any intraspecific competition which might have further
necessitated a more precise utilization of available resources.

The mean foraging height of Yellow-throated Vireos is considerably above that of the Solitary Vireo, and corresponds more
closely to higher mean nest heights. However, mean nest heights
are above mean foraging heights (Fig. 8) further indicating a
preference for foraging below the leafy canopies where nests are
better concealed.

In the overall analysis, there is considerable similarity in
the foraging behaviour of Solitary and Yellow-throated vireos,
and one would not expect them to occupy the same habitat. Indeed,
populations which are locally sympatric are confined to separate
habitats. Hamilton (1958) postulates the origin of Yellow-throated
Vireos from Solitary Vireo stock. He further postulates that
secondary contact between these two populations initially produced selective pressure on Yellow-throated Vireos to restrict
its foraging to crown layers of broadleaf trees and subsequent
spatial separation into different habitats to complete the
isolation.

However, as the Yellow-throated Vireo is not restricted to
crown layers, and as they still forage in a very similar manner,
initial habitat separation would appear to have been more important in isolating these two species.

Since Solitary Vireos use deciduous as much as coniferous
trees, such an initial shift to purely deciduous trees would
not be a difficult step (Hinde, 1959). If the Yellow-throated
Vireo had initially been forced to occupy the upper layers of
deciduous trees, one would expect it to have modified its foraging
to glean from leaves. As this is not so, I consider that it never
confined foraging to the upper canopy, nor would this be necessary
when in sympatry with the Red-eyed Vireo, as long as modes of
foraging differed between them. Conversely, the change to foliage
gleaning would have brought the Yellow-throated Vireo into
increasing conflict with the more abundant Red-eyed Vireo, which
does forage on leaves (James, 1976).

ACKNOWLEDGMENTS

I wish to thank Dr. Jon C. Barlow for guidance during initial phases of the study. The National Research Council of Canada assisted with grants to Dr. Barlow for the study of vireos. Mrs. S. Merson typed various drafts of the manuscript, the photography department of the Royal Ontario Museum assisted in preparation of the figures, and the library of the ROM provided reference material. To them I am grateful for their assistance.

REFERENCES

Balda, R. (1969). Condor 71, 339.
Bent, A. (1950). "Life histories of North American wagtails, shrikes, vireos and their allies". U.S. Nat. Mus. Bull. 197
Bond, R. (1957). Ecological Monographs 27, 351.
Chapin, E. (1925). "Food habits of vireos, a family of insectivorous birds." U.S. Dept. Agric. Bull. 1355.
Cottam, G. and Curtis, J. (1956). Ecology 37, 451
Emlen, J. (1967). Ecology 48, 158.
Hamilton, T. (1958). Wilson Bull. 70, 307.
Hamilton, T. (1962). Condor 64, 40.
Hinde, R. (1959). Biol. Rev. 34, 85.
James, R. (1973). "Ethological and ecological relationships of the Yellow-throated and Solitary Vireos (Aves: Vireonidae) in Ontario." Ph.D. thesis, Univ. of Toronto, Toronto.
James, R. (1976). Wilson Bull. 88, 62.
Kendeigh, S. (1947). "Bird population studies in the coniferous forest biome during a spruce budworm outbreak." Ont. Dept. Lands and Forests Biol. Bull. 1.
MacArthur, R. (1958). Ecology 39, 599.
Morse, D. (1968). Ecology 49, 279.
Orenstein, R. (1970). "Variation in the jaw musculature of some members of the avian family Vireonidae." M.S. thesis, Univ. of Toronto, Toronto.
Park, O. (1931). Ecol. Monogr. 1, 189.
Schoener, W. (1967). Science 155, 474.
Selander, R. (1966). Condor 68, 113.
Sokal, R. and Rohlf, F. (1969). "Biometry". W. H. Freeman & Co., San Francisco.
Sutton, G. (1949). "Studies of the nesting birds of the Edwin S. George Reserve. Part I. The Vireos." Misc. Publ. Mus. Zool., Univ. Mich. 74.
Williamson, P. (1971). Ecol. Monogr. 41, 129.

FORAGING ECOLOGY AND HABITAT UTILIZATION
IN THE GENUS SIALIA

Benedict C. Pinkowski[1]

Ecological Sciences Division
NUS Corporation
Pittsburgh, Pennsylvania

Factors influencing foraging behavior are examined for the
three species of Sialia in a variety of habitats. Searching for
ground-dwelling prey from a perch is the fundamental predatory
tactic of each species, but the proportion of insects in the diet
and the frequency of aerial foraging vary interspecifically with
wing-loading. S. currucoides occurs in several vegetation
communities and is the most ecologically tolerant species, partly
because it may forage by hovering in areas having low perch
densities. Bioenergetic advantages of hover-foraging vis-a-vis
perch-foraging include increased capture opportunities in areas
away from perches and an enlarged perceptual field afforded by
the relatively great heights at which hovering occurs. S.
sialis hybridizes with S. currucoides but does not hover-forage;
it occurs at low elevations and is frequently found in oak
(Quercus sp.) woodlands. S. mexicana, rather limited in ecological tolerance, is usually found in ponderosa pine (Pinus
ponderosa) forests and captures more of its prey above-ground
than the other species. Two different species may occur in
intermediate habitats, in which case the lack of major differences in foraging behavior and the importance of prey located
in a two-dimensional (ground) space apparently make interspecific
territoriality adaptive. Prey abundance, perch distribution, and
density of ground cover are important habitat variables readily
identifiable by the forest manager; changes in these variables
have different effects on the various species and may alter the
relative proportions of two coexisting species.

---

[1]Research supported in part by Frank M. Chapman Memorial
Fund of the American Museum of Natural History.

I. INTRODUCTION

A. Secondary Cavity Nesting Strategy

Probably no group of birds has been as affected by forest
management practices as have secondary (= non-excavating) cavity-
nesting species, at least insofar as numbers of individual birds
are concerned. Secondary cavity nesters depend on other agents
for their nesting places and thus are influenced by snag management practices recommended by Balda (1975a) and Conner (1978) at
recent U.S.F.S. symposia. Densities of many secondary cavity
nesters are reportedly increased by events that increase the
number of suitable nest sites, including forest fires (Lloyd,
1938; Ahlgren and Ahlgren, 1960; Bock and Lynch, 1970), selective
logging (Warbach, 1958; Conner and Adkisson, 1975; McClelland,
1975; Scott, 1978), and the provision of artificial nest sites
(Hesselschwerdt, 1942; Hamerstrom et al., 1973). Most avian
species, however, respond to habitats on the basis of physio-
gnomic stimuli such as foliage profile height and foliage density
(MacArthur and MacArthur, 1961; Wiens, 1969; James, 1971; Verner,
1975), and a suitable nest site represents only one structural
cue in a constellation of features that render a particular
locale acceptable or not acceptable to a secondary cavity nesting
species.

Secondary cavity nesters evidently evolved their nesting
habit later than primary (= excavating) cavity nesters (Haartman,
1957). For reasons probably not yet fully appreciated, virtually
all secondary cavity nesting passerines of the southwestern
ponderosa pine forests are insectivores (Balda, 1975b). A
similar situation exists in the eastern hardwood and coniferous
forests, where 16 native passerine species representing seven
families regularly use cavities excavated by woodpeckers or
caused by natural decay (Scott et al., 1977). All of these
species are primarily insectivorous, as are the families to which
they belong, which suggests that most of the species began nesting in cavities at an evolutionarily later date than they
acquired their insectivorous habits. In some cases the original
transition may have been attributable to the fact that the young
of insectivorous species require greater ability on the wing for
prey capture during the post-fledging period than do those of
gramnivorous, frugivorous, or omnivorous species. Greater flight
skills by recently fledged young can be achieved by prolonging
the nestling period; the young of cavity nesters, being
relatively safe from predators, can leave the nest in a more
developed state than their open-nesting relatives (Haartman,
1957; Pinkowski, 1975a).

North American passerines that are primarily or totally
insectivorous also tend to be migratory (Morse, 1971a). Only a
limited number of nest sites are available to secondary cavity

nesters, however, and natural selection may favor individuals of these species remaining on the breeding grounds year-round (Haartman, 1968) or migrating early in spring because of intraspecific competition (Johnston and Hardy, 1962; cf. Rohwer and Niles, 1977) or interspecific competition (Brown, 1978) for nest sites. Nine of the 16 native secondary cavity nesters occurring in the eastern deciduous forests are permanent residents in at least part of their ranges (Scott et al., 1977), but seven of these (four Parids and three Sittids) feed on relatively inactive insects or consume seeds during winter. The various selective pressures involving diet, nesting habit, and migration strategy have no doubt resulted in very delicate balances in the remaining two species, the Carolina Wren (Thryothorus ludovicianus) and Eastern Bluebird (Sialia sialis), as indicated by the sensitivity of both species to severe winters (James, 1962; Robbins and Erskine, 1975; Van Velzen, 1978).

B.   The Avian Genus Sialia

In addition to the Eastern Bluebird, the genus Sialia consists of two other morphologically and ecologically similar species endemic to the New World. The Mountain Bluebird (S. currucoides) hybridizes locally with the Eastern Bluebird (Lane, 1968, 1969) and both may occupy the same nest site in consecutive seasons (Pettingill and Whitney, 1965). The Western Bluebird (S. mexicana) and Eastern Bluebird are also close in their habitat requirements and reportedly exhibit interspecific territoriality (Marshall, 1957), as do Western and Mountain Bluebirds and Mountain and Eastern Bluebirds in areas of sympatry (pers. obs.). All three species build loose grass nests in tree cavities although S. currucoides also nests in crevices in cliffs and mesas (Oberholser, 1974). One or another species occurs nearly everywhere in the continental United States (Fig. 1) and all three are present in the southwest, a likely place of origin for the genus (J. D. Webster, pers. comm.).

In this report I examine the feeding ecology of Sialia and attempt to translate the predatory tactics and foraging requirements of these birds into habitat needs identifiable by the forest manager.

II.   STUDY AREAS AND METHODS

Eight study areas that I visited during 1968-1978 provided a sample of the habitats utilized by the three species; S. sialis was studied in Macomb and Oscoda Cos., Michigan, and Leon Co., Florida; S. mexicana in Pima and Coconino Cos., Arizona; S. currucoides in Custer and Pennington Cos., South Dakota; and S.

FIGURE 1.   Approximate breeding ranges of <u>Sialia</u> <u>currucoides</u>
(Sc), <u>S</u>. <u>mexicana</u> (Sm), and <u>S</u>. <u>sialis</u> (Ss).   Figure is based on
personal observations and Miller (1951), American Ornithologists'
Union (1957), Bailey and Niedrach (1965), Pettingill and Whitney
(1965), Miller (1970), Oberholser (1974), Herlugson (1975), and
Witzeman <u>et</u> <u>al</u>. (1975, 1976).

<u>currucoides</u> and <u>S</u>. <u>mexicana</u> in Fremont Co., Colorado. <u>S</u>. <u>sialis</u>,
the principal species studied, was sympatric with <u>S</u>. <u>mexicana</u> in
Pima Co., Arizona, and <u>S</u>. <u>currucoides</u> in Custer Co., South
Dakota. Observations of <u>S</u>. <u>currucoides</u> in the relatively treeless terrain of the Badlands National Monument, Pennington Co.,
South Dakota, offered a marked contrast to the ponderosa pine
forests used by this species in the Black Hills, Custer Co.
Similarly, the two study areas in Michigan, abandoned old fields
in Macomb Co. and lumbered and burned pine-oak woodlands in
Oscoda Co., offered contrasting habitats for <u>S</u>. <u>sialis</u>.
     Spatial arrangement of nests was examined for birds using
nest boxes in the southeast Michigan (Macomb Co.) study area; all
other birds were nesting in natural cavities. Territory sizes
and related measurements were assessed by following the birds'
movements for a continuous period and recording these on a series
of maps (Kendeigh, 1944; Odum and Kuenzler, 1955). Areas were
calculated from maps of known scale using a HP 9821 calculator

and HP 9864A digitizer. Only nest boxes in the southeast Michigan study area that were used for spring nests in at least one season were considered in a comparison of random and observed site selection for the spring nesting period. Random sites were assigned to the same number of sites that were used each spring in 1969-1976 by employing the random function (RND) on a HP 2000A computer. "Spacing pressure" (Sp) is defined as:

$$Sp = -(\text{observed} - \text{expected})/\text{expected}^{\frac{1}{2}}$$

where observed and expected are the actual and random frequencies for the nearest neighbor distance frequency categories of the actual and random distributions, respectively (Krebs, 1971). Percentages are compared by a t-test for percentage equality (Sokal and Rohlf, 1969:607).

III. RESULTS AND DISCUSSION

A. Diet, Foraging Methods, and Wing Morphology

Studies by Beal (1915) indicate that S. currucoides consumes more animal foods on an annual frequency basis (91.6%, n = 66 birds examined) than does S. mexicana (81.9%, n = 217) or especially S. sialis (68.0%, n = 855). Little difference exists in the frequencies of the various invertebrate groups among the species, however; Orthoptera, Lepidoptera, Arachnida, and Coleoptera collectively comprise over half of the diet of each species. Plant foods in the diet consist primarily of dry or succulent fruits, rarely hard seeds. Mistletoe (Phoradendron californicum), a favorite food of S. mexicana, and the sumacs (Rhus typhina and R. glabra), favorites of S. sialis, are frequently taken in the non-breeding period, especially in early morning before temperatures rise and insects become active (pers. obs.).

The most complete data on the diet of Sialia nestlings were obtained for S. sialis in southeast Michigan (Pinkowski, 1978). Principal foods of nestlings were: Lepidoptera larvae (32.4% on a frequency basis), with Noctuidae the predominant family; Orthoptera (25.6%), mostly field crickets (Gryllus sp.) and grasshoppers; and Arachnida (11.3%), all spiders. Succulent, early-maturing fruits such as Morus sp., Rubus sp., Cornus sp., Prunus sp., and Lonicera sp. are a short-lived food supply (Snow, 1971) fed to older nestlings and fledglings.

The presence of fruit as well as aerial, geophilous, and phytophilous invertebrates in the diets of bluebirds suggests a diversity of foraging tactics. Seven different types of predatory behavior have been described:

Dropping, a ground-directed foraging tactic, occurs when a bird searches the ground from a perch and, having located prey, drops to the ground and seizes the prey with the bill. Small prey is often swallowed on the ground, but large prey is usually brought up to a perch for preparation (Pinkowski, 1977). Foraging birds change positions on the same perch or move to a new perch when food is not located; they may search the ground monocularly, particularly when on a low perch, or binocularly, especially when on a high perch.

Hopping, also directed at prey on the ground, occurs when a bird hops along the ground and feeds upon items encountered after it lands on the ground. Hopping is most common when the birds are consuming small prey or foraging in areas containing few perches. Bluebirds searching for prey while on the ground are not limited to prey sighted at close range, do not search areas having much leaf litter, and do not flip aside debris with the bill, behaviors typical of many other North American Turdidae (Bent, 1949; Dilger, 1956).

Hovering, another ground-directed tactic, almost always occurs in areas of low perch densities. A hover-foraging bird assumes an airborne position and, on flapping wings and spread tail, it maintains zero ground speed while searching for previously undetected prey. Hover-foraging birds, like perch-foraging birds, may change positions by short, swooping flights from one location to another; they may also land on a perch and begin perch-foraging or land on the ground and forage by hopping.

Gleaning, a foliage-directed foraging tactic, occurs when a bird lands on and removes prey from the foliage and branches of trees or shrubs, or tree trunks. Gleaning is a modification of the dropping mode, with the bird landing on and obtaining prey from a surface above ground level.

Flight-gleaning (= drop-gleaning; Pinkowski, 1977), like gleaning, is directed at foliage-inhabiting prey. The bird descends toward the ground after sighting prey from a perch, but instead of landing it remains airborne while plucking the item from the foliage. A flight-gleaning bird may "flutter" briefly before seizing its prey.

Flycatching, an aerially-directed predatory tactic, is diverse in bluebirds, as is the case in some tyrannid flycatchers that frequently forage by this technique (Leck, 1971; Verbeek, 1975). Typically, aerial prey is captured by short excursions into the air from a perch but on rare occasions the birds capture aerial insects without taking flight or capture prey sighted while flying. Usually only one item is obtained per flycatching sortie. Power (1974:40) thought that flycatching by Mountain Bluebirds was totally opportunistic; this is evidently true for all three species, any of which may forage exclusively by this tactic when aerial prey is abundant.

Soaring, though not observed in this study, was noted in Western Bluebirds by Pitelka (1941). The birds remained airborne for 6-8 seconds in an up-draft created about 2 m above the ground along a slope, apparently in response to local conditions of wind and terrain.

A comparison of frequencies of the various foraging tactics employed by the three Sialia species feeding in similar habitats (Table I) indicates the following: 1) the relative importance of the dropping tactic and hence prey located on the ground stratum in all species; 2) the occurrence of hovering in S. currucoides but not in the other two species; and 3) the greater frequency of flycatching employed by S mexicana, which therefore obtains less of its food on the ground than the other species. Differences in foraging tactics of the species, however, are not great compared with those reported by Verbeek (1975) and James (1976) for other sympatric congeners.

Length of the wing and hence wing-loading in gm/cm^2 (Clark, 1971) are known to vary interspecifically among birds of similar size, being least for strong-flying species that are required routinely to fly long distances and greatest for ground feeding species, including some thrushes, that occur in more closed habitats (Poole, 1938; Temple, 1972). Wing-loading for the three Sialia species (Table II), calculated as suggested by Temple (1972), reflects these trends and is closely correlated with feeding habits, specifically the amount of foraging that each species does while airborne. S. currucoides engaged in more aerial feeding (hovering + flycatching + flight-gleaning = 42.6% for all areas studied; n = 345) than S. mexicana (25.7%; n = 175), which engaged in more aerial feeding than S. sialis (12.1%; n = 3,123). Thus increased wing-loading is correlated with an interspecific decline in the proportion of insects in the diet as well as a decline in the frequency of aerial feeding.

TABLE I. Frequencies of foraging tactics of the three species of Sialia feeding in similar habitats.[a]

States observed No. birds	S. sialis MI, AZ (12)		S. mexicana CO, AZ (11)		S. currucoides CO, SD (10)	
	No.	%	No.	%	No.	%
Dropping	363	74.8	111	63.4	121	85.2
Flycatching	49	10.1	41	23.4	5	3.5
Gleaning	47	9.7	13	7.4	5	3.5
Flight-gleaning	19	3.9	4	2.3	3	2.1
Hopping	7	1.4	6	3.4	5	3.5
Hovering	0	0.0	0	0.0	3	2.1

[a] Open pine and pine-oak woodlands.

TABLE II. Wing-loading in males of the three species of
Sialia.

| Species | Wing Chord (mm) | | | Weight | Wing-loading |
	(n)	Mean	Range	(g)	(g/cm^2)
S. s. sialis	10^a	100.9	97–105	30.3_d	0.298
S. mexicana	$62^{b,c}$	107.8	102–114	28.0^d	0.282
S. currucoides	45^b	114.6	110–125	31.9^e	0.276

[a]Based on Pinkowski (1975a); birds weighed and measured in
Macomb Co., Michigan.
[b]Specimens examined at the University of Michigan Museum of
Zoology.
[c]Distributed by subspecies as follows: S. m. bairdi (45);
S. m. occidentalis (13); S. m. anabelae and S. m. australis
(2 each).
[d]Includes one University of Michigan specimen, three records
given by Paynter (1952), and one record given in Davis
(1945).
[e]Several weighings at various times of the year for four
captive birds (Pinkowski, 1975b).

B. Energetics of Foraging

Hovering imposes greater energy costs per unit time on foraging bluebirds than other feeding methods because a hovering bird
is in flight for a long search time during a foraging sequence.
Tarboton (1978) found that hover-foraging by the Black-shouldered
Kite (Elanus caeruleus) is 6.9 times more expensive than perch-
foraging, but the difference is probably less in smaller species
like Sialia because of a direct relationship between body weight
and the cost of flight (Utter and LeFebvre, 1970). The energetic
cost of flight in the Purple Martin (Progne subis), a species
similar in size to Sialia, has been estimated as 4.6–6.1 times
greater than its standard metabolism (Utter and LeFebvre, 1970)
and 4.3 times the cost of perching (Utter and LeFebvre, 1973;
Ricklefs, 1974).
    Flycatching, like hovering, may sometimes involve aerial
searching for prey and thus can be a bioenergetically expensive
foraging tactic. In Sialia, however, most prey caught by either
flycatching or flight-gleaning is located from a perch so that
the bird is in flight only during pursuit and capture. Gleaning
and dropping are less expensive foraging methods than flycatching
or flight-gleaning because the bird is normally in flight only
during pursuit. Hopping entails no flight for search, pursuit,
or capture and consequently is the least expensive tactic.

Among the reported advantages of hover-foraging are increased opportunities for prey capture (Grubb, 1977), particularly in foraging situations away from perches (McNicholl, 1972). In Colorado I recorded a mean (+ SD) height of 2.6 (+ 1.4) m for 64 Mountain Bluebird hovering sorties; all but two of these were at greater heights than those of the surrounding vegetation (grazed rabbitbrush, Chrysothamnus sp., and sagebrush, Artemisia sp.). The hover-foraging presumably permitted the birds to search a larger area because the perceptual field increases with foraging height (Pinkowski, 1977). By contrast, although hopping is not an expensive foraging tactic, birds using this hunting method have a very small perceptual field.

C. Foraging Perch Utilization

Perch-foraging permits avian predators to search for prey inexpensively while alternating searching with maintenance activities (e.g., preening and sunning) and watchful viewing of their territories for predators or interlopers (Craig, 1978). To achieve these advantages, bluebirds employ a variety of natural objects as foraging perches (tree branches, dirt mounds, boulders, cliff peaks and ledges, shrubs, and coarse weedstalks such as Yucca sp., Verbascum sp., Oenothera sp., Daucus sp., and Aster sp.) as well as many types of artificial objects (fences and fence posts, buildings and roofs, picnic tables, refuse cans, nesting boxes, highway signs, utility wires, stakes and utility poles, and ground debris). Some types of perches such as fences are preferred to others such as utility wires, probably because height is critical (Power, 1974:48) and fences closely approximate the height used by bluebirds foraging from "natural" perches (Pinkowski, 1977).

All three Sialia species prefer dead branches to living branches; this preference was displayed by S. currucoides perch-foraging on sagebrush, S. sialis on oak limbs, and S. mexicana on ponderosa pines. Dead, leafless branches offer the birds an unobstructed view of their surroundings, and dead branches are often abundant in the same types of ecological situations (e.g., burned woodlands) that contain dead snags with cavities for nesting. Bluebirds foraging in living pines obtain ideal perches from the natural death of the lower branches; they may also perch horizontally on tree trunks to obtain an unobstructed view of the ground from an optimum height.

Eastern Bluebirds foraging in old fields use higher perches in summer than in spring (Pinkowski, 1977). Because the perceptual field is larger at greater heights, relatively high perch densities are needed for a bird to search a given area early in the season, but perch density can be lower in summer.

D. Ground Cover and Vegetation

Some avian predators that feed on ground-dwelling prey require low ground vegetation because tall vegetation may interfere with hunting success (Maher, 1974). Perch-foraging bluebirds evidently prefer well-lighted areas containing a low, sparse vegetation and little understory, probably for the same reason. Experiments by Power (pers. comm.) and Pinkowski (1974) on S. currucoides and S. sialis, respectively, have shown that both species prefer mowed to non-mowed areas where the natural vegetation is tall. Dry, non-fertile soils with bare ground and patches of low vegetation are also preferred, as are park-like situations where ground cover is kept short by mowing or grazing.

Relatively poor soils existed in the southeast Michigan study area, where the structural ground cover requirements of Eastern Bluebirds were satisfied by the floristic elements of an old field flora typified by such species as Poa sp., Hieracium sp., Vicia sp., Rumex acetosella, Tragopogon major, Potentilla sp., Erigeron philadelphicus, and Chrysanthemum leucanthemum. This low-growing, mixed herbaceous perennial sere appears late in field succession following disturbance and, on poor soils, it occurs concurrently with woody plants (Beckwith, 1954) that bluebirds use as foraging perches. At high elevations in the montane areas of the west, leaching, erosion, and burning are among factors that result in poor soil capable of supporting only limited plant growth.

Mountain Bluebirds evidently prefer a sparser ground cover than do Western Bluebirds. In the Colorado study area Mountain Bluebirds often foraged by hopping or hovering in grazed valleys that contained little vegetation; dropping by this species was more common on flat to gently sloping (0-15°) terrain (76% of 125 foraging sorties recorded as to slope) than was the case for the Western Bluebird (52% of 143 sorties). The steep slopes that were frequently used by Western Bluebirds had experienced less grazing and contained denser vegetation so that the two species appeared to be differentially responding to vegetation physiognomy rather than to slope.

Bluebirds are opportunistic foragers that greatly alter their foraging strategies according to habitat and ground vegetation. In old fields in southeast Michigan, dropping (78.8%) and flycatching (10.7%) were the most common feeding tactics of Eastern Bluebirds (n = 2,638; Pinkowski, 1977), their frequencies not differing significantly from those observed in open woodlands (Table I; P>0.05 in each case). Hopping (2.6%) and flight-gleaning (1.1%) were not important in old fields, as was the case in woodlands, but gleaning was significantly more common in woodlands than in old fields (6.8%, t = 2.1, P<0.05) because in wooded areas the birds commonly exploited insects dwelling on trees.

Goldman (1975), observing Eastern Bluebirds feeding on lawns in Ohio, found the dropping mode more frequent (87.4%, n = 508) than I observed in either woodlands (t = 5.1, P<0.001) or old fields (t = 4.8, P<0.001); this difference reflects increased exploitation of ground-dwelling invertebrates by birds foraging in the short grass. Goldman did not report birds feeding on lawns by hopping but I found this common around residences in southeast Michigan, especially in late summer and autumn when vegetation elsewhere was relatively tall and dense.

Although the Mountain Bluebird typically feeds by dropping in areas containing perches (Table I), this species' most common foraging tactic in treeless areas of Pennington Co., South Dakota, was hovering (61.6%, n = 203). The percentage of food obtained there by drop-foraging (24.6%) was lower than I observed in any <u>Sialia</u> species in any other habitat type. Many drops were undertaken from boulders and the tops of hills and thus closely resembled hopping, which was comparatively uncommon on flat substrates (2.5%). Some drops were also directed at the sides of cliffs and bluffs and thus were analogous to gleaning. Flight-gleaning and gleaning were expectedly uncommon in the used sense that prey was removed from vegetation (5.9% combined).

Mountain Bluebirds evidently increase the frequency of hover-foraging during periods of high nutritional needs; such is the case during inclement weather or while adults are feeding large broods. There appears to be an upper limit of absolute use (Power 1974:175), and this limit was probably approached by the hover-foraging birds that I observed in Pennington Co. because most of these birds were feeding nestlings. Many of the Mountain Bluebirds foraging from perches in open woodlands (Table I) were also feeding nestlings, however, so that the foraging differences in the two habitat types appeared due to differences in perch availability and not to differences in work load.

Orthopterans, especially field crickets, are a preferred food of the Mountain Bluebird (Bent, 1949); in Pennington Co. they accounted for 87% of 61 items observed and dietary diversity was low (diversity index H' = 0.88, four orders represented). In Custer Co. Orthopterans accounted for only 25% of 42 items observed and diversity was greater (H' = 1.88, seven orders represented). Hover-foraging evidently increases food availability, as suggested by several foraging models that predict an inverse relationship between food availability and the range of items taken by a predator (MacArthur and Pianka, 1966; Emlen, 1966; Levins and MacArthur, 1969; Schoener, 1971).

In South Dakota the size of individual items fed to the young was smaller for drop-foraging Mountain Bluebirds in Custer Co. than for hover-foraging birds in nearby Pennington Co.; also, the drop-foraging birds often brought several items to the nest at a time (Table III). Thus hover-foraging may also represent an adaptive strategy permitting the birds to obtain large prey on proportionately fewer attacks in relatively perchless areas

TABLE III. Prey characteristics for Mountain Bluebirds using two foraging tactics.[a]

Principal feeding mode	No. items per trip			Size of items[b]		
	1	2	>3	S	M	L
Hovering	65 (89%)	8 (11%)	0 (0%)	0 (0%)	7 (9%)	68 (91%)
Dropping	17 (49%)	8 (23%)	10 (29%)	10 (24%)	21 (50%)	11 (26%)

[a] Unpublished data obtained in Pennington (principally hover-foraging) and Custer (principally perch-foraging) Cos., South Dakota, June 1972.
[b] Small (S) = <1 cm; medium (M) = 1-2 cm; large (L) = >2 cm. Size of items could not be estimated for six items fed by hover-foraging birds and 21 items fed by drop-foraging birds.

rather than forage for the same biomass of small prey by making more attacks from perches. Consequently, part of the greater cost of hover-foraging is apparently offset by a greater caloric yield per hovering sortie.

E. Seasonal and Weather Effects on Foraging

The opportunistic nature of avian foraging behavior is sometimes reflected in seasonal changes in foraging behavior as vegetation physiognomy changes (Best, 1977). In southeast Michigan vegetation height increased late in the nesting season; Eastern Bluebirds progressively attempted more aerial hunting forays (flycatching) and foliage-directed forays (gleaning and flight-gleaning) as the season progressed and ground-directed forays (hopping and dropping) became less frequent (Table IV). Interestingly, vegetation height in some parts of Montana decreases late in the season because of grazing, and Power (1974:62) found that Mountain Bluebirds generally feed more on the ground late in the season.
Seasonal changes in the foraging methods of Eastern Bluebirds are reflected in the types of prey fed to the young. In southeast Michigan geophilous Noctuidae larvae were a principal component of the nestling diet in May (26.4%), but became less common in June (15.2%) and were not important in July (7.0%) and August (3.3%). Among spiders there is a notable trend from geophilous species (Trochosa terricola, Thanatus formicinus,

TABLE IV.   Seasonal variation in direction of Eastern Bluebird
hunting forays.[a]

Direction	Percentage Occurrence			
	March n = 584	April n = 595	May n = 770	June n = 689
Ground	99.5	90.3	92.5	46.2
Foliage	0.1	0.4	2.0	27.6
Aerial	0.4	9.4	5.6	26.3

[a]Data based on Pinkowski (1977) for Macomb Co., Michigan.

Lycosa *frondicola*, and *Phidippus* *princeps*) in spring to phytophilous species (*Xysticus* *elegans*, *Schizocosa* *avida*, and
Tibellus *oblongus*) in summer (Pinkowski, 1978).   Obviously the
birds are responding to the increased vegetation height, which
brings about a new resource of available prey, by altering their
foraging tactics.   Tall summer vegetation, however, evidently
renders capture of some geophilous forms difficult or impossible
because late in the season some of these are available but are
not taken (Pinkowski, 1978).

Quantitative data on the effects of weather on foraging
behavior of Eastern Bluebirds are presented elsewhere (Pinkowski,
1977).   Conclusions related to habitat utilization, probably
applicable to perch-foraging by all three *Sialia* species, are:
   1.   Foraging height of birds feeding by the dropping mode
decreases with decreasing temperature and percentage of
sunshine, apparently because insects become less active and
more difficult to locate.
   2.   As foraging height decreases, less time is spent on
unsuccessful perches, presumably because by foraging at lower
heights during inclement weather the birds are able to search
a smaller perceptual field more rapidly.
   3.   The birds move shorter distances after leaving a successful perch (one at which prey was sighted and an attack was
executed) and greater distances when leaving an unsuccessful
perch (no attack executed).
   4.   Frequency of flycatching decreases and frequency of
gleaning increases during increasingly inclement weather,
when some aerial insects evidently remain on vegetation
instead of flying.
   5.   Frequency of flycatching decreases during very windy
conditions, probably because aerial insects are not active at
these times (Freeman, 1945).
   6.   Fruit consumption occurs when insects cannot be obtained
but is uncommon at other times.

The rigorous weather conditions encountered by bluebirds establishing territories during late winter and early spring make some potential nest sites unsuitable for early nestings. Consequently, Eastern Bluebirds use nest sites located in a greater variety of habitats in summer than in spring (Pinkowski, 1977). Also, in spring it is not uncommon for birds to abandon previously established territories and move elsewhere during cold weather, when areas acceptable during favorable weather become unsuitable; this rarely occurs in summer because food is more abundant.

Eastern Bluebirds foraging in early spring follow a regular and quite predictable sequence to obtain food (Fig. 2). The sequence is based on the behavioral patterns noted above as well as the tendency of the birds to continue feeding by hopping after consuming small food items on the ground and to rest or perform maintenance activities after consuming fruit. The entire sequence is usually not performed in summer, when more favorable conditions and greater insect abundance make fruit consumption unnecessary and foraging behavior is more opportunistic and hence less predictable.

F. Space Use, Territory, and Habitat Relationships

Bluebirds, like many other passerines, maintain a type "A" territory in that the defended area is used for mating, nesting, and feeding (Nice, 1941). Although some workers (Brown and Orians, 1970; Archibald, 1975) have suggested that one of the essential characteristics of a territory is that it changes little over a period of time, Eastern Bluebird territories vary seasonally (Fig. 3). During 1972 and 1973 territories in the southeast Michigan study area were largest in March (mean = 11.5 and 16.2 ha, respectively) but became progressively smaller in April (8.5 and 8.8), May (5.5 and 6.8), and June (5.0 and 4.4). Thus exceptionally large territories were maintained during the period preceding the general arrival of birds in early April.

The trend toward large territories early in the season has been noted in other passerines (Thompson and Nolan, 1973) and non-passerines (Meyerriecks, 1960); in bluebirds it is probably related to a need for large foraging areas when insects are not abundant and a great diversity of feeding situations (woodlands with many perches, south facing slopes, and fruit) are required to satisfy nutritional demands. Later in the season peripheral parts of the territory are no longer economically defensible as insects become more abundant and other males arrive; consequently, territories become smaller, reflecting their "compressibility" (Beer et al., 1956). The trend toward smaller territories late in the season occurs despite the fact that parts of a given territory become unusable for ground-directed foraging in summer due to the increased vegetation height.

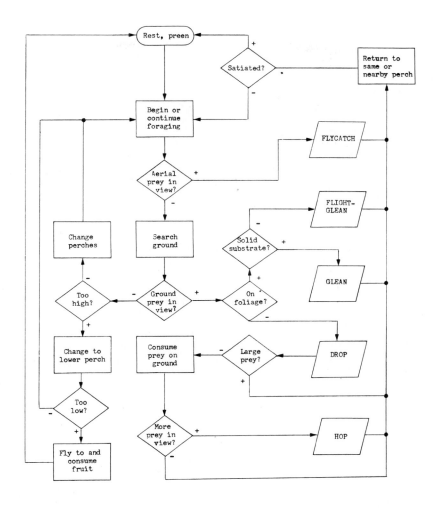

FIGURE 2. A generalized regime showing the major foraging
options available to perch-foraging bluebirds. Successful
foraging tends to the upper right, unsuccessful to the lower
left. Changes to lower perches are accompanied by an increase
in the rate of perch changes (see text). "Too high" and "too
low" refer to the adequacy of perch height (and hence the
perceptual field) relative to existing temperature and prey
movement so that a minimum height below which perch-foraging
cannot succeed is predicted for a given set of foraging
conditions.

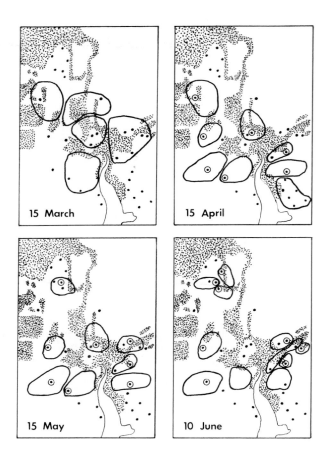

FIGURE 3. Monthly variation in Eastern Bluebird territories of various males in the southeast Michigan study area in 1973. Points represent nest boxes, circled points represent active nest boxes. The study area measures approximately 1.6 x 2.4 km. Stipuled areas represent woodlands and a large lake occurs in the south-central region.

Natural selection should favor circular territories with nest sites located close to the center of the territory, thus minimizing traveling time (Cody, 1974a; Covich, 1976). Some Eastern Bluebird territories deviate considerably from circularity and nest sites are often near the periphery, not the center, of the territory (Fig. 3). Deviations such as these occur at the expense of energetic efficiency and are caused by the relative distribution of suitable perches, nest sites, and ground cover. Some nests may be remarkably close to one another, but these are often at different stages of the nesting cycle (Pinkowski, 1974), which probably reduces competition for food resources (Kilham, 1973).

A plot of bluebird foraging locations on a map showing an entire (defended and advertised) territory permits an estimate of the size of the area used for foraging (A_f), the total advertised and defended area (A_t), and the usability percentage (UP = A_f/A_t). Although A_t values of ten S. sialis nesting territories mapped in southeast Michigan in 1972 ranged from 4.5-38.9 ha, A_f values were fairly constant at 3.9-8.4 ha and had less variance (F = 41.5, P<0.01; Pinkowski, 1977). Thus a wide range of UP values (22%-98%) was largely attributable to a wide range of A_t values.

Two contrasting situations involving very different A_t and UP values for Eastern Bluebird territories are shown in Fig. 4. A_t, A_f, and UP values for both territories varied little in other seasons during 1968-1977, and both nest sites had high and nearly identical success rates: Territory 1 had 14 nests, 9 successful, 42 young fledged; and Territory 2 had 14 nests, 10 successful, 34 young fledged. Thus territory size is generally correlated with food supply (Fig. 3), as is the case in other species (Schoener, 1968), but the presence of the requisites to gain access to the food supply is also important in determining territory size (Fig. 4); territory size and shape may be adjusted as foraging requisites vary.

Eastern Bluebird nests in boxes in the southeast Michigan study area were not distributed randomly. Twenty-three different boxes were used for 5-12 spring nests during each season in 1969-1976. When distances between each nest and its nearest neighbor are compared with nearest neighbor distances between nest boxes assigned randomly for the same number of nests per year, the observed and random distributions of nests differ significantly (Table V; χ^2 = 21.2, df = 5, P<0.01). Fewer nests are located as close as would be theoretically expected as spacing pressure is greatest at sites nearest a nest (<201 m). Pressure is lowest at sites 201-600 m apart but also occurs at very distant sites (>600 m) because nests were rarely placed far apart. Territorial behavior apparently limits the number of birds using the available nest sites because of the paucity of nests located very close to one another (Patterson, 1965; Krebs, 1971). Pressure at

distant sites apparently results from the tendency of the species
to form aggregations or loose colonies, as noted by Marshall
(1957).

    Eastern Bluebird territories in the lumbered and burned study
area in Oscoda Co., Michigan, ranged in size from 0.8 to 2.9 ha
(mean = 1.4 ha) and thus were much smaller than those observed in
old fields in southeast Michigan. Consequently, during June 1974
the Oscoda Co. area had a higher density of birds (7.9 nests/100
ha) than the maximum density in southeast Michigan in 1973 (2.1
nests/100 ha; Fig. 3). The difference is probably due to the
optimum habitat conditions, particularly ground cover and
perches, that exist after a burn; in every case the portions of
territories suitable for drop-foraging in the Oscoda Co. study
area exceeded 90% whereas such high values were noted for only
one of the ten territories in old fields of southeast Michigan.

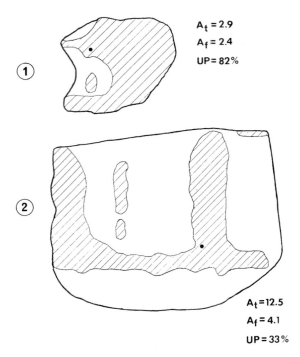

$A_t = 2.9$
$A_f = 2.4$
UP = 82%

$A_t = 12.5$
$A_f = 4.1$
UP = 33%

    FIGURE 4. Total area of territory (A_t, outer border), area
used for foraging (A_f, shaded), and usability percentage (UP =
A_f/A_t) of two contrasting Eastern Bluebird territories in the
southeast Michigan study area in May 1974. Areas are in ha;
solid circles represent nest boxes, both of which contained
nestlings when territories were mapped.

TABLE V. Frequencies of observed and expected inter-nest
distances of Eastern Bluebirds using nest boxes in southeast
Michigan, 1969-1976.

Distance Interval (m)	Observed	Expected[a]	Spacing Pressure[b]
0-100	0	6	+2.45
101-200	14	21	+1.53
201-400	20	17	-0.73
401-600	31	18	-3.07
601-800	5	6	+0.41
801-1000	1	3	+1.16

[a]Obtained by computer-simulated occupancy of the number of
nest sites used in each year.
[b]Defined in text.

Balda (1967, 1975b) obtained comparative data on territory
size for all three species of Sialia in Arizona. S. sialis
territories in oak-juniper woodlands were larger (mean = 1.5 ha)
than territories of S. mexicana (0.4 ha) or S. currucoides (0.6
ha) in ponderosa pine forests. These data together with my
observations indicate that S. sialis requires a larger nesting
territory than the other species. Capture of prey in areas
having few perches and prey consumption in a third, vertical
dimension are foraging options that reduce the size of foraging
territories of S. currucoides and S. mexicana, respectively.
These options are not available to S. sialis, which therefore
requires a larger nesting territory.

G. Competition and Niche Overlap

Of the three species of Sialia, S. currucoides is found in
the greatest variety of habitats. In Colorado, Arizona, and
South Dakota this species was most common in the Upper Sonoran
and Transition Life Zones; it also occurred near grassy openings
in the Canadian and Hudsonian Zones, often in association with
aspen (Populus tremuloides) (cf. Kermott et al., 1974; Flack,
1976). S. sialis occasionally occurred in the Transition Zone of
montane regions of Arizona and South Dakota but was more
characteristic of lower elevations, especially oak or "encinal"
(Shreve, 1915) portions of the Upper Sonoran Zone; in the eastern
U.S. the natural nests of this species are frequently found in
oaks (Pinkowski, 1976). S. mexicana was most limited in its

habitat distribution; I nearly always found it in the Transition
Zone forests of Arizona and Colorado though elsewhere it
occasionally occurs at lower elevations (Miller, 1951) or in the
Canadian Zone (Marshall, 1957; Herlugson, 1975). Thus interspecific habitat segregation is incomplete, partly because of
marked differences in ecological tolerance among the species.
Cooccupancy and interspecific competition for resources are most
likely to occur in the Transition Zone, where all three species
have reportedly nested close together (Bailey and Niedrach,
1965).

    S. currucoides, the largest species of Sialia, forages
farthest from perches and may occur in open areas with low perch
densities because of its ability to hover. Neither S. sialis nor
S. mexicana hover-forages and both are restricted to areas containing perches. S. mexicana, the smallest species, obtains less
food on the ground than the others and may occur in relatively
dense pine woodlands (Marshall, 1957; pers obs.). S. sialis is
intermediate in size and prefers intermediate perch and vegetation densities. Thus body size varies inversely with preferred
vegetation density; this trend has been noted in some tyrannid
flycatchers (Cody, 1974b:79) and, in Sialia, probably functions
to maximize foraging efficiency and reduce interspecific competition.

    Because each Sialia species can vary its foraging strategy
from one place to another, there is more variation in foraging
tactics of a given species between habitats (Table III and text)
than between different species in the same habitat (Table I).
Interspecific competition in intermediate habitats must be keen
because in such situations different species forage mainly by the
dropping mode, presumably because other foraging methods are less
efficient and therefore energetically more costly (Tramer and
Tramer, 1977). Each of two different species cooccupying the
same habitat would be under selective pressure to optimize
foraging behavior (Rice, 1978), and evidently the optimal foraging patterns of different species are similar in intermediate
habitats where there is often little difference in the vegetation
on the different species' territories. Very open or forested
areas sometimes inhabited by S. currucoides and S. mexicana,
respectively, are "environmental refugia" (Morse, 1971b) that lack
interspecific competition because they contain only one species.

    Although subtle behavioral differences may reduce competition
for feeding resources and thus permit overlapping territories in
closely related sympatric species, the maintenance of mutually
exclusive territories by different Sialia species suggests that
the usual options of interspecific divergence in resource use may
not be advantageous to these species (Cody, 1973). Many other
perch-foraging avian predators (e.g., the genera Lanius, Otus,
Empidonax, and Buteo) that feed on prey concentrated at a given
stratum maintain interspecific territories (Orians and Willson,

1964). Despite overall differences in resource utilization by the three *Sialia* species, the similarity of foraging tactics in intermediate habitats results in similar ecological requirements, especially the need for prey located on the ground stratum near perches. Territorial overlap in intermediate habitats would mean excessive competition for prey dwelling on the ground near perches so that it seems unlikely that the interspecific territories are nonadaptive, as suggested by Murray (1971). In intermediate habitats different *Sialia* species behave ecologically like a single species, and mutually exclusive territories are a logical consequence of their similar behavior.

H. Human Activities and Management

Marked changes in the relative abundance of two coexisting *Sialia* species are not uncommon and may be brought about by human activities. Around 1900 S. sialis and S. currucoides both expanded their ranges into the grasslands of the Great Plains (Bent, 1949) and now compete for nest sites in this region (Criddle, 1927; Miller, 1970). These range expansions are similar to those noted for other arboreal species in that they were probably caused by habitat changes attributable to human activities (Anderson, 1971). More recently S. sialis has expanded its range in the southwestern United States (Witzeman et al., 1975, 1976), possibly as a response to lumbering in that region, and S. currucoides has evidently replaced S. mexicana as the predominant species in the northwest (Herlugson, 1975). Changes in relative numbers of S. mexicana and S. currucoides in their broad area of sympatry are also probably caused by changes in the openness of the terrain. Natural ponderosa pine woodlands, though open and park-like, have dense grasses in the understory (Cooper, 1960; Weaver, 1961); grazing reduces the density of the herbaceous understory of these forests (Rummell, 1951) and thus may benefit S. currucoides at the expense of S. mexicana because the former forages more on open ground.

Management practices for any one *Sialia* species would obviously be different with and without the presence of a competing congener. All three species are abundant in disturbed areas containing a few mature or dead trees and all reportedly benefit from selective cutting of mature timber stands (Hagar, 1960; Conner and Adkisson, 1974; Davis, 1976). S. currucoides probably benefits more from logging than the other species; at least one study (Szaro, 1976:174) recorded the replacement of S. mexicana by S. currucoides following clear-cutting. Optimum conditions will exist after cutting if large snags are left as nesting places, if a variety of smaller saplings and snags are left as foraging perches, and if native or planted fruit-bearing trees or shrubs are present. Overmature trees should be left to allow for adequate snag recruitment (Szaro, 1976:250).

Savannas and open stands of mature pines are natural bluebird
habitats that often require recurrent fires for their maintenance
(Komarek, 1976; Dickman, 1978). Prescribed burning is usually
beneficial to Sialia species, especially if it controls shrubs
and hardwood trees in the understory (Biswell, 1973) or maintains
forest openings (Gartner and Thompson, 1973). Suitable habitats
must be extensive or clearings frequent, however, because of the
tendency of bluebirds to occur in loose colonies.

I.   Future Research

Studies of avian foraging behavior and competition in
closely-related species have been very much in vogue since the
pioneering work of MacArthur (1958). Many studies, however, are
applicable to two or more species in one area only and thus have
been criticized (Wiens, 1977) because they ignore important
geographic and environmental variables. This study shows that
avian predators are very capable of modifying their foraging
behavior from one time or place to another. The extent to which
local foraging strategies are fixed as a result of selective
pressures on sympatric species or flexible and capable of modification whenever competition becomes excessive is not known
(Diamond, 1978). The role of inherited behavior as well as the
importance of habitat selection and philopatry (Mayr, 1963:568)
in resource partitioning by sympatric congeners can only be
determined if future studies are directed at different species
co-inhabiting different geographical areas. Where possible,
experimental manipulation of environmental variables may also
prove instructive (Wiens, 1977).

ACKNOWLEDGMENTS

Discussions with Drs. Kenneth Parkes, William Thompson, and
John Wiens were very helpful in developing some of the ideas
contained in this report. Drs. Ernst Mayr, Robert Storer, Gary
Friday, and Melvin Weisbart offered many helpful criticisms of
this and earlier versions of the manuscript. James Stevens and
Phyllis Pinkowski assisted with the field work, Ken Hough cal-
culated the territory sizes, and Vera Percy typed the manuscript.
I thank all of these persons for their assistance.

REFERENCES

Ahlgren, I., and Ahlgren, C. (1960). Bot. Rev. 26, 483.
American Ornithologists' Union. (1957). "Check-list of North
    American birds." Port City Press, Baltimore, Maryland.
Anderson, B. (1971). Condor 73, 342.
Archibald, H. (1975). J. Wildl. Manage. 39, 472.
Bailey, A., and Niedrach, R. (1965). "Birds of Colorado." Vol.
    2. Denver Museum of Natural History, Denver, Colorado.
Balda, R. (1967). "Ecological relationships of the breeding
    birds of the Chiricahua Mountains, Arizona." Ph.D. Thesis,
    Univ. of Illinois, Urbana.
Balda, R. (1975a). In Proc. Symp. Management of Forest and
    Range Habitats for Nongame Birds, U.S. For. Ser., Tucson,
    Arizona.
Balda, R. (1975b). "The relationship of secondary cavity
    nesters to snag densities in western coniferous forests."
    U.S. For. Ser. Wildl. Habitat Technical Bull. No. 1,
    Albuquerque, New Mexico.
Beal, F. (1915). "Food of the robins and bluebirds of the
    United States." U.S.D.A. Biol. Surv. Bull. 171.
Beckwith, S. (1954). Ecol. Monogr. 24, 349.
Beer, J., Frezel, L., and Hansen, N. (1956). Wilson Bull. 68,
    200.
Bent, A. (1949). "Life histories of North American thrushes,
    kinglets, and their allies." U.S. Natl. Mus. Bull. 196.
Best, L. (1977). Condor 79, 192.
Biswell, H. (1973). Proc. Tall Timbers Fire Ecology Conf.
    12, 69.
Bock, C., and Lynch, J. (1970). Condor 72, 182.
Brown, C., (1978). Bird-Banding 49, 130.
Brown, J., and Orians, G. (1970). Ann. Rev. Ecol. Syst. 1, 239.
Clark, R. (1971). Auk 88, 927.
Cody, M. (1973). Ann. Rev. Ecol. Syst. 4, 189.
Cody, M. (1974a). Science 183, 1156.
Cody, M. (1974b). "Competition and the structure of bird
    communities." Princeton Univ. Press, Princeton, New Jersey.
Connor, R. (1978). In Proc. U.S.F.S. workshop management of
    southern forests for non-game birds. U.S.D.A. For. Ser.
    Gen. Tech. Rep. SE-14.
Conner, R., and Adkisson, C. (1974). J. Wildl. Manage. 38, 934.
Conner, R., and Adkisson, C. (1975). J. Forestry 73, 781.
Cooper, C. (1960). Ecol. Monogr. 30, 129.
Covich, A. (1976). Ann. Rev. Ecol. Syst. 7, 235.
Craig, R. (1978). Auk 95, 221.
Criddle, N. (1927). Can. Field-Nat. 41, 40.
Davis, P. (1976). "Response of vertebrate fauna to forest fire
    and clearcutting in south central Wyoming." Ph.D. Thesis,
    Univ. of Wyoming, Laramie.

Davis, W.   (1945).   Auk 62, 272.
Diamond, J.   (1978).   Amer. Scientist 66, 322.
Dickman, A.   (1978).   J. Forestry 76, 24.
Dilger, W.   (1956).   Wilson Bull. 68, 171.
Emlen, J.   (1966).   Amer. Natur. 100, 611.
Flack, J.   (1976).   "Bird populations of aspen forests in western
    North America." Ornithol. Monogr. 19, American Ornitholo-
    gists' Union.
Freeman, J.   (1945).   J. Anim. Ecol. 14, 128.
Gartner, F., and Thompson, W.   (1973).   Proc. Tall Timbers Fire
    Ecology Conf. 12, 37.
Goldman, P.   (1975).   Auk 92, 798.
Grubb, T., Jr.   (1977).   Wilson Bull. 89, 149.
von Haartman, L.   (1957).   Evolution 11, 339.
von Haartman, L.   (1968).   Ornis Fenn. 45, 1.
Hagar, D.   (1960).   Ecology 41, 116.
Hamerstrom, F., Hamerstrom, F., and Hart, J.   (1973).   J. Wildl.
    Manage. 37, 400.
Herlugson, C.   (1975).   "Status and distribution of the Western
    Bluebird and the Mountain Bluebird in the state of
    Washington." Master Thesis, Washington State Univ., Pullman.
Hesselschwerdt, R.   (1942).   J. Wildl. Manage. 6, 31.
James, D.   (1962).   Audubon Field Notes 16, 306.
James, F.   (1971).   Wilson Bull. 83, 215.
James, R.   (1976).   Wilson Bull. 88, 62.
Johnston, R., and Hardy, J.   (1962).   Wilson Bull. 74, 243.
Kendeigh, S.   (1944).   Ecol. Monogr. 14, 68.
Kermott, H., Fields, R., and Trout, A.   (1974).   Wilson Bull.
    86, 83.
Kilham, L.   (1973).   Bird-Banding 44, 317.
Komarek, E., Sr.   (1976).   Proc. Tall Timbers Fire Ecology Conf.
    14, 201.
Krebs, J.   (1971).   Ecology 52, 2.
Lane, J.   (1968).   Auk 85, 684.
Lane, J.   (1969).   Blue Jay 27, 18.
Leck, C.   (1971).   Wilson Bull. 83, 310.
Levins, R., and MacArthur, R.   (1969).   Ecology 50, 910.
Lloyd, H.   (1938).   J. Forestry 36, 1051.
MacArthur, R.   (1958).   Ecology 39, 599.
MacArthur, R., and MacArthur, J.   (1961).   Ecology 42, 594.
MacArthur, R., and Pianka, E.   (1966).   Amer. Natur. 100, 603.
Maher, W.   (1974).   "Ecology of Pomarine, Parasitic, and
    Long-tailed Jaegers in northern Alaska." Pacific Coast
    Avifauna No. 37, Cooper Ornithological Society, Los Angeles,
    California.
Marshall, J., Jr.   (1957).   "Birds of pine-oak woodland in
    southern Arizona and adjacent Mexico." Pacific Coast Avifauna
    No. 32, Cooper Ornithological Society, Berkeley, California.
Mayr, E.   (1963).   "Animal species and evolution." The Bleknap
    Press, Harvard Univ., Cambridge, Massachusetts.

McClelland, B.   (1975).   J. Forestry 73, 414.
McNicholl, M.   (1972).   Blue Jay 30, 96.
Meyerriecks, A.   (1960).   "Comparative breeding behavior of four
    species of North American herons." Publ. Nuttall Ornith.
    Club, No. 2, Cambridge, Massachusetts.
Miller, A.   (1951).   Univ. Calif. Pub. Zool. 50, 531.
Miller, W.   (1970).   Blue Jay 28, 38.
Morse, D.   (1971a).   Ann. Rev. Ecol. Syst. 2, 177.
Morse, D.   (1971b).   Wilson Bull. 83, 57.
Murray, B., Jr.   (1971).   Ecology 52, 414.
Nice, M.   (1941).   Amer. Midl. Nat. 26, 441.
Oberholser, H.   (1974).   "The bird life of Texas." Vol. 2.
    Univ. of Texas Press, Austin, Texas.
Odum, E., and Kuenzler, E.   (1955).   Auk 72, 128.
Orians, G., and Willson, M.   (1964).   Ecology 45, 736.
Patterson, I.   (1965).   Ibis 107, 433.
Paynter, R., Jr.   (1952).   Auk 69, 293.
Pettingill, O., Jr. and Whitney, N., Jr.   (1965).   "Birds of
    the Black Hills." Lab. of Ornithology, Cornell Univ.,
    Ithaca, New York.
Pinkowski, B.   (1974).   "A comparative study of the behavioral
    and breeding ecology of the Eastern Bluebird (Sialia sialis)."
    Ph.D. Thesis, Wayne State Univ., Detroit.
Pinkowski, B.   (1975a).   Bird-Banding 46, 273.
Pinkowski, B.   (1975b).   Avicultural Mag. 81:14.
Pinkowski, B.   (1976).   J. Wildl. Manage. 40, 556.
Pinkowski, B.   (1977).   Wilson Bull. 89, 404.
Pinkowski, B.   (1978).   Wilson Bull. 90, 84.
Pitelka, F.   (1941).   Condor 43, 198.
Poole, E.   (1938).   Auk 55, 511.
Power, H.   (1974).   "The Mountain Bluebird: sex and the evolu-
    tion of foraging behavior." Ph.D. Thesis, Univ. of Michigan,
    Ann Arbor.
Rice, J.   (1978).   Ecology 59, 526.
Ricklefs, R.   (1974).   In "Avian energetics." (R. Paynter, Jr.,
    ed.) p. 152. Publ. Nuttall Ornith. Club, No. 15,
    Cambridge, Massachusetts.
Robbins, C., and Erskine, A.   (1975).   Trans. 40th N.A. Wildl.
    Conf., 288.
Rohwer, S., and Niles, D.   (1977).   Bird-Banding 48, 162.
Rummell, R.   (1951).   Ecology 32, 594.
Schoener, T.   (1968).   Ecology 49, 123.
Schoener, T.   (1971).   Ann. Rev. Ecol. Syst. 2, 369.
Scott, V.   (1978).   J. Forestry 76, 26.
Scott, V., Evans, K., Patton, D., and Stone, C.   (1977).
    "Cavity-nesting birds of North American forests." U.S. Dept.
    Agric., Agric. Handbook 511.
Shreve, F.   (1915).   "The vegetation of a desert mountain range
    as conditioned by climatic factors." Carnegie Inst.
    Washington, Publ. 217.

Snow, D.   (1971).   Ibis 113, 194.
Sokal, R., and Rohlf, F.   (1969).   "Biometry." W. H. Freeman
    and Co., San Francisco, California.
Szaro, R.   (1976).   "Population densities, habitat selection,
    and foliage use by the birds of selected ponderosa pine
    forest areas in the Beaver Creek watershed." Ph.D. Thesis,
    Northern Arizona Univ., Flagstaff.
Tarboton, W.   (1978).   Condor 80, 88.
Temple, S.   (1972).   Auk 89, 325.
Thompson, C., and Nolan, V., Jr.   (1973).   Ecol. Monogr. 43, 145.
Tramer, E., and Tramer, F.   (1977).   Wilson Bull. 89, 166.
Utter, J., and LeFebvre, E.   (1970).   Comp. Biochem. Physiol.
    35, 713.
Utter, J., and LeFebvre, E.   (1973).   Ecology 54, 597.
Van Velzen, W.   (1978).   Amer. Birds 32, 49.
Verbeek, N.   (1975).   Wilson Bull. 87, 231.
Verner, J.   (1975).   In Proc. Symp. Management of Forest and
    Range Habitats for nongame birds, U.S. For. Ser., Tucson,
    Arizona.
Warbach, O.   (1958).   J. Wildl. Manage. 22, 23.
Weaver, H.   (1961).   Ecology 42, 416.
Wiens, J.   (1969).   Ornithol. Monogr. 8, 1.
Wiens, J.   (1977).   Amer. Scientist 65, 590.
Witzeman, J., Hubbard, J., and Kaufman, K.   (1975).   Amer. Birds
    29, 890.
Witzeman, J., Hubbard, J., and Kaufman, K.   (1976).   Amer. Birds
    30, 985.

FORAGING TACTICS OF FLYCATCHERS IN
SOUTHWESTERN VIRGINIA

Jerry W. Via

Department of Biology
Virginia Polytechnic Institute and State University
Blacksburg, Virginia

Foraging characteristics of seven species of flycatchers
were studied in the ridge and valley province of southwestern Virginia. For each species the percentage of use for
various substrates was recorded and an index of substrate
diversity was calculated. Foraging perch heights within
trees were recorded and compared to the canopy height. Three
characteristics of the foraging flights were also recorded;
length of the flight, method of capture, and whether the
flycatcher returned to the same foraging perch after the
flight.

There was a pronounced vertical stratification relative
to canopy height for the eastern forest flycatchers. The
Great Creasted Flycatcher foraged significantly higher in the
forest canopy than the Eastern Wood Pewee, Least Flycatcher,
Acadian Flycatcher, and Eastern Phoebe. The Eastern Phoebe
had the highest index of substrate diversity since it was
found in both woodland and open country habitats. The
Eastern Kingbird had the longest median foraging flight distance which was significantly longer than similar measurements for the two other open country flycatchers, the Willow
Flycatcher and the Eastern Phoebe. For the forest flycatchers, the Great Crested Flycatcher had the longest median
flight distance. Most species in this study showed an equal
number of returns versus non returns to the foraging perch.
The Great Crested Flycatcher returned to the perch in only

about one third of the observations. Most all species foraged primarily by hawking insects, but gleaning of insects
from vegetation was an important means of foraging for the
Willow Flycatcher and the Great Crested Flycatcher. Differences in these measurements between species may act to reduce
overlap in resource use.

I. INTRODUCTION

The tyrant flycatchers, family Tyrannidae, derive their name
from the habit of aerial pursuit and capture of insect prey.
These "sorties" are also performed by species other than flycatchers, but in North America this trait is most definative of the
Tyrannidae. Within the family, flycatching is not a stereotyped
behavior for there is considerable variation in many of the
foraging tactics (Lederer, 1972; Verbeek, 1975).

In areas of several sympatric species of flycatchers, there
may be overlap in the use of food resources. This overlap may
be lessened by differences in prey type and size, foraging site,
and foraging tactics. Several studies have compared the ecological relationships between sympatric species of North American
flycatchers (Johnson, 1963; Hespenheide, 1964, 1969, 1971;
Crowell, 1968; Smith, 1966; Johnston, 1971; Lederer, 1972; and
Verbeek, 1975).

In the Southern Appalachians there are seven common breeding
species of tyrannids: Eastern Kingbird (Tyrannus tyrannus), Great
Crested Flycatcher (Myiarchus crinitus), Eastern Phoebe (Sayornis
phoebe), Acadian Flycatcher (Empidonax virescens), Willow Flycatcher (E. traillii), Least Flycatcher (E. minimus), and Eastern
Wood Pewee (Contopus virens). While all of these species are
sympatric over a large area, many species are not syntopic (Rivas,
1964) because of habitat requirements or altitudinal differences
(Johnston, 1971).

II. METHODS

The ridge and valley province of southwestern Virginia varies
in elevation between 300 and 700m. Most of the land below 750m
elevation has been developed for agriculture while areas above
750m are generally forested. Observations for this study were
made in Montgomery, Botetourt, Giles, Craig, Roanoke, Smyth and
Grayson counties. Data were collected from 26 May to 15 August
1977. Most observations were made between 6:00 and 11:00 am, but
observations were recorded throughout the day whenever a fly-

catcher species was encountered.

Foraging substrates and foraging tactics were recorded using the methods of Verbeek (1975). Substrates included any perch from which the bird made a sortie. This method arbitrarily divides used trees into an inner core and peripheral shell. Each of these areas is then divided into a lower, middle and upper region plus the apex. This partitions the tree into seven regions. For this study, trees were defined as any woody plant ≥ 4 meters high. Perch height was measured with an Abney level. If the perch was in a tree, the height of the tree was also measured. Note was also made if the perch was living or dead vegetation. The flight distance between the perch and point of prey capture was estimated to the nearest foot and later transformed into meters. Foraging behavior was classified as either hawking (aerial prey capture) or gleaning (removal of prey from a substrate). Note was made after each sortie as to whether the bird returned to the same perch, within a few centimeters or flew to a new perch.

Substrate diversity was calculated using the Shannon Information Formula (Shannon, 1948). In calculating the substrate diversity index, all values within the seven partitions of the tree were lumped as one observation. The different aged forest stands of southwest Virginia made comparisons of the different species impossible without relating them to canopy height. Use of a canopy affinity value facilitates comparisons of the same species from stands of different canopy heights as well as comparing different species from the same area. Canopy affinity was calculated by dividing the height of the foraging perch by the height of the tree containing the perch. Observations were made on as many individuals as possible. Between two and four observations were usually recorded for each individual, but in each case, no more than ten observations were made on any one individual.

III. RESULTS

A survey of the ridge and valley province of southwest Virginia revealed two basic flycatcher assemblages; the woodland species and the open country species. In forested areas above 1070m, the woodland assemblage often consisted of four syntopic species; Great Crested Flycatcher, Eastern Wood Pewee, Least Flycatcher, and Eastern Phoebe. In lowland forested areas the assemblage included; Great Crested Flycatcher, Eastern Wood Pewee, Acadian Flycatcher, and Eastern Phoebe. In the lowland open farmlands and waste areas, the Eastern Kingbird, Willow Flycatcher, and Eastern Phoebe were often syntopic. The distinction between the high elevation forest and the lowland forest flycatchers was solely based on the presence of the Least Fly-

TABLE I. Characteristics of Foraging Perches

	Substrate Diversity	Canopy Affinity	Foraging Perch Substrate		
	Index	Median	Living	Dead	Inanimate
Crested	.720	.84	31.3%	62.5%	6.3%
Pewee	.567	.69	8.6	82.8	8.6
Least	.302	.59	18.6	76.7	4.7
Acadian	.284	.41	50.8	49.2	0.0
Phoebe	1.727	.40	36.5	27.0	36.5
Kingbird	1.279	.93	60.0	18.0	22.0
Willow	1.563	.67	58.1	33.8	8.1

catcher in modified forests at higher elevations and its absence
at lower elevations. Only in rare circumstances were open country species associated with woodland flycatchers.

A. Substrate Diversity and Perch Site Selection

Within the forest flycatcher assemblage, the Great Crested
Flycatcher had the second highest substrate diversity (Table I).
It foraged primarily from the dead perches in the canopy and outer
periphery of trees (Fig. 1). Approximately 79% of the foraging
perches of crested flycatchers were located in trees, but they
also frequented utility lines and short shrub like growth in
recent clearcuts. The crested flycatcher also had the highest
canopy affinity (.84) of the forest flycatchers. A median test
(Siegel, 1956) shows that the median canopy affinity of the crested flycatcher was significantly greater than that of the pewee
($P<0.01$), Least Flycatcher ($P<0.0005$), Acadian Flycatcher
($P<0.0005$), and phoebe ($P<0.0005$).

The pewee had the third highest substrate diversity index of
the forest flycatchers (Table I). Pewees foraged primarily from
dead twigs on the outer periphery of trees (Fig. 2) and did not
frequent the tops of trees as often as the crested flycatcher,
hence they had a significantly lower (.69) canopy affinity. This
value is not significantly different from the median value for the
Least Flycatcher, but was significantly greater than the median
value for the Acadian Flycatcher ($P<0.005$) and phoebes ($P<0.01$)
when compared with a median test. The pewee used dead perches
more than any other species (82.8%).

Least Flycatchers foraged primarily from dead branches in the
mid and lower portions of trees (Fig. 3) and had the highest use
of trees (93%) for foraging perches. When compared to other
flycatchers, it has a low index of substrate diversity (.302).
The Least Flycatcher's median value for canopy affinity (.59) was
not significantly different from that of any other forest flycatcher except for the crested flycatcher.

The Acadian Flycatcher also foraged in the mid and lower parts
of trees and understory and was rarely found in the upper portions
of trees (Fig. 4). It used dead and living perches with equal
frequency and had the lowest substrate diversity of the forest
species. The median canopy affinity (.41) was significantly less
than that of the crested flycatcher and pewee.

The Eastern Phoebe was probably the most adaptable of the
flycatchers with regard to substrate since it had the highest
substrate diversity index and was found in both forest and open
habitats. Most foraging perches were in trees (42.3%), but
phoebes regularly foraged from man made structures (36.5%) annual
herbs and from the ground (Fig. 5). When compared to other- forest
flycatchers, the median canopy affinity (.40) was significantly
less than that of the crested flycatcher and pewee. In open
country situations, the median canopy affinity of the phoebe was

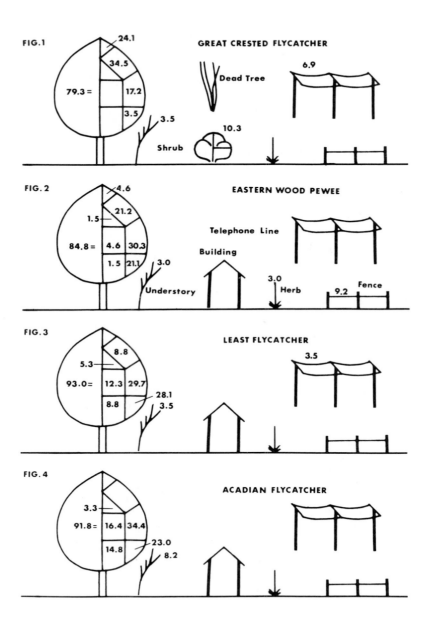

FIG.1   24.1   GREAT CRESTED FLYCATCHER
             34.5
79.3 =   17.2   Dead Tree   6.9
             3.5   3.5
                      Shrub   10.3

FIG. 2   4.6   EASTERN WOOD PEWEE
              21.2
1.5
84.8 =   4.6 30.3   Telephone Line
         1.5 21.1 3.0 Building
             Understory   3.0   Herb   9.2   Fence

FIG. 3   8.8   LEAST FLYCATCHER
5.3   3.5
93.0 = 12.3 29.7
        8.8   28.1
                   3.5

FIG. 4   ACADIAN FLYCATCHER
3.3
91.8 = 16.4 34.4
        14.8   23.0
                   8.2

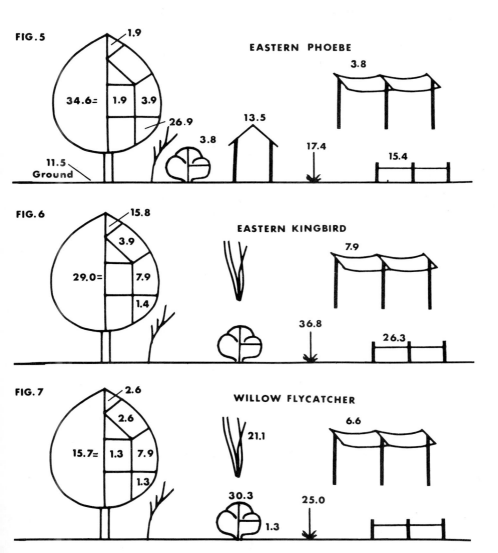

FIGURE 1-7. Percent use of various substrates as foraging perches for each species studied. (after Verbeek, 1975).

TABLE II. Characteristics of Foraging Flights of Seven Species of Tyrannid Flycatchers

	Sortie Flight Distance		Returns to Perch		Methods of Capture	
	N	Median Distance(m)	Return	No Return	Hawking	Gleaning
Crested	44	2.44	29.2%	70.8%	67.9%	32.1%
Pewee	43	1.83	50.8	49.2	93.5	6.5
Least	42	1.22	48.7	51.3	81.0	19.0
Acadian	42	.91	56.9	43.1	83.3	16.7
Phoebe	44	1.52	43.5	56.5	78.0	22.0
Kingbird	68	3.66	48.8	51.2	92.0	8.0
Willow	50	.91	60.0	40.0	56.8	43.2

only significantly less than that of the kingbird (P<0.005).

The remaining two open country species, the Eastern Kingbird and the Willow Flycatcher primarily foraged from living perches (Fig. 6, Fig. 7) such as trees, shrubs and annual herbs, but kingbirds frequently used fences and utility wires. While the substrate diversity indices of these species were not as high as that of the phoebe (Table I), they were considerably higher than values for any of the forest flycatchers. When foraging in trees, the median canopy affinity for the kingbird (.94) was higher than that of any of the flycatchers.This value was significantly higher than the value for the phoebe, but was not statistically different from that of the Willow Flycatcher.

B. Sortie Characteristics.

Comparison of the forest flycatchers (Table II) shows that the crested flycatcher had the longest median flight distance while the Acadian Flycatcher had the shortest. There was no significant difference in median flight distances of the forest flycatchers except between the crested and Acadian flycatchers (P<0.01) and pewee and Acadian Flycatchers (P<0.005). The kingbird had the longest median flight distance of the open country flycatchers. This value is significantly greater than the median flight distance for both the Willow Flycatcher (P<0.005) and the phoebe (P<0.01).

There was essentially equal frequency of returns and non returns to foraging perches for all species except for the crested flycatcher which returned to the foraging perch only 29.2% of the observations. All species foraged primarily by hawking insects. The pewee and kingbird foraged almost exclusively by hawking (93.5% and 92% respectively). The Willow Flycatcher gleaned more than any other species (43.2%).

IV. DISCUSSION

Ecological separation between members of the eastern Tyrannidae is best accomplished by differences in habitat, Hespenheide (1971a) Johnston (1971) and Lederer (1972). While many of the flycatchers considered in this study were separated from one another by habitat differences, the potential for unique assemblages in southwest Virginia was present. The varied stands of forest maturity, differences in elevation, along with other disturbances and edge effects resulted in more resident species and unique assemblages which may have contributed to resource overlap.

Other studies on the eastern members of the Tyrannidae have primarily discussed habitat differences of woodland flycatchers.

Hespendeide (1971a) measured nesting habitats of the Least Flycatcher, Acadian Flycatcher and Eastern Wood Pewee with regards to
the vegetation density between 0.6m and 6.1m. He concluded on the
basis of vertical stratification that in the eastern forest, only
two species of flycatchers could exist in the same habitat, the
crested flycatcher and one small bodied flycatcher such as a
pewee, acadian or least. He failed to sample nonuniform habitats
which could allow for another potential species such as an edge
species like the pewee. This fact was pointed out by Johnston
(1971) who demonstrated in southwest Virginia a large bodied,
medium size bodied and a small bodied species of flycatcher may
coexist. His assemblages for this study were either crested-
pewee-least or crested-phoebe least. Observations made during
the present study indicate that three species may occupy the same
habitat and in unusual edge circumstances, four species may be
present. Johnston (1971) noted no interspecific agression among
the flycatchers he observed, and only rarely have I observed such
agressive encounters and these have usually occurred in edge
situations.
     Several studies have related significant differences in perch
height to vertical habitat partitioning by the species involved
(MacArthur, 1958; Morse, 1968; Pearson, 1971; Williamson, 1971;
Johnston, 1971; Verbeek, 1975). Within the flycatchers, Johnston
(1971) reports vertical stratification of perches between the
Least Flycatcher and pewee in southwest Virginia. Verbeek (1975)
found a significant median perch height difference between the
Black Phoebe (Sayornis nigricans) and Western Flycatcher
(Empidonax difficilis) and the Black Phoebe and Western Wood Pewee
(Contopus sordidulus). Other studies of flycatchers also show
trends of feeding at different heights: Crowell (1968) with
Elaenia martinica and E. flavogaster; Johnson (1963) with
Empidonax oberholseri and E. wrightii: Lederer (1972) with five
species of flycatchers including the Great Crested Flycatcher,
Eastern Phoebe, Eastern Wood Pewee, Acadian Flycatcher and Alder
Flycatcher (Empidonax alnorum). In the present study, selection
of foraging perches with regard to height revealed significant
differences in vertical stratification between most of the forest
flycatchers and between two of the open country species.
     Increased substrate diversity is another method which species
may use to decrease resource overlap as demonstrated with flycatchers by Verbeek (1975) and woodpeckers by Conner (1977). The
phoebe's high index of substrate diversity may have reduced
resource overlap with other syntopic species, particularly Willow
Flycatchers and kingbirds. The lower substrate diversity of the
forest species may reflect a lack of different substrates in many
cases, but these species probably rely more on vertical habitat
partitioning to reduce resource overlap than do the open country
species.
     The length and angle of the foraging flight are other methods
which a species may use to exploit the available resources

(Williamson, 1971; Leck, 1971; Lederer, 1972; Verbeek, 1975).
Leck (1971) believes that the longer foraging flights of the
Eastern Kingbird are associated with pursuit and capture of larger
prey. In the present study, the kingbird and crested flycatcher
displayed the longest median flight distances for their respective
habitats. Hespendheide (1971b) found that these two species do
capture on average larger insect prey items when compared to other
flycatchers.

The length of the flight may affect the frequency of return
to the original perch since after long foraging flights, kingbirds
often did not return to the original perch (Leck, 1971). In this
study, returns versus non returns to the original perch were
virtually equal for all species except the crested flycatcher.
Other reasons for non returns may be due to the need for dispatch-
ing large items of prey at the nearest perch, prey densities and
availability of alternative foraging perches (Leck, 1971; Verbeek,
1975).

Values of canopy affinity (Table I) and median flight length
(Table II) decrease in the same order: crested flycatcher, pewee,
Least Flycatcher and Acadian Flycatcher (phoebe excluded). This
trend probably indicates an increasing density of vegetation from
the canopy to the forest floor. The longer unhampered flights of
the crested flycatcher and pewee occurred in the more open areas
of the upper canopy, while the least and acadian flycatchers
exhibited shorter flights in the denser subcanopy and understory
layers.

Most species foraged by hawking insects with the pewee being
the most consistent in regard to this habit. The Willow Flycatcher and crested flycatcher also showed a high percentage of
gleaning sorties. The practice of gleaning may be one method by
which the Willow Flycatcher, a recent addition to the Virginia
avifauna reduces resource overlap with the kingbird and phoebe. In
other comparative studies of flycatchers, Empidonax flycatchers
demonstrated a greater affinity for gleaning than other species.
(Lederer, 1972; Verbeek, 1975).

Schoener (1965) included tyrant flycatchers in a group of
sympatric congeners that show little character divergence. In
times of resource shortage, these species could either partition
the habitat or adopt differences in foraging tactics. Similar
ecological pressures may also exist at the familial level, but to
a lesser degree. Many past studies of the eastern Tyrannidae
defined habitat requirements of several or individual species,
but frequently did not consider differences in foraging tactics.
Studies such as Crowell (1968), Williamson (1971), Lederer (1972),
Verbeek (1975), Conner (1977) and this one demonstrate that while
habitat separation is probably the most important means of
ecological separation, differences in foraging tactics may be yet
another means by which similar species avoid resource overlap.

REFERENCES

Conner, R. (1977). "Seasonal Changes in the Foraging Methods and
    Habitats of Six Sympatric Woodpecker Species in Southwestern
    Virginia" Unpublished PhD Dissertation, Blacksburg, Virginia,
    Virginia Polytechnic Institute and State University.
Crowell, K. (1968). Auk 85,265.
Hespenheide, H. (1964). Wilson Bull. 76,265.
Hespenheide, H. (1969). "Niche Overlap and the Exploitation of
    Flying Insects as Food by Birds with Special Reference to the
    Tyrannidae" Unpublished PhD Dissertation, Philadelphia Univ,
    Pennsylvania.
Hespenheide, H. (1971a). Auk 88,61.
Hespenheide, H. (1971b). Ibis 113,59.
Johnson, N. (1963). Univ. Calif. Publ. Zool. 66,79.
Johnston, D. (1971). Auk 88,796.
Leck,c. (1971). Wilson Bull. 83,310.
Lederer, R. (1972). "Foraging Behavior and Niche Overlap in Seven
    Species of North American Flycatchers (Tyrannidae). Unpubl.
    PhD Dissertation" Univ. of Illinois at Urbana-Champagin, Ill.
MacArthur, R.(1958). Ecology 39,599.
Morse,D. (1968). Ecology 49,779.
Pearson, D. (1971). Condor 73,46.
Rivas, L. (1964). Syst. Zool. 13,42.
Schoener, T. (1965). Evolution 19,189.
Shannon, C. (1948). The Bell Sys. Tech. Journ. 27,379 and 623.
Siegel, S. (1956). "Nonparametric Statistics for the Behavioral
    Sciences." McGraw-Hill, New York.
Smith, W. (1966). Publ. Nuttall Ornithol. Club 6,1.
Verbeek, N. (1975). Wilson Bull. 87,231.
Williamson, P. (1971). Ecol. Monographs 41,129.

VERTICAL AND TEMPORAL HABITAT UTILIZATION WITHIN A BREEDING BIRD COMMUNITY[1,2]

Stanley H. Anderson[3]
Herman H. Shugart, Jr.
Thomas M. Smith

Environmental Sciences Division
Oak Ridge National Laboratory
Oak Ridge, Tennessee

Vertical and temporal stratification of birds in an eastern deciduous forest are discussed. It is shown that habitat use varies among species within the vertical strata and on a temporal basis in the forest. Subdivision of the habitat by means of behavioral differences in time and in the vertical is shown as an additional means of resource partitioning. Bark gleaning birds show a vertical stratification while canopy feeders are active at different times. Vertical and temporal stratification and stratification of behavior are shown to be forms of community subdivisions which allow birds to use the forest habitat.

I. INTRODUCTION

Several avian habitat-selection studies have related the spatial heterogeneity of vegetation structure to spatial heterogeneity in bird distribution (James, 1971; Shugart and Patten, 1972). Other studies (Hildén, 1965; Wiens, 1969) have concluded that birds select habitats largely on the basis of vegetation structure. Presumably, species tend to use specific resources and exclude other species from these resources (MacArthur and Levins, 1964) and the high degree of specialization related to food exploitation in tropical communities (Karr, 1971) tends to

---

[1]Research sponsored by the United States Department of Energy under contract with Union Carbide Corporation.
[2]Publication No. 1261, Environmental Sciences Division, Oak Ridge National Laboratory.
[3]Present address: Migratory Bird and Habitat Research Laboratory, U.S. Fish and Wildlife Service, Laurel, Maryland 20811.

support this generalization. Cody (1968) shows that in grassland communities, species segregation (implying resource
division) may be divided into horizontal, vertical, or temporal
dimensions.

    The nature of habitat selection in the horizonal dimension
was previously documented for breeding birds of Walker Branch
Watershed, a predominantly deciduous forest in eastern Tennessee
(Anderson and Shugart, 1974). Here we investigate the temporal
and vertical aspects of resource division in this same bird
assemblage. Our specific objectives are: (1) to determine the
vertical distribution of birds; (2) to document vertical and
temporal activity patterns; and (3) to examine vertical and
temporal differences in behavior.

                        II. STUDY AREA

    Walker Branch Watershed is located on the U.S. Department
of Energy Reservation in eastern Tennessee. The Walker Branch
Watershed project is a U.S. Department of Energy ecosystem study
conducted in collaboration with the Eastern Deciduous Forest
Biome Project of the United States International Biological
Program. The watershed occupies a 97.5 ha area of steeply
sloping ridges and narrow valleys which range in elevation from
285 m to 375 m. It is described in detail by Auerbach et al.
(1971).

    Vegetation composition and structure were determined from
298 permanent 0.08 ha sample plots (Curlin and Nelson, 1968).
The basic vegetational data used were tree species, DBH
(diameter breast height) and tree height. A combination of
ordination and classification techniques (Grigal and Goldstein,
1971) was used to define the structure of the watershed using
the basal area of the 12 most common tree species as variables.
Four distinguishable forest types were classified using 131 of
the 298 stands. These 4 types were pine (predominantly Pinus
echinata), yellow poplar (dominated by Liriodendron tulipifera),
oak-hickory (mixed Quercus spp., Carya spp.) and chestnut oak
(Quercus prinus). Six plots, referred to as "core plots"
(Gridal and Goldstein, 1971), of each of the 4 types were used
in this study.

III. METHODS

A. Sampling Methods

Birds were observed on the 24 core plots between 30 May and
20 July 1972. The daylight hours were divided (for sampling
purposes) into morning (6:00-10:00 EST), midday (10:00-5:00)
and evening (5:00-9:00). Three 1-h midday and two 1-h. evening
periods were used to observe birds on each of the 24 plots.
Data were collected by the observer sitting near the edge of
a core plot where a view of the entire plot was possible. At
the end of each 5-min interval of observation the total number
of birds on the plot at that time was recorded. The following
information was noted for each individual: species, sex, age
(juvenile or adult), height, activity. Activity patterns were
designated as: feeding (trunk gleaning, branch gleaning,
foliage gleaning, hawking, pecking); moving (hopping--on
ground, flying--between branches, flying--overhead); singing;
perching.

B. Analytical Methods

A height distribution curve for each species and for the
total bird population on the watershed was plotted by combin-
ing data from all plots at 1.5-m intervals. Temporal curves
were prepared for each species and for the total bird population by tabulating the number of individuals observed during
each 5-min interval. Since the behavior of each individual
was recorded at each height for each time, the following comparisons were possible: (1) variation in bird species composition with height; (2) variation in species behavior with
height; (3) variation of species observed with time of day;
(4) variation in species behavior with time of day. Since the
24 study plots represented four vegetation types, the vertical,
temporal, and behavioral variations examined for each of the
four vegetation types were combined to obtain a description
of avian behavior in the watershed. The Kolmogorov-Smirnov
two-sample test (Siegel, 1956) was used to determine if species
or groups of species showed a significant difference within
behavioral patterns at different heights and time periods.

IV. RESULTS

A. Vertical Distribution of Birds

The cumulative percentage of individual birds (regardless
of species) observed by height in each of the four vegetation
types (Fig. 1) was shown to differ significantly in three of
the four vegetation types using the Kolmogorov-Smirnov sta-
tistic. A difference ($\alpha \leq 0.05$) was found to exist between the
pine and the yellow poplar plots. Likewise the pine and yellow
poplar plots differed significantly ($\alpha \leq 0.05$) from the chestnut oak and oak-hickory but no significant difference was found
between chestnut oak and oak-hickory plots. When the data from
all height observations on the watershed were graphed (all
plots, Fig. 1), it was found that total activity of birds by
height was different ($\alpha < 0.01$) between the combined date plot
and both the pine and yellow poplar stands. The significant
differences in the distributions of vertical activity patterns
can be attributed to differences in the total heights typical
of each of the forest types. The activity patterns are shifted
upward in the yellow poplar plots (which tend to have taller
trees), and are shifted downward in the pine plots (which typically have shorter trees).

In the analyses to follow, we will present data on the common species breeding on Walker Branch Watershed. These species
all can occur on any of the different types of plots and any
tendency for activity patterns to be shifted upward (on yellow
poplar plots) or downward (on pine plots) would tend to increase the variance in the results presented below. We have
chosen not to "correct" the results that follow for plot differences in vertical patterns but note that such corrections
would tend to increase the significance levels presented.

Although there are significant differences in the vertical
distributions of activity of the total breeding bird communities associated with certain of the vegetation types, the general patterns of activity are very similar in shape. About 40%
of the activity is at ground level, and the remainder of the
activity is distributed more or less uniformly through the forests (Fig. 1). There are no sharp zonations in the vertical
pattern nor are there any layers (other than the forest floor)
of concentrated activities.

B. Vertical and Temporal Activity Patterns

To consider in more detail the overall patterns of
vertical-temporal differences between the common bird species
on Walker Branch Watershed, the temporal and vertical activity patterns of all possible pairs of 22 common (more than 19

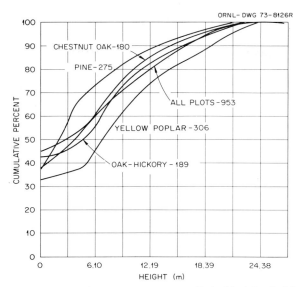

FIGURE 1.   Cumulative percentage of individual birds at each height on watershed (all plots) and on four vegetation types.   Numbers in parentheses indicate degrees of freedom.

observations) bird species are tested for statistical significance using the Kolmogorov-Smirnov test (Siegel, 1956).
Pairs of bird species showing a significant difference in vertical stratification ($\alpha \leq 0.05$) are indicated with a H while
pairs of species showing a significant difference in the time
of activity ($\alpha \leq 0.05$) are shown with T (Table I).   Bird species which concentrate their activity in one or two strata of
the forest yet are occasionally observed over a broad height
range may be distinct from many other species in the table.
    Groups of species that exploit the same class of environ-
mental resources in a similar way, without regard to taxonomic
position, are called guilds (Root, 1967).   The shrub foraging
birds, including Carolina Wren (<u>Thryothorus ludovicianus</u>),
Ovenbird (<u>Seiurus aurocapillus</u>), Wood Thrush (<u>Hylocichla
mustelina</u>), Kentucky Warbler (<u>Oporonis formosus</u>), and Hooded
Warbler (<u>Wilsonia citrina</u>), form one such group and do not
show a significant difference in vertical distribution of activities from one another (Table I).   Similarly, a canopy foraging guild--Blue-gray Gnatcatcher (<u>Polioptila caerulea</u>),
Cerulean Warbler (<u>Dendroica cerulea</u>) and Pine Warbler (<u>Dendro-
ica pinus</u>) do not show vertical stratification differences;
however, comparisons of all warbler species do reveal other
differences among species.   Members of the bark foraging
guild--Red-bellied Woodpecker (<u>Melanerpes carolinus</u>), Downy

TABLE I.   Comparisons of vertical and temporal activity patterns between 22 common bird species using the Kolmogorov-Smirnov test.   Pairs of bird species showing·a significant difference in vertical stratification ($\alpha \leq 0.05$) are indicated with H; pairs of bird species showing a significant difference in time of activity ($\alpha \leq 0.05$) are indicated with T.   Number following species name on left is degrees of freedom (N-1).

	Yellow-billed Cuckoo	Red-bellied Woodpecker	Downy Woodpecker	Eastern Wood Pewee	Blue Jay	Crow	Carolina Chickadee	Tufted Titmouse	White-breasted Nuthatch
Yellow-billed Cuckoo-40		T		T	T		T		T
Red-bellied Woodpecker-26	H			T				T	
Downy Woodpecker-35	H	H		T	T		T		
Eastern Wood Pewee-89	H	H	H		T	T	T	T	T
Blue Jay-54	H					T	T	T	T
Crow-13		H	H	H	H				
Carolina Chickadee-243	H.	H		H		H		T	T
Tufted Titmouse-63	H	H	H	H	H		H		T
White-breasted Nuthatch-23		H	H			H			
Carolina Wren-15	H	H	H	H	H	H	H	H	H
Wood Thrush-14	H	H	H	H	H	H	H	H	H
Blue-gray Gnatcatcher-24	H	H	H	H	H		H	H	H
White-eyed Vireo-5		H	H	H					
Red-eyed-Vireo-138	H	H	H	H	H	H	H	H	
Cerulean Warbler-8		H	H	H	H		H	H	H
Pine Warbler-21	H	H	H	H	H		H	H	H
Ovenbird-4	H					H		H	H
Kentucky Warbler-4	H					H			
Hooded-Warbler-38	H	H	H	H	H	H	H	H	H
Scarlet Tanager-20			H			H			
Summer Tanager-14	H					H		H	
Cardinal-19	H			H		H		H	H

TABLE I. (continued)

Carolina Wren	Wood Thrush	Blue-gray Gnatcatcher	White-eyed Viero	Red-eyed Vireo	Cerulean Warbler	Pine Warbler	Ovenbird	Kentucky Warbler	Hooded Warbler	Scarlet Tanager	Summer Tanager	Cardinal
T	T		T		T	T			T	T	T	T
T			T		T							
T	T		T		T				T	T		
T		T		T		T			T	T		T
T		T		T		T			T		T	T
	T		T		T			T	T	T		
T		T	T	T	T	T			T			
T	T	T	T	T	T	T	T	T	T	T	T	T
T	T		T		T			T	T	T		
	T	T	T	T	T	T			T	T	T	T
H		T			T		T					
H	H		T	T	T	T		T	T	T	T	
H	H	H		T		T	T				T	T
H	H				T				T	T		T
H	H			H		T					T	T
H	H			H					T	T		
H		H	H	H	H	H						
	H	H			H	H			T			
H		H	H	H	H	H						
H	H	H			H	H					T	T
H	H	H	H	H	H	H			H			
H	H	H	H	H	H				H	H		

Woodpecker (<u>Picoides</u> <u>pubescens</u>), and White-breasted Nuthatch
(<u>Sitta</u> <u>carolinensis</u>)--differ from one another in their vertical
stratification.

Temporal variation in activity seems an integral component
of avian community organization and many species differ significantly in their temporal activity patterns (Table I). Concentration of activity of all species in the morning hours is
apparent during data collection. While some species become
active at sunrise, others do not become active until an hour
to an hour and a half after sunrise. Most species are active
until about 11:00. Comparison of the difference in activity
periods of the birds is summarized in Table I. Canopy foraging birds are active at different times while shrub foraging
birds typically did not exhibit either vertical or temporal
stratification.

C. Vertical Behavioral Differences

Since activity sometimes varies with height, a comparison
of the major activity patterns with height of observation was
made to further examine habitat division by the bird species.
Selected results of these analyses are shown in Fig. 2. If
all individuals engaged in the indicated behavior pattern are
distributed equally among the different heights, a horizontal
line at zero on the abscissa would result. The figure on the
y-axis indicates the cumulative percentage difference from an
equal distribution. When fewer individuals are observed in the
behavior pattern than would be expected from an equal distribution among all heights, a negative slope results. Each
curve (Fig. 2) differs significantly ($\alpha \leq 0.05$) from a uniform
distribution.

From examination of 4 behavior patterns of the total breeding bird population on the watershed, it is seen that branch
and foliage gleaning is concentrated between 6 and 20 m, while
trunk gleaning is largely restricted between 6 and 10 m (Fig.
2). Singing birds are most commonly observed between 6 and
12 m. Branch gleaning, foliage gleaning, and singing all
follow similar patterns in the study area and do not differ
significantly among plots. Trunk gleaning occurs at significantly different heights ($\alpha \leq 0.05$) from the other three
gleaning activities.

By examining species which display behavioral differences
with height, we have documented vertical stratification of
behavior patterns. For example, the Yellow-billed Cuckoo
(<u>Coccyzus</u> <u>americanus</u>) was frequently observed in the middle
vegetation layer (Fig. 2). Singing was concentrated between
11 and 17 m while foliage gleaning occurred most commonly
between 6 and 14 m. Other feeding behavior activities
of the cuckoo was not significantly different from foliage

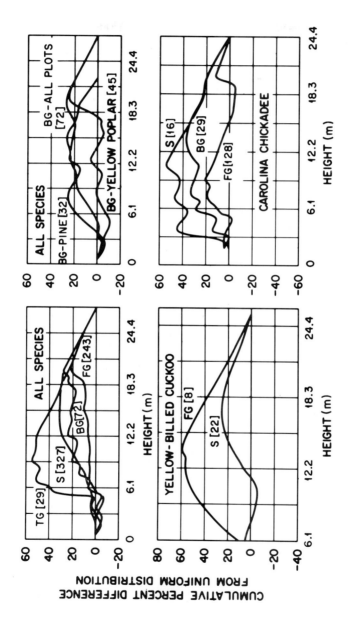

FIGURE 2. Vertical changes in behavior patterns of selected bird species. Numbers following behavior pattern indicate degrees of freedom. (FG = foliage gleaning, BC = branch gleaning, TG = trunk gleaning, PE = pecking, P = perched, FB = flying between branches, S = singing, H = hawking)

gleaning pattern and there was a pattern of stratification of
singing and feeding behavior in this species.

Examination of the behavior of the Carolina Chickadee
(Parus carolinensis), one of the more common species in the
study area, shows a significant difference ($\alpha \leq 0.05$) in the
height at which foliage gleaning and branch gleaning occur.
Foliage gleaning is concentrated in the lower canopy (6-9 m)
and the upper canopy (18-21 m) while branch gleaning is most
common between 3 and 14 m. These 2 behavior patterns change in
the mid-canopy where fewer birds would be expected to foliage
glean. Singing is common in the 2-11 m range.

D. Differences in Temporal Behavior

Whereas activity in general is concentrated in the morning
hours, different forms of behavior predominate at different
time periods. Fig. 3 is developed in the same manner as Fig. 2.
If activity is equally distributed throughout the day a horizontal line at zero on the y-axis would result. A positive
slope during a particular time period indicates that a larger
than expected number of individuals (based on a uniform distribution of behaviors during all time periods) exhibit a given
behavior during that time. A negative slope indicates that a
smaller than expected number of individuals engage in a given
behavior. When the activity of all individuals is plotted
(Fig. 3), feeding behavior patterns of foliage gleaning and
trunk gleaning do not differ at the $\alpha \leq 0.05$ level of significance from one and another. Foliage gleaning is shown as an
example (Fig. 3) and is common in the 07:00-11:00 period.
Singing is concentrated in the 07:00 to 10:30 period. Trunk
gleaning peaks during 09:30-11:00 period while pecking is most
common between 06:00 and 08:30. This may reflect a temporal
adjustment in bark-foraging birds to accommodate differences
in insect activity as more insects emerge and alight on the
bark in the later hours of the morning.

Comparisons of the 4 vegetation types on the watershed
show that 3 types of vegetation differ in temporal as well as
vertical patterns. Branch gleaning, for example, on the com-
posite plot curve occurs throughout the morning hours (Fig. 3).
Branch gleaning is concentrated in the 10:00-11:00 period on
yellow poplar plots but is common in the 08:00-11:30 periods
on the pine plots.

The Yellow-billed Cuckoo shows little temporal variation
in feeding behavior as all feeding behavior patterns are
similar to foliage gleaning behavior (Fig. 3). The feeding
behavior patterns are different ($\alpha \leq 0.05$) from singing.
There is no significant temporal stratification of feeding
behavior in the Carolina Chickadee. Branch gleaning occurs

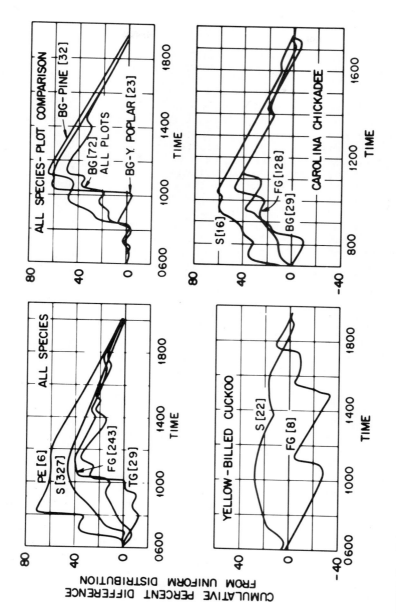

FIGURE 3. Temporal changes in behavior patterns of selected bird species (code as in Fig. 2).

between 07:00 and 11:30 hours, while foliage gleaning occurs
between 8:00 and 10:30. Singing occurs during the 07:00 to
10:30 time period in the chickadee.

V. DISCUSSION

Habitat selection studies show that the presence of bird
species in habitats are correlated with features of the vegetation structure (James, 1971; Anderson and Shugart, 1974).
Bird species composition does not show sharp zones of demarca-
tion but varies on a continuum with vegetation structure and
physical features of the habitat (Bond, 1957; Beals, 1960).
Habitat use at any point along the continuum, therefore, is
variable and related to the habitat features which allow the
species to successfully compete and establish itself in the
community or not to compete if other species can make better
use of the habitat.

Analysis of avian communities requires the determination
of habitat use in relation to the other bird species present.
One form of community subdivision by bird species results in
a form of zonation (MacArthur, 1958; Stallcup, 1968). Such
zonation can be seen as a method of reducing competition by
specialization. In Maine habitat use by warblers varies with
the particular species considered (Morse, 1971). Some species
have stereotyped foraging patterns, requiring specific forms
of habitat structure. Other species have a greater plasticity
in habitat use, thereby being able to live in a larger variety
of habitats.

Karr (1971) shows that bird species feed at preferred
heights, while Cody (1974) finds this form of habitat segregation to be particularly common in foliage insectivores and
sallying flycatchers. Further, birds seem to respond to vegetation in three horizontal layers (0-2', 2-20', and >20') in
temperate forests, and four layers in tropical forests (0-2',
2-10', 10-25' and >25') (Cody, 1974). Colquhoun and Morley
(1943) indicate that birds in an English woodland segregate
into three separate vertical strata each with its own dominant
individuals. Our results agree in part with those of Colquhoun
and Morley (1943) and Cody (1974) in that community subdivision
by height strata appears much finer near the ground than in the
forest canopy.

The behavior of a species may vary within a vertical strata
to provide a finer subdivision of the habitat. For example,
our results indicate that the Yellow-billed Cuckoo ranges from
6 to 26 m, but foliage gleaning in the species predominates
between 6 and 15 m and singing occurs commonly between 11 and
17 m. Vertical stratification of behavioral patterns represent
a further source of community subdivision of bird species.

Additional subdivision on a temporal basis can serve as a method of resource partitioning. Cody (1974) indicates that time is not a large factor in the organization of terrestrial bird communities. Our results show that most of the species are most active during the morning hours. During this period many species show significant temporal variations in behavior (Table I).

The coexistence of bird species depends on habitat use patterns which evolve in each population in the community. These patterns include vertical and temporal changes in behavior which allow different forms of habitat exploitation. The actual sequence of these patterns may affect exploitation patterns (Williamson, 1971). Pearson (1971) shows a direct correlation between relative foliage density and use of each level for foraging by birds in a tropical dry forest. Some bird species require several types of habitat in order to exist. Root (1967) indicates how differences in habitat use occur in feeding young and self-maintenance feeding. Territory selection by birds may be based on the presence of several essential structural features. Forest communities, in particular the eastern deciduous forest, are mosaics of structural forms which become important determinants of the avifauna present. The present study and earlier work (Anderson and Shugart, 1974) indicate that in the breeding bird community considered, species differ significantly in habitat, vertical, and temporal dimensions.

ACKNOWLEDGMENTS

We thank Carolyn Gard for the programming assistance and W. Frank Harris for the vegetation data on Walker Branch Watershed.

REFERENCES

American Ornithologists' Union. (1957). "Check-list of North American Birds" and Supplements. Lord Baltimore Press, Baltimore.
Anderson, S., and Shugart, H. (1974). *Ecology 55*, 828.
Auerbach, S., and others. (1971). "Ecological Sciences Division Annual Progress Report." ORNL-4759, Oak Ridge, Tennessee.
Beals, E. (1960). *Wilson Bull. 72*, 156.
Bond, R. (1957). *Ecol. Monogr. 27*, 351.
Cody, M. (1968). *Am. Nat. 102*, 107.
Cody, M. (1974). "Competition and the Structure of Bird Communities." Princeton University Press, N.J.
Colquhoun, M., and Morley, A. (1943). *J. Anim. Ecol. 12*, 75.

Curlin, J., and Nelson, D. (1968). "Walker Branch Watershed
    Project: Objectives, Facilities, and Ecological
    Characteristics." ORNL-TM-2271, Oak Ridge, Tennessee.
Grigal, D., and Goldstein, R. (1971). *J. Ecol. 59*, 481.
Hildén, O. (1965). *Ann. Zool. Fennica 2*, 53.
.James, F. (1971). *Wilson Bull. 83*, 215.
Karr, J. (1971). *Ecol. Monogr. 41*, 207.
MacArthur, R. (1958). *Ecology 39*, 599.
MacArthur, R., and Levins, R. (1964). "Competition, habitat
    selection and character displacement in a patchy environ-
    ment." *Proc. N.A.S. Zoology 51*, 1207.
Morse, D. (1971). *Ecology 52*, 216.
Pearson, D. (1971). *Condor 73*, 46.
Root, R. (1967). *Ecol. Monogr. 37*, 317.
Shugart, H., and Patten, B. (1972). *In* "Systems Analysis and
    Simulation in Ecology II" (B. Patten, ed.). Academic Press,
    New York.
Siegel, S. (1956). "Nonparametric Statistics for the Behavioral
    Sciences." McGraw-Hill, New York.
Stallcup, P. (1968). *Ecology 49*, 831.
Wiens, J. (1969). "An approach to the study of ecological
    relationships among grassland birds." *Ornithol. Monogr. #8*.
    American Ornithologists' Union, Ithaca, New York.
Williamson, P. (1971). *Ecol. Monogr. 41*, 129.

THE USE OF NESTING BOXES TO STUDY THE BIOLOGY OF THE MOUNTAIN CHICKADEE (PARUS GAMBELI)[1] AND ITS IMPACT ON SELECTED FOREST INSECTS

Donald L. Dahlsten
William A. Copper

Division of Biological Control
University of California
Berkeley, California

Nesting boxes were placed on trees in five plots consisting of 50 boxes each in northeastern California in 1966. The boxes were spaced at 100 m intervals in a grid. The Mountain Chickadee was the most common occupant and the habits of these birds during nesting are described.

The 1966-1975 occupancy rate ranged from 38 to 63%. There were significantly more chickadees in areas with nest boxes than in areas without. Mean clutch size varied between 5.8 and 8.6 and was smaller in second broods when they occurred.

Feeding habits of chickadees were studied by taking stomach samples. Chickadees feed primarily on insects, but ate considerable plant material at certain times of the year.

There was no discernable relationship between pair density, clutch size or nestling density and mortality. The most common cause of mortality was predation due to weasels and snakes. Mean annual fledging success on all plots varied between 56-85%.

---

[1]Studies on predation of different life stages of Douglas-fir tussock moth were supported by the USDA Douglas-fir Tussock Moth Expanded Research and Development Program, Forest Service, Portland, Oregon, FS-PNW-Grant No. 46. Portions of other studies were supported by National Institute of Health Grant #CC0283.

217

Adults and nestlings in some nest box plots were banded
annually. Approximately 50% of the adults, but less than
1% of the fledglings, were recaptured the following year.
The age structure of the breeding population was determined.
About 50% of the nesting pairs stayed with the same mate
during their lifetime. The annual fledgling survival rate
was calculated to be around 15%.

In 1977 predation on several life stages of Douglas-fir
tussock moth was studied. Tussock moth egg masses were
glued on branches at 3 densities and then were preyed upon.
Birds were thought to be responsible for much of the destruction of egg masses.

Nest boxes with cameras showed chickadees fed their
nestlings a variety of arthropods but no tussock moth
larvae. Over one half of the prey consisted of hymenop-
terous and lepidopterous larvae.

In late summer tussock moth cocoons were glued on the
branches and boles of white fir at different densites in
areas of high, low, and no nest boxes. There were
significant differences between areas suggesting that nest
box or chickadee density may be important. Ant predation
was higher than bird predation on the bole of the tree but
the reverse was the case on the branches.

I.   INTRODUCTION

The role of natural enemies in the dynamics of forest insect
populations has been the major focus of our laboratory for 15
years.   In 1965, at the conclusion of a Douglas-fir tussock moth
outbreak in northeastern California we initiated a study of one
of the most common birds in the forests of California, the
Mountain Chickadee (Parus gambeli). We chose this bird because
it was insectivorous, abundant and nested in cavities. We
rationalized that their nesting habit would permit the use of
nesting boxes and allow us the greatest degree of flexibility in
terms of manipulation, as well as facilitate studies of feeding
habits and general biology of the Mountain Chickadee. From the
wealth of literature on European parids we felt that the potential for success with American parids was high.

The overall objectives of our studies were: 1) to evaluate
the influence of increased nesting sites on chickadee populations, 2) to study the biology, ecology and dynamics of
Mountain Chickadee populations in northeastern California, and
3) to investigate the impact of chickadees and other predators
on selected forest insects, particularly the Douglas-fir tussock
moth (Orgyia pseudotsugata (McDunnough)).

II. METHODS

The areas chosen for nest box plots were all similar in that
each had been the site of an outbreak of Douglas-fir tussock
moth during 1964-65. In addition, intensive sampling of white
fir for tussock moth was conducted in these plots in 1976-77.
Plots were in areas between 1340-2800 m in elevation that were
"high graded" (most of the pine removed) from the late 1800's
to mid 1940's, resulting in second growth stands of primarily
white fir.

All of the plots were named from the immediate locale in
which they occurred (Fig. 1). Roney Flat (6), Yellowjacket
Springs (1) and Tom's Creek (4) were studied for the duration of
the study, 1966-1975. Lower Fredonyer (7) and Upper Fredonyer
(9) were studied from 1966-1971 and Middle Fredonyer (8) from
1970-1971. (Middle Fredonyer was used as a non-box census plot
from 1967-1969.) Hilton Spike (2), Rush Creek (3) and Adin Pass
(5) were used as non-box census plots.

A. Plot Design

All nest box plots and census plots were set up in the same
manner with a staff compass and metal tape measure. In other
studies (Kluyver, 1951; Lack, 1955; Campbell, 1968; Ortiz
de la Toree, 1970) nest boxes were not placed at equal intervals. In this study, in order that different areas could be
compared, boxes were placed in a grid pattern at 100 m intervals.
The rectangular plots were laid out so that ten rows of 5 boxes
each ran from north to south as determined by a compass reading.
Check plots were set out in a similar manner with lines being
cut and paint and flagging used to mark the lanes and the 100 m
intervals. The markings in the check plots were used for the
breeding census. No attempt was made to remove natural nesting
sites from any of the plots.

The plots when completed measured 400 x 900 m using the outer
boxes to delineate the area and a 50 m buffer around the entire
plot gave an area of 50 ha. Box density was 0.56/ac or 1.4/ha.

Nesting boxes used were constructed of cement and sawdust
with removable front (Fig. 2). These were purchased from
Schwegler and Sons® in Munich, Germany. In most other nest box
studies a variety of hole sizes and shapes were used, perhaps
due to the large compliment of parids in Europe, but in this
study a front with a standard round 33 mm diameter hole was
used. Boxes were 25 cm high with an outside diameter of 11.5
cm.

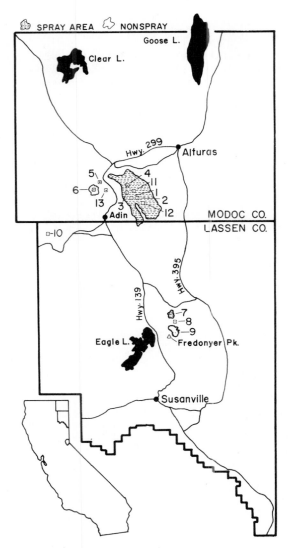

FIGURE 1. Map of northeastern California showing the location of the study plots (1, 4 and 6 were 10-year nest box plots,
7 and 9 were 6-year nest box plots, and 8 was a 2-year nest box
plot; 2, 3, 5, 8 were used as non-nest box census areas; and 10
to 13 were used to collect stomachs). The black solid areas are
lakes and the outlined areas are location of previous Douglas-fir
tussock moth activity.

FIGURE 2.   Commercially produced nesting box used in this
            study.

B. Census Techniques

To investigate the effect of artificial nesting sites a
breeding census was conducted. In both nest box and non-nest
box plots two investigators would start on line three and walk
25 m intervals to the end of the line (400 m). Census takers
walked 50 m apart with one worker walking the cut line, pacing
and keeping time. This procedure was repeated up through line
seven. Counts were made on the basis of sightings and singing.
The censuses were normally taken in the early morning hours. A
census was made on each plot between four and six times,
beginning usually, with early nest building activity in April.
The counting was concluded once egg incubation commenced.

C. Feeding Habits

Four shooting areas (10, 11, 12, 13) were established outside of the nest box and census plots (Fig. 1) and stomach
samples collected from 1966-1972. Specimens were collected in
late summer (September), fall (October-November), winter
(January-February), and early spring (April) during the morning
hours only (7 a.m.-12:00 noon). Stomach samples were analyzed
for percent by volume of animal matter (entirely arthropods),
vegetative material (primarily seed and buds) and grit. A
limit of two hours hunting time was allowed for each area.
Five birds (when available)from each of four areas were taken.
No birds were shot during the breeding season (mid April-
August). However, observations at nest-box sites and some
nestling stomachs allowed identification of some of their food.
A .410 guage shot gun with #9 (bird) shot and .22 (smooth bore)
rifle was used to obtain specimens.

D. Life History Studies

Nesting boxes not only facilitated observation of
chickadees during the nesting period but also during the
remainder of the year since most breeding birds were permanent
residents and tended to stay in or near the plot areas throughout the year. Adult behavior before and during mating,
nest building and nestling development were recorded.
Mortality of eggs and nestlings could be documented and parasites in the nest were identified as were nest inquilines.
Boxes also facilitated the study of interspecific competition
for nesting sites.

Wing chord measurements (Van Balen, 1967) were made of all adults shot for stomach analyses over a four year period. Adults in the nesting boxes were also measured annually and compared. Also, weights of all adults collected and those using the nesting boxes were taken using a 30 g Pesola® spring scale.

Adults and nestlings in the nesting boxes were banded on five plots (Yellowjacket Springs, Tom's Creek, Roney Flat, Upper Fredonyer, and Lower Fredonyer) in 1968 and 1969 but birds on only three plots (Yellowjacket Springs, Tom's Creek, and Roney Flat) were banded from 1970 to 1975. The bands (size #0) were supplied by the U. S. Fish and Wildlife Service.

E.   Tussock Moth Predation Studies

In 1977, a study was initiated to try to determine the impact of chickadees and other predators on various stages of tussock moth. There were three types of areas available for study--areas without bird boxes, areas with boxes spaced at 100 m intervals and an area with boxes at 50 m intervals. In 1975, extra boxes were placed in the eastern half of the Yellowjacket Springs plot so that half of the plot had boxes at 50 m intervals and the other half at 100 m intervals.

The tussock moth has a one-year life cycle. The larvae hatch from overwintering egg masses in early June and develop through five or six instars. Pupation occurs in July and August. Adults begin emerging in late July or early August and continue to emerge into September and October. Wingless females emerge from cocoons, mate and lay their eggs on the top of the cocoon. Egg masses, larvae and pupae within the cocoon are all subject to predation by the Mountain Chickadee as well as other birds.

1) Three separate studies were undertaken. The first was a spring egg mass study. Egg masses overwinter and are therefore subject to predation for more than six months. Since populations were very low, egg masses were obtained from a laboratory colony that is maintained by the USDA, Forest Service, Pacific Southwest Forest and Range Experiment Station, in Berkeley. Only 100 egg masses were available so this study was conducted in the high density nest box plot (50 m intervals).

The egg masses were placed on the undersides of branches in the lower crown of white firs that were chosen randomly. There were 18 egg mass placement sites, six replicates of three

densities (1, 5 and 10 egg masses per branch). Three treatments
were done per row and   repeated until 96 egg masses were placed
on 18 trees. The egg masses were glued on with Weldwood® contact
cement as were the cocoons that are discussed below.

2) Since nest boxes were already present in areas known to
have tussock moth larvae we decided to concentrate on the
Mountain Chickadee as a predator of DFTM larvae. The hatching
of chickadee eggs coincide with DFTM egg hatch, and the birds
feed commonly in areas on the tree where encounters with DFTM
larvae would be expected--the freshly open buds and new foliage
of white fir.

To determine the extent of DFTM in the diet of nestling
chickadees a method developed by Royama (1957) was used. Three
special nest boxes, equipped with Minolta® super 8 mm movie
cameras, were constructed to photograph adult birds returning to
the nest with food (Fig. 3). Each time a bird enters, a single
frame picture of the bird's head and beak contents is taken. In
addition a small watch on the inside near the entrance is also
photographed. In this way time of day, date and amount and kind
of insects can be recorded automatically for the entire nesting
period. Nests, containing nestling chickadees, are easily
transferred to nest boxes with cameras.

3) The cocoon-pupal stage is relatively short compared with
the other stages but based on previous observation it is one of
the most often preyed upon. Cocoons in the numbers that we
needed for this study could not be collected readily in the field.
In addition we wanted to work with known densities of cocoons on
the branches and bole of white fir. In order to do this we field
collected larvae. Approximately 5000 fourth, fifth and sixth
instar DFTM larvae were collected by beating white fir foliage.
These DFTM larvae were placed in large (1 m x 1 m x .5 m) card-
board boxes with muslin tops. The larvae were fed fresh white
fir foliage every three days and those larvae that had spun
their cocoons and pupated on the old foliage were collected.
Cocoons were clipped off along with the small portion of foliage
on which they had spun. All cocoons at each collection time were
divided into lots of 26. This was the number of cocoons needed
to stock one four-tree block with cocoons and was done to
standardize the age of cocoons in each treatment block.

Assuming that the nest box plots might have an influence on
Mountain Chickadee density, three areas were selected for the
establishment of treatment blocks: 1) an area without nest
boxes, 2) an area with a nest box density of 1 box per hectare
(100 m interval), and 3) an area with a density of 4 boxes per
hectare (50 m interval).

FIGURE 3. A nesting box with a movie camera attached as used in this study for evaluation of the prey brought to Mountain Chickadee nestlings.

Sites throughout each study area were selected by finding four open grown, white fir trees approximately 10-15 m in height and five to 15 m apart. In this manner 14 blocks were chosen in each study area. Each tree in the four tree block was flagged and numbered one through four and the treatment assigned randomly to each tree.

In each treatment cocoons were placed at a distance of 1.5-2 m from the ground. The four treatments were as follows:

1) one cocoon near the tip of a branch.
2) five cocoons on the bole of the tree in an area 0.3 x 0.6 m.
3) ten cocoons, placed near the tip of the foliage, spaced 0.5-1.0 m apart, around the tree.
4) ten cocoons, on a single branch, in an area approximately 0.36 m².

Cocoons were glued to the underside of the foliage or on the
bole with contact cement which sticks quickly, is waterproof, and
does not become hard or brittle. Only the foliage on which the
cocoon had spun was glued to the tree's foliage. The cocoons
were observed and examined periodically for evidence of predation
and to determine the type of predation.

III. RESULTS AND DISCUSSION

In addition to the Mountain Chickadee other cavity nesters
used the boxes such as the Red-breasted Nuthatch (Sitta cana-
densis Linnaeus) and the White-breasted Nuthatch (Sitta carolinensis Latham). Nuthatches nested one to two weeks earlier
than chickadees. House wrens (Troglodytes aedon Vieillot) were
relatively common in certain years on the Upper Fredonyer plot,
and only nested occasionally on two of the other plots. The
wren, a migratory bird, usually began nest building a week to
two weeks later than the chickadees and actively competed for
nest boxes. Chipmunks (Eutamias spp.) and white-footed deer
mice (Peromyscus spp.) frequently use the nest boxes and sometimes produce offspring in them. The chipmunk is suspected of
preying on nesting chickadees, eggs, and nestlings. Bees and
vespids used the boxes for nesting on several occasions.
Pairing of Mountain Chickadees on the Modoc Plateau was in
the first part of April. Nest building began during the latter
part of April and the first eggs were usually laid in early May
depending upon weather and elevation. In some years second
broods were produced and in these cases egg laying commenced
during July. Second broods seem to be a result of cold, wet
weather conditions during the early first brood incubation
period.
Nests are usually constructed of small woodchips and fur
(Fig. 4). The woodchips are often taken from soft rotting logs
and arranged in a circular fashion to form a base. In some
cases, the birds will use grass, moss, lichen, thistles or a
combination of the above materials to form the base and this
presumably depends upon availability. Above and into this
funnel shaped base is placed fur from animal scats, mainly coy-
ote, (probably raptor pellets too) as small bones of rodents
can be seen throughout the nesting material. A center depression is developed which is at first and through egg-laying
rather loose and covered with a cap of fur. Once incubation
starts the nest becomes more compact with a well defined egg-cup.
Both the male and female participate in bringing material to
the nest. However, a number of boxes each year are found with
only a few chips and/or small clumps of fur indicating a possible

FIGURE 4. A typical Mountain Chickadee nest showing the fur cup on top of a wood chip foundation.

nesting site choice, presumably on the part of the male. The female is the only one that incubates the eggs with the male bringing her food; at times she leaves to forage for herself. The female lays one egg each day and incubation begins within 24 h after the last egg is laid.

Incubation usually lasts 12–13 days, however, in several years (1967, '68, '71, '74) and especially in 1971 there was evidence of extended incubation of up to 15 days in some nests

due to adverse weather. Hatching of a clutch of eggs usually
took place within 24 h. Brooding of nestlings by the female
continues during the first part of nestling life with fledging
occurring within 18-21 days.

A. Nest Box Occupancy

Nesting boxes were used almost exclusively by the Mountain
Chickadee, however, Red-breasted Nuthatches and White-breasted
Nuthatches occasionally occupied the boxes. It cannot be
ascertained whether or not the nuthatches actually competed with
the chickadees for the nesting boxes due to the low numbers of
these birds. The total number of Red-breasted Nuthatch occu-
pancies in the ten year study on all plots was 21 and they were
distributed through the years as follows: 1966 (1), 1967 (1),
1968 (2), 1969 (11), 1970 (3), 1972 (2) and 1975 (1).

White-breasted Nuthatches were even more uncommon and only
used the boxes in the Lower and Middle Fredonyer plots (1 in
1969; 4 in 1970; and 2 in 1971). House Wrens were only common at
Upper Fredonyer and the number of nests by year was as follows:
1966 (10), 1967 (21), 1968 (19), 1969 (13), 1970 (12), and 1971
(5 + 2 second broods). There were only 7 nests during the 10
year study on all of the other plots combined.

Studies of European parids have shown that the number of
nesting pairs could be increased by increasing the number of
nest boxes per hectare (Bruns, 1960). As a part of another
study (Dahlsten and Copper, unpubl. data), we found that the
number of Mountain Chickadee nesting pairs was two to three
times greater in the half of the Yellowjacket plot with the
boxes spaced at 50 m intervals (1.76 boxes/ha) than in the half
with the standard 100 m spacing (0.54/ha) in three years, 1976-
1978. In the plots established in this study we felt that the
occupancy rate of the boxes was a good general indicator of
Mountain Chickadee population size (Table I). The lowest overall
occupancy was in 1967 and 1968 and there was a noticeable drop
(25%) in 1971 from 1970. The factors responsible for these low
years or apparent drops in chickadee populations remain unknown.
Whether or not it is due to abundance of total food at a certain
time of year, to weather, or to fluctuations in insect populations cannot be ascertained. An attempt to demonstrate that
insect abundance was critical would be most difficult. The
methods for sampling the diverse numbers of insects that would
be included in the diet preclude any meaningful statement from
such data. The occupancy data for each plot are fairly consistent in that the low years in each area usually coincide.
However, there are some exceptions (Table I).

TABLE I. Percent Occupancy of Nesting Boxes by Mountain Chickadees in Two Plots (1966-1971) and Three Plots (1966-1975) in Northeastern California (Modoc and Lassen Counties)

Percent Occupancy

Areas	1966	1967	1968	1969	1970	1971	1972	1973	1974	1975	Ten Year $\overline{x}$
Yellow-Jacket	44	34(2)[a]	50(4)	62	86	64(40)	76	62	58(4)	92	63%
Tom's Creek	42	20(4)	26(8)	34	50	30(30)	60	56	52(2)	80	45%
Roney Flat	22	16(8)	28(8)	50	54(6)	20(18)	44	44	34(2)	64	38%
Upper Fredonyer	66	18(6)	18(6)	50(2)	72	54(28)					46%[b]
Lower Fredonyer	56	20	29(10)	52	64	34(24)					41%[b]
$\overline{x}$	46	22(4)	28(7.2)	50(.4)	65(1.2)	40(28)	60	54	48(3)	79	

[a] ( ) indicates percent of boxes with 2nd broods
[b] Six year mean.

B.   Nest Boxes and Chickadee Abundance

The results of the spring census showed that there were more
birds (not pairs) in nest box plots than in those plots without
nesting boxes in every year (Fig. 5). There was only one year
when the number of birds in a check plot exceeded those in any
of the nest box plots and that was 1969 (Table II) when Rush
Creek exceeded Tom's Creek. Because of the variability between
years in chickadee populations, plots are compared within years
only. A two-way analysis of variance showed significant differences between nest box and non-nest box areas (P < .05) and
between years (P < .05).

There is no doubt that taking a census within a nest box
plot introduces an artifact but the degree to which the boxes
are accepted (Table I) by the Mountain Chickadee indicates that
addition of nesting sites can indeed increase the number of
birds in a given area.

Kluyver (1951) demonstrated a four-fold increase of the Great
Tit, (Parus major) in Europe by using nest boxes. Tinbergen
(1946) has also shown that in areas with few natural nesting
sites the population of titmice could be doubled by the addition
of nesting boxes. In both these studies a singing male census
was used to evaluate the test. Many other authors (e.g.,
Mackenzie, 1952; Lack, 1954; Franz, 1961; Tichy, 1963; Staude,
1968) have also stated that cavity nesters such as the parids
can be increased by the addition of nesting sites. Our results
show that the same generality is applicable to the Mountain
Chickadee in northern California.

C.   Stomach Analyses

Samples were taken four times per year and there were
definite peaks in the abundance of arthropods in the stomachs
(Fig. 6). This pattern was similar in each of the areas where
stomachs were collected regularly so the data were pooled.
Whether or not the same patterns occurred in the nest box areas
cannot be stated, however it is well known that insect species
are not equally abundant everywhere. It is possible, on the
other hand, that insect biomass is generally higher in some
years than in others. This might be a reflection of the climate
during the winter, overall moisture, etc. It can be speculated
that insect abundance may influence Mountain Chickadee occupancy
rates in the nest boxes, clutch size and possibly egg and
nestling mortality.

Since stomachs were not taken during the nesting period,
samples prior to and after the nesting period were used as a
reflection of insect abundance during nesting. From our data

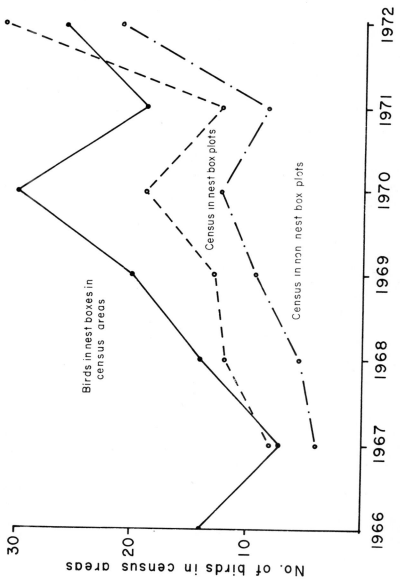

FIGURE 5. Mean annual census figures showing the number of birds in nest box and non-nest box plots and the known number of birds on the nest box plots.

TABLE II. Census Data from Study Areas With and Without Nesting Boxes

			Nest box areas				Non box areas		
Area	Year	#[a]	x̄ # birds observed	Birds/ha	# birds[b] in area	Area	#[a]	x̄ # birds observed	#/ha
Yellow-jacket Sp.	1967	4	12.3	0.61	14	Hilton Spike			
	1968	4	17.3	0.85	20				
	1969	4	16.5	0.81	26				
	1970	4	25.5	1.26	40		4	14.5	0.72
	1971	4	15.0	0.74	32		4	9.0	0.44
	1972	5	31.6	1.56	28		5	21.2	1.05
			x̄=23.7		x̄= 33.0			x̄=14.9	
Tom's Creek	1967	4	7.9	0.39	4	Rush Creek	4	5.3	0.26
	1968	4	11.7	0.58	10		4	6.6	0.33
	1969	4	10.5	0.52	10		4	14.2	0.70
	1970	4	17.7	0.87	24		3	10.0	0.49
	1971	1	11.0	0.54	16		2	7.5	0.37
			x̄=11.3		x̄=12.2			x̄= 8.7	
Roney Flat	1967	4	7.7	0.38	6	Adin Pass	3	8.3	0.41
	1968	4	12.0	0.59	14				
	1969	4	14.5	0.72	26				
	1970	4	15.6	0.77	26				
			x̄=12.6		x̄=17.2			x̄ = 8.3	

TABLE II. (Continued)

Area	Year	#	x̄ # birds observed	Birds/ha	# birds in area[b]	Area	#[a]	x̄ # birds observed	#/ha
Lower Fredonyer	1967	3	6.0	0.30	12	Middle Fredonyer	4	2.5	0.12
	1968	3	13.0	0.64	10		3	5.3	0.26
	1969	4	19.5	0.96	22		4	7.2	0.35
			x̄ = 9.5		x̄=14.7			x̄=5.0	
Upper Fredonyer	1967	3	3.6	0.18	10				
	1968	3	10.0	0.49	12				
	1969	4	16.0	0.79	22				
			x̄=10.3		x̄=14.7				
x̄ all years			13.5		22.4			9.2	

[a]Indicates number of times plot was censused.
[b]Indicates number of birds in nest boxes in census area.

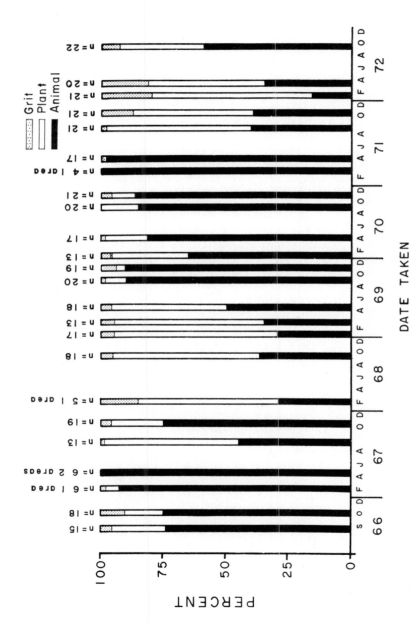

FIGURE 6. Average percent of animal and vegetable matter, and grit in Mountain Chickadee stomachs at different times of the year from four study areas.

there was no obvious relationship between contents of the
stomachs and overall mortality. For example, fledging success
was high (84%) in 1968 but the arthropod content was around 30%
both before and after nesting. In 1970, the arthropod content
was over 75% but fledging success was only 68%. In addition,
each area must be considered individually as mortality was shown
to vary considerably from year to year by plot (Dahlsten and
Copper, unpubl. data).

Occupancy rates were also not affected by arthropod abundance
(Table I). However, clutch size may have been influenced. Clutch
sizes were higher in 1967, 1970 and 1971 and lower in the other
three years that stomachs were taken (Table III). This is only
circumstantial evidence but arthropod abundance, as reflected by
analysis of chickadee stomachs, in the months preceding nesting
may well have an influence on clutch size.

Lack (1955) in studies on British tits noted that clutch
sizes tend to be larger in years that caterpillars are more
abundant. His reference was to insect abundance closer to the
time of nesting than ours as we were looking at arthropod
abundance several months prior to the onset of nesting.

Mountain Chickadees are opportunistic feeders and feed on a
wide variety of insects. While they are primarily insectivorous,
they do feed extensively on seeds and other vegetative material
at certain times of the year (Fig. 6). There does not appear to
be any pattern to this switch to plant materials but it may be
related to arthropod availability. Nestlings were always fed
arthropods but on several occasions seeds were found in the guts
of dead nestlings. This may have been an indication of arthropod shortage and that the nestlings had starved.

Most of the plant material in the stomachs could not be
identified, but the birds did feed on conifer seeds, probably
white fir. The arthropods were identified from stomachs and by
observation at the nesting boxes in each plot (Table IV). Many
times insects were found on the edge of the nest or when banding,
in the adult bird's beak.

Generally, during nesting in late spring and early summer,
chickadees feed on large larvae of geometrids (Melanolophia and
others), sawflies (Neodiprion spp.), web-spinning sawflies
(Acantholyda spp.) and the smaller bud-mining sawflies (Pleroneura
spp.). Most of these occur on white fir except for Acantholyda
which is on pine. An adult scarab, Dichelonyx sp., which is
abundant on pine in certain years was brought to the nest boxes
frequently. The metallic green elytra were easily recognized.
In addition, almost every nestling stomach examined had head cap-
sules of this scarab.

During the late summer (September, Table IV) food consists
primarily of lepidopterous larvae and some adults, spiders,
aphids, cercopids and cicadellids. It should be noted that

TABLE III. The Number, Mean, and Standard Deviation of Clutch Sizes[a]

1st Broods

Area	1966			1967			1968			1969			1970		
	N	X̄	S.d.	N	X̄	S.d.	N	X̄	S.d.	N	X̄	S.d.	N	X̄	S.d.
YJ	22	6.1	.89	10	8.3	1.17	21	7.6	1.01	31	7.3	1.28	43	7.3	1.15
TC	21	5.8	.60	6	7.5	.50	10	6.6	.66	16	7.2	.99	24	7.5	1.28
RF	10	5.9	.73	8	8.1	.76	9	7.4	1.01	25	6.7	.94	25	7.3	.94
UF	33	5.9	.93	8	7.9	1.13	6	7.8	.41	17	6.6	1.27	30	7.3	1.17
LF	27	6.1	.78	6	7.5	1.38	7	8.0	.82	23	6.3	.85	32	6.8	.96
MF													15	7.5	.91

Area	1971			1972			1973			1974			1975		
	N	X̄	S.d.	N	X̄	S.d.	N	X̄	S.d.	N	X̄	S.d.	N	X̄	S.d.
YJ	31	8.3	1.24	30	6.4	1.24	30	7.6	.89	24	7.7	.91	41	7.2	.99
TC	14	8.6	1.11	28	6.7	.71	27	7.4	1.01	25	7.8	.90	35	7.5	1.18
RF	9	8.6	.53	18	6.7	.95	20	7.2	.73	17	7.6	.93	31	7.1	1.02
UF	22	7.7	1.10												
LF	16	7.8	.75												
MF	14	7.5	1.45												

TABLE III. (Continued)

							2nd Broods									
	1967			1968			1969			1970			1971		
Area	N	X̄	S.d.	N	X̄	S.d.	N	X̄	S.d.	N	X̄	S.d.	N	X̄	S.d.
YJ	1	6.0	--	2	6.0	--							18	6.0	1.33
TC	2	5.5	.71	4	5.5	.50							13	5.8	.77
RF	4	5.7	.50	4	5.0	.82							9	6.5	.73
UF	2	5.0	--	3	7.0	--				3	5.6	.58	13	5.8	.90
LF				5	5.4	.55							12	5.2	1.03
MF							1	6.0	--				9	6.1	.93

	1974		
Area	N	X̄	S.d.
YJ	2	5	1.0
TC	1	5	--
RF	1	4	--

[a] X̄ clutch size calculated from nests from which no eggs were taken for analysis and from those in which no disruption occurred. Obvious renests or extremely late nests were also excluded.

TABLE IV. Percent Occurrence of Arthropod Prey in the Stomachs of the Mountain Chickadee at Four Different Time Periods during the Year from Five Areas

SEPTEMBER   N = 139 stomachs

	1966					1967					1968					1969					1970					1971					1972					% Occurrence
	A	B	C	D	E	A	B	C	D	E	A	B	C	D	E	A	B	C	D	E	A	B	C	D	E	A	B	C	D	E	A	B	C	D	E	
	5	6	5	0	5[b]	4	4	3	1	0	5	4	5	4	4	5	7	5	5	0	5	5	5	5	0	4	7	5	5	0	5	7	4	5	0	
Lepidoptera lar. & ad.	x	x	x			x[c]	x	x			x	x	x	x	x	x	x	x	x	x	x		x	x	x	x		x	x	x			x			38.2
Spiders	x	x	x			x	x	x			x					x					x	x	x			x	x		x		x					10.5
Aphids	x					x					x						x	x			x					x									x	9.4
Cercopidae																		x	x			x	x	x												8.0
Cicadellidae																		x	x				x	x	x			x	x	x			x	x		6.5
Psocidae	x	x	x			x	x				x	x						x																x		4.5
Hymenoptera (wasps)	x	x	x			x	x	x				x	x																x				x			4.5
Hymenoptera (ants)													x	x														x				x	x			3.5
Hemiptera	x						x					x																				x				3.0
Diptera																		x					x	x									x			2.5
Chrysopidae		x																					x	x									x			1.5
Pentatomidae eggs	x																																			1.0
Tussock moth eggs		x																																		1.0
Pentatomidae adults			x																															x		1.0

TABLE IV. (Continued)

SEPTEMBER N = 139 stomachs

	1966					1967					1968					1969					1970					1971					1972					% Occurrence
	A	B	C	D	E	A	B	C	D	E	A	B	C	D	E	A	B	C	D	E	A	B	C	D	E	A	B	C	D	E	A	B	C	D	E	
	5	6	5	0	5[b]	4	4	3	1	0	5	4	5	4	4	5	7	5	5	0	5	5	5	5	0	4	7	5	5	0	5	7	4	5	0	
Sawfly larvae																										x										0.5
Coleoptera	x																																			0.5
Pseudo-scorpions	x																																			0.5

NOVEMBER N = 98

	1966					1967					1968					1969					1970					1971					1972					% Occurrence
	A	B	C	D	E	A	B	C	D	E	A	B	C	D	E	A	B	C	D	E	A	B	C	D	E	A	B	C	D	E	A	B	C	D	E	
	5	5	5			4	5	5	5		NS[d]					5	5	4	5		5	7	4	0		4	6	5	6		NS					
Aphids	x	x	x			x					x					x	x	x	x	x	x	x	x			x										19.7
Lepidoptera lar. & ad.	x	x				x										x	x	x	x		x	x	x			x										15.4
Sawfly adults- Neodiprion						x	x				x					x					x	x	x			x	x	x								14.8
Spiders	x	x	x	x		x	x									x					x	x				x	x	x								10.9
Diptera (lar. pup. & ad.)	x	x	x	x							x					x										x										6.8
Sawfly larvae	x	x	x	x		x	x																													3.1
Hymenoptera (wasps)											x					x	x	x																		3.1
Cercopidae	x																									x										3.1
Psocidae	x																				x															2.5
Cicadellidae						x															x					x										2.5
Homoptera											x					x	x									x										2.5

239

TABLE IV. (Continued)

NOVEMBER N = 98 (cont.)

	1966 (A B C D E)	1967 (A B C D E)	1968 (A B C D E)	1969 (A B C D E)	1970 (A B C D E)	1971 (A B C D E)	1972 (A B C D E)	% Occurrence
(sample size)	5 5 5	4 5 5 5	NS[d]	5 5 4 5	5 7 4 0	4 6 5 6	NS	
Chrysopidae	x x							1.8
Coleoptera	x	x						1.8
Curculionidae				x	x			1.3
Hymenoptera (ants)	x			x				1.3
Matsucoccus bisetosus		x						0.6
Pentatomidae eggs						x		0.6
Tussock moth eggs	x							0.6

FEB., JAN., MARCH N = 79

	1966 (A B C D E)	1967 (A B C D E)	1968 (A B C D E)	1969 (A B C D E)	1970 (A B C D E)	1971 (A B C D E)	1972 (A B C D E)	% Occurrence
(sample size)	NS	6 0 0 0 0	5 0 0 0 0	8 0 6 10 5	5 5 0 3 0	5 0 0 0 0	4 7 5 5 0	
Matsucoccus bisetosus preadults								48.0
Lepidoptera lar. & ad.		x	x	x	x	x	x	12.6
Coleoptera		x	x	x x x	x	x		6.3
Curculionidae		x		x				6.3
Spiders				x			x	5.2
Pentatomidae eggs		x	x	x		x		5.2

240

TABLE IV. (Continued)

FEB., JAN., MARCH N = 79 (cont.)

	1966					1967					1968					1969					1970					1971					1972					% Occurrence
	A	B	C	D	E	A	B	C	D	E	A	B	C	D	E	A	B	C	D	E	A	B	C	D	E	A	B	C	D	E	A	B	C	D	E	
	NS					6	0	0	0	0	5	0	0	0	0	8	0	6	10	5	5	5	0	3	0	5	0	0	0	0	4	7	5	5	0	
Sawfly ad.																																x				4.2
Aphids																					x	x														3.1
Hymenoptera (ants)																					x										x					3.1
Hymenoptera (wasps)																					x	x														2.1
Diptera																					x	x														2.1
Coccinellidae						x																														1.0
Hemiptera						x										x																				1.0

APRIL-MAY N = 79

	1966					1967					1968					1969					1970					1971					1972					% Occurrence
	A	B	C	D	E	A	B	C	D	E	A	B	C	D	E	A	B	C	D	E	A	B	C	D	E	A	B	C	D	E	A	B	C	D	E	
	NS					2	0	0	4	0	NS					5	5	5	4	0	4	4	5	4	0	4	6	3	4	0	4	7	4	5	0	
Curculionidae (Scythropsus sp)						x										x					x	x	x	x		x	x	x	x		x	x	x			38.6
Matsucoccus bisetosus ♀ ad.						x										x	x	x	x		x	x				x	x	x	x		x	x				28.5
Lepidoptera larvae						x										x	x	x			x	x				x	x				x	x				24.0
Hymenoptera	x																									x					x					3.2
Spiders	x																				x					x					x					3.2
Cercopidae	x																																			0.8

TABLE IV. (Continued)

Observations of adults feeding nestlings June-July

Dichylonyx
Acantholyda
Neodiprion
Xyelidae--Pleroneura spp.

Geometridae--particularly <u>Melanolophia</u> sp.
Cicadidae (abdomens)
Ants (<u>Camponotus</u>)
Tupulidae

[a]Letters refer to the following areas: A = Brown's Canyon, B = Likely Mill, C = Head of Rush Creek, D = Roney Flat (not the nest box plot), E = Fredonyer Canyon.

[b]Number refers to the number of stomachs examined per area.

[c](x) indicates presence.

[d]NS = no samples taken.

Note: We are trying to show similarities between areas and seasons.

nearly 25% of the diet at this time of the year consists of
sucking insects or homopterans. The presence of insect eggs
(pentatomids and Douglas-fir tussock moth) is also interesting.
Tussock moth egg masses in the study areas also showed signs of
predation.

In the fall (November, Table IV) aphids predominate in the
diet, followed by Neodiprion sawfly adults. Sawfly adults oviposit on the tips of fir or pine foliage and after oviposition
remain on the tips. Lepidopterous larvae and adults are much
less abundant in the diet while spiders remained about the same
as in September. Insect eggs were found in the stomachs during
this period also. The pre-adult of a mealy bug, Matsucoccus bisetosus, also began to show up in the stomachs. The overwintering
form of this mealybug is unique and easily recognized.

The most common insect in the stomach during the winter
(Table IV) was the overwintering preadult of M. bisetosus
followed by lepidopterous larvae and adults. Aphids were much
less abundant and spiders were 1/2 of what they were in the previous two periods.

In the early spring the predominant insect in the diet was a
small brown adult weevil, Scythropus sp. Members in this genus
feed on old foliage of pines and Douglas-fir in spring and early
summer and are most often seen on young trees but also occur on
older trees (Furniss and Carolin, 1977). Matsucoccus female
adults and lepidopterous larvae were also common (Table IV).

Much can be learned of the chickadee's feeding habits in the
trees during the various seasons, however, there is a possibility
that the chickadees store food. The seasonal habits could be
misleading since Haftorn (1954) found European parids to store
food and noted similar habits with the Mountain Chickadee in
Colorado (Haftorn, 1974).

D. Chickadee Mortality

The nesting boxes permitted a close examination of egg and
nestling mortality. Mortality in the various stages of development was variable between plots in any given year as was the
variation in a given plot from year to year. Pooling all plots,
mortality was greater in the egg stage than apparent mortality
in the nestling stage for six of the years and nestling mortality exceeded that in the egg stage in four of the years.
Total mean values for the ten year study showed apparent
nestling mortality to be slightly higher than egg mortality
(15.7 - 14.1%).

Egg mortality in second broods during the study period was
slightly higher than that in first broods (17.3%) but nestling
mortality was lower (9.1%). Lack's (1955) data on second brood

mortality of the Great Tit varied by plot and was inconclusive,
but Perrins (1965) stated that for the Great Tit in Britain
second broods were highly unsuccessful in most years as compared
with earlier broods. In our studies with the Mountain Chickadee
we found second brood mortality to be variable.

Determination of the cause of mortality was difficult as the
boxes were checked at three to seven day intervals. In cases
where nests were abandoned we could not be certain if the adult
had died or had left the nest for some other reason. Nestling
mortality was particularly troublesome as we did not know if it
was disease, starvation, weather, or a combination of these.
As we did not have a scheme for sampling all the prey of the
chickadee it was impossible to relate mortality to insect
abundance. There was no relationship between pair density, clutch
size, or nestling density and mortality. In addition we could
not find any definite relationship between cold, wet weather and
mortality.

Since nests were not checked daily, the activity of carpenter
ants (Camponotus modoc Wheeler) further complicated accurate
determination of mortality because of their nest cleaning
activities. Ants were observed removing insect parts left by
the birds and portions of dead nestlings. Ants also tunnel
through the nesting material, which makes further mortality
analysis difficult. The ants, however, do not disturb the
larvae and puparia of a fly, Protocalliphora sp., that is parasitic on chickadees. Other animals, white footed deer mice
(Peromycsus sp.) or chipmunks (Eutamias sp.), may also remove
unhatched eggs or dead nestlings.

Predation was probably the most common cause of mortality
but not a single predator was caught in the act on the plots in
northeastern California. In other studies we have film of a
weasel, Mustela sp., preying on nestlings as well as observations
of 2 gopher snakes, Pituophis catenifer, investigating a camera
unit from which a nest had just been removed (Dahlsten and
Copper, unpubl. data). Predation of a rubber boa, Charina
bottae, on Chestnut-backed Chickadees (9 day old nestlings) has
been observed in California (Copper et al., 1978).

Snakes apparently remove the contents of the nest but do not
disrupt the nest structure itself. This was found by Copper et
al. (1978) with the rubber boa as well as by Laskey (1946) who
found 23-40% of eastern blue birds (Sialis sialis) in nest
boxes were preyed on by the pilot snake (Elaphe obsoleta).

Mammal predation was easy to recognize since parts of the
nestlings or adults were present and the nest disrupted. Parts
of brooding females and broken eggs have been found as well as
dead nestlings. With mammal predation the entire contents of
the nest are rarely removed.

Four small mammals were suspected as predators of chickadee eggs and nestlings. The long-tailed weasel (<u>Mustela frenata</u>) and the least or short-tailed weasel (<u>M. erminea</u>) are known to occur in the area and the long-tailed weasel was observed on the Yellowjacket plot. White-footed deer mice occasionally nested in the boxes and may have competed for nest sites with chickadees. They may also have fed on remnants left by predators or acted as predators themselves as in some situations they were found in boxes that had chickadee nests in them. The most common mammal inhabitants of the nest boxes were chipmunks, but they usually did not occupy the boxes until fledging was completed. Often times they would occupy boxes that had old chickadee nests in them and incorporate the chickadee nesting material into their own nest. On occasion, chipmunks would use the boxes coincident with the chickadee nesting period and it was at these times that predation was suspected.

Lack (1966) and Perrins (1965) list three predators of the Great Tit in Europe, Great Spotted Woodpeckers (<u>Dendrocopos major</u>), gray squirrels (<u>Sciurus carolinensis</u>), and weasels (<u>Mustela nivalis</u>). Squirrels made the holes larger in the wooden boxes in order to get at the nestlings, and the woodpeckers made holes through the sides at the level of the nest due to noises made by the nestlings. Occasionally, the cement sawdust boxes used in this study would show signs of gnawing around the hole as if to enlarge it. Also, a box or two had holes bored through the sides. However, these activities couldn't be associated with a specific predator.

The weasel was considered to be the most important predator of the Great Tit by both Lack (1966) and Perrins (1965). Lack contended that the weasel learned to associate the nesting boxes with eggs and that is why adjacent boxes would have the clutches destroyed. Lack (1966) noted that this pattern did not exist with the nestlings and felt that weasels used noises made by the nestlings to locate their prey. Perrins (1965) stated that there was little doubt that the large broods of the Great Tit were taken more frequently by predators because the nestlings were hungrier and calling more. He artificially increased brood sizes of some nests by adding nestlings at hatching time. Perrins stated further that the disadvantage of raising a larger brood is further increased by the fact that 20% of the females were preyed on along with these broods.

In our studies we found predation to be highly variable between plots and years. The peak was reached at the Yellow-jacket Springs plot in 1975 when 75% of the nestlings were destroyed. The reason for this variability in northeastern California is unknown. Dunn (1977) studied weasel predation on breeding tits in Great Britain and found weasels to destroy an

average of 20% of the nests annually. He, too, found marked
annual fluctuations and showed that predation was proportional
to nesting density. Dunn (1977) also found an interesting relationship between rodent density and predation on Blue Tit and
Great Tit populations. High nest density hastened the onset of
predation but if rodent density was high it tended to retard the
rate of predation on tit nests by providing an alternate food
source.

An example of mortality during developmental stages of
Mountain Chickadees in northeastern California is shown for all
plots combined for the year, 1969, of least mortality (Table V)
and the year, 1975, of the greatest mortality (Table VI).

It is certainly apparent from our work in California as well
as from the many European workers noted above, that predation,
particularly by weasels, must be taken into consideration in
any attempt to increase cavity nesters by the use of nesting
boxes.

E. Banding Studies

During the eight years of banding, 584 adults (269 males
and 315 females) were banded. Counting adults that were recovered a total of 1,007 adults (499 males and 508 females) were
banded and/or recovered. Almost all of the adults that nested
in the boxes were banded annually (78.8% of 1,278 nesting birds
over the entire study period).

Since adults tend to stay in the same area or return each
year for nesting, a good estimate of survival can be obtained
and age structure of the population determined. The assumption
then is that a bird banded or recovered in one year that is not
recorded in the following year has died. The return rate can
then be assumed to be the minimum survivorship rate from an area
since some birds may move from the area and others are not retrapped due to nest failure. The percent return of adults
banded the previous year on the three plots was as follows:

	1969	'70	'71	'72	'73	'74	'75	$\bar{x}$ annual survival
Males	55	80	36.5	66	51.9	49.2	67.2	58.0
Females	65.2	65	30.6	45.6	45.8	42	58.5	50.2

This shows that about one-half of the birds return each year to
the same plot and that male survival is higher than female.
There were differences between plots and this could have been
due to predator activity prior to the banding period. Mean
annual return rate was highest at Tom's Creek (63% male and 62%

TABLE V. Partial Life Table for the First Brood of Mountain
          Chickadees in 1969

x Age interval	lx^a	DxF Mortality factor		dx^b	$100qx^c$	$100rx^d$
Eggs	807	addled or infertile		18	2.2	2.2
		broken		14	1.7	1.7
		abandoned		26	3.2	3.2
		predation (?)		13	1.6	1.6
		unknown		18	2.2	2.2
			Total	89	11.0	11.0
Nestlings 1-7 days	718	predation		0	--	--
		unknown		0	--	--
			Total	0	--	--
Nestlings 8-14 days	718	predation		27	3.8	3.3
		unknown		2	0.3	0.2
			Total	29	4.0	3.6
Nestlings 15-21 days	689	predation		0	--	--
		unknown		1	0.1	0.1
			Total	1	0.1	0.1
Fledglings	688					
% Fledged	85.2	% Total Mortality				14.7

[a] lx = number alive at the beginning of x.

[b] dx = number dying during x.

[c] $100qx$ = dx as a percentage of lx, apparent mortality.

[d] $100rx$ = dx as a percentage of the number at the beginning
(eggs), real mortality.

female) and lower at Yellowjacket (54% male and 40% female) and
Roney Flat (56% male and 47% female). Mountain Chickadee adults
therefore tend to stay in the same area every year for nesting.
Kluyver (1951) found this to be true for the Great Tit, and the
annual return rates for the Mountain Chickadee are very similar.
Kluyver found a differential between male and female returns

TABLE VI. Partial Life Table for the First Brood of Mountain
           Chickadees on Three Plots in Northeastern California
           in 1975

x	lx	DxF		dx	100qx	100rx
Eggs	825	addled or infertile		10	1.2	1.2
		broken		4	0.5	0.5
		abandoned		5	0.6	0.6
		predation		24	2.9	2.9
		unknown		41	5.0	5.0
			Total	84	10.2	10.2
Nestlings 1-7 days	741	predation		21	2.8	2.5
		unknown		19	2.7	2.3
			Total	40	5.4	4.8
Nestlings 8-14 days	701	predation		111	15.8	13.4
		unknown		99	14.1	12.0
			Total	210	29.9	25.4
Nestlings 15-21 days	491	predation		9	1.8	1.1
		unknown		18	3.7	2.2
			Total	27	5.5	3.3
Fledgings	464					
% Fledged	56.2	% Total mortality				43.8

(49% male and 43% female) over a nine year period ($\bar{x}$ 46%).
Kluyver (1951) also used a correction factor for his studies
on the Great Tit based on birds that were missed one year and
then retrapped in the next.

A total of 4512 nestlings fledged from the boxes over the
eight year period. Only 138 of these birds were not banded
(3.1%). In these cases birds fledged prior to banding. This
happened primarily during the first two years of the study and
in 1974 with second broods.

The recapture of birds banded as nestlings in the following
year was extremely low. Only 28 of 4374 nestlings known to
fledge were recaptured (0.6%). This is considerably lower than
the recapture rates recorded by Kluyver (1951) for the Great Tit
(approximately 6%). We assume that Mountain Chickadee fledglings
move from nest box areas but that they move much greater

distances than those recorded for the Great Tit. Kluyver (1951)
stated that 1/3 of the Great Tit fledglings moved 200 to 800 m
from their nests and the majority of the birds not more than
two kilometers. This could not have been the case with the
Mountain Chickadee as the plot dimensions were 400 by 900 m.
Surely a higher recovery rate would have been achieved if the
habits of the two birds were similar. The alternate   hypothesis
is that the mortality rate of the mountain chickadee fledglings
was unusually high. We believe that the recapture rate was low
due to movement. The vast contiguous forested areas in northeastern California where this study was conducted would permit
movement of far greater distances than the areas where Kluyver
did his work, namely, isolated islands of wooded areas which
were not conducive to extensive movement of Great Tit fledglings.

For most birds mortality of first year individuals is 80%
or higher (Welty, 1975). Kluyver (1951) estimated that 86.8% of
the Great Tits died in the first year. In our study we assumed
that all unbanded birds in any given year were first year birds
from outside the plot. If we consider these birds as replacements for the banded fledglings that left, then the difference
can be assumed to be mortality. Overall we calculated a 15%
survival rate of first year birds, which is almost the same as
Kluyver's (1951) figure for the Great Tit. The annual fledgling
survival rate is shown in Table VII.

Most of the birds recaptured were one or two years old, however, one male was recaptured seven years after banding (Table
VIII).   This male was assumed to be at least one year old in
1968. Approximately one-half of the birds stay with the same
mate until one or the other dies.

F.   Douglas-fir Tussock Moth Predation Studies

1. Spring Egg Masses. The number of egg masses still intact at each observation period is shown in Table IX. This
shows that there was little activity during the first week (27
April to 2 May). Unfortunately the egg masses were not observed
for 3 weeks from 2 May to 23 May due to weather, so it is not
certain when the 23 egg masses were preyed on. The period of
greatest activity was between 23 May and 8 June. It is likely
that most of the egg masses were preyed on during the 3 week
period preceding DFTM egg hatch. In the Yellowjacket Springs
area natural egg hatch occurred between June 8 and 18. Only 5
egg masses were preyed on after 8 June. None of the laboratory-
produced egg masses produced larvae, presumably due to
laboratory handling, but this should not have affected the
predators.

Donald L. Dahlsten and William A. Copper

TABLE VII. Mean Annual Mountain Chickadee Fledgling
            Survival Rate

Year	No. fledglings banded	No. unbanded adults captured in following year	Year	Assumed fledgling survival - %[b]
1968[a]	481	61	1969	12.7
1969	688	107	1970	15.5
1970[a]	819	56	1971	6.8
1971[a]	827	86	1972	10.4
1972	357	61	1973	17.1
1973	444	66	1974	14.9
1974[a]	432	104	1975	24.1
1975	464	94	1976	20.2
	4512	635		

    - 138 not banded   8 Year Mean = 14.8%
    4374 = Total fledglings

[a] Indicates that second brood occurred in these years.
[b] Survival rate calculation explained on page 33.

TABLE VIII. Average Age Structure of Mountain Chickadee
             Breeding Populations

Age of birds - years	Banding returns males	Banding returns females	Total %
8	1	--	0.1
7	3	--	0.3
6	7	3	1.0
5	11	10	2.1
4	29	21	5.0
3	59	47	10.5
2	120	112	23.0
1	269	315	58.4
Total	499	508	100.0

TABLE IX. Tussock Moth Egg Masses Remaining on Branches Stocked at 3 Different Densities[a]

Tree No.	# Egg masses	Observation date									
		IV-27	V-2	V-23	V-30	VI-8	VI-12	VI-18	VI-22	VII-5	VII-26
13	1	1	1	1	1	1	1	1	1	1	1
32	1	1	1	1	1	1	1	1	1	1	1
49	1	1	1	1	1	1	1	1	1	1	1
59	1	1	1	1	1						
79	1	1	1	1	1	1	1	1	1	1	1
86	$\frac{1}{6}$	$\frac{1}{6}$	$\frac{1}{6}$	$\frac{1}{6}$	$\frac{1}{6}$	$\frac{1}{5}$	$\frac{1}{5}$	$\frac{1}{5}$	$\frac{1}{5}$	$\frac{1}{5}$	$\frac{1}{5}$
17	5	5	5	5	5	3	1	1	1	1	1
31	5	5	5								
45	5	5	5	5	5	5	5	5	5	5	5
57	5	5	5	2	1						
72	5	5	5								
89	$\frac{5}{30}$	$\frac{5}{30}$	$\frac{5}{30}$	$\frac{5}{17}$	$\frac{5}{16}$	$\frac{5}{13}$	$\frac{5}{11}$	$\frac{5}{11}$	$\frac{5}{11}$	$\frac{5}{11}$	$\frac{5}{11}$
19	10	10	10	10	8	8	7	6	6	6	6
36	10	10	10	8	6	5	5	4	4	4	4
46	10	10	9	9	6	3	3	3	3	3	3
56	10	10	10	2	1	1	1	1	1	1	1
78	10	10	10	10	8	1	1	1	1	1	1
88	$\frac{10}{60}$	$\frac{10}{60}$	$\frac{10}{59}$	$\frac{10}{49}$	$\frac{2}{31}$	$\frac{2}{20}$	$\frac{2}{19}$	$\frac{2}{17}$	$\frac{2}{17}$	$\frac{2}{17}$	$\frac{2}{17}$
Total remaining	96	96	95	72	53	38	35	33	33	33	33

[a] It is assumed that the missing egg masses were destroyed or removed by birds.

The concentration of egg masses had an influence on predators. Of the single egg masses only one of six egg masses was preyed on (83.3% survived), 33.3% of the egg masses in groups of 5 survived, and 28.3% of the egg masses in groups of 10 survived. This suggests that the predators will continue to search in an area if they are successful.

Attempts to observe the predators feeding on the egg masses were made on numerous occasions but without success. It is almost certain that the egg masses were preyed upon by birds, particularly those that forage in a manner similar to that of the chickadee. Carpenter ants (Camponotus sp.) were noted on the study branches on several occasions but they were never seen on an egg mass or carrying eggs.

2. Nest Box Camera Units. During late June and early July, 1977, five chickadee nests were transferred to camera-box units. The nests were only partially recorded since nestlings were not transferred until at least 5 days old. In addition, there were a number of technical difficulties since the equipment was still being perfected.

Seven rolls of film were exposed during this study and approximately 10,000 frames of birds with prey in their beaks have been developed for analysis. All of the film has been examined in a cursory manner but to date only the data for two nests with 5 and 7 days of feeding activity have been analyzed.

Adult bird activity began about 0530 h and stopped at 2050 h, amounting to almost continuous activity during daylight hours. A total of 2,582 trips in which 4,737 food items were brought to nests were analyzed (Table X). Unknown insect larvae and Lepidoptera larvae accounted for 43.5% of the total food items brought to the nest. These were mostly Lepidoptera (geometrids, tortricids), and Acantholyda larvae, but they could not be distinguished for certain and so were placed in a more general category. Positively identified Acantholyda larvae accounted for 22.9% of the total and tortricid larvae and pupae for an additional 10.1%. Spiders were also evident and accounted for 4.5% of the total of which Pelloctanes sp. (1.9%) was the most abundant. The remaining film, which includes some 5,000 to 7,000 frames, did not reveal any tussock moth larvae.

Preliminary analysis indicates that Mountain Chickadees rely heavily on a number of potential pest insects in Modoc County's forests and in particular the Modoc budworm, Neodiprion sawflies, geometrids, etc.

The camera units were useful in recording the predators of the chickadee, too, as in one instance a weasel was photographed preying on the nestlings.

TABLE X. Food Items Brought to the Nest by Chickadees (Recorded as a Percent of Total Food)

Food item	Julian Date	Box 22					Box 8							Total
		182	187	188	189	190	181	182	183	184	185	187	188	
Insect larva		4.7	3.6	15.8	16.9	3.4	26.8	33.3	29.8	43.8	48.2	30.4	19.8	24.4
Insect pupa		--	--	1.0	.7	.3	.4	--	.1	--	--	--	--	.2
Insect adult		1.8	3.6	.7	--	--	1.3	.2	.3	.5	.7	--	.5	2.0
Insect unknown		8.3	--	7.9	4.1	3.3	2.2	1.0	.9	.2	--	.2	--	.4
Lepidoptera larva		45.0	21.4	9.5	19.3	41.5	24.1	18.6	13.8	5.8	.7	8.3	18.2	19.1
Lepidoptera pupa		3.6	7.1	.3	1.7	1.8	.9	1.0	.9	.7	1.4	.5	.7	1.1
Lepidoptera adult		2.4	7.1	4.6	4.7	3.4	5.4	2.2	.7	1.8	--	10.4	5.5	3.7
Tortricidae larva		--	10.7	17.1	14.2	10.1	--	.5	--	--	--	--	22.7	6.4
Tortricidae pupa		5.9	--	4.3	.7	3.4	--	.2	.3	.2	--	.5	.2	1.2
Noctuidae pupa		--	--	--	--	--	--	--	--	--	--	--	--	.0
Geometridae larva		1.2	3.6	.3	--	--	--	--	--	--	--	--	.4	.1
Dioryctria larva		1.8	--	.7	--	--	--	--	--	--	--	--	--	.1
Hymenoptera larva		--	--	.3	1.7	.1	7.1	11.6	15.1	8.1	--	2.5	.5	5.4
Hymenoptera adult		--	--	--	--	.1	.4	--	--	--	--	--	--	.0
Acantholyda larva		8.3	14.3	18.1	14.5	18.0	15.2	12.1	24.0	30.4	37.6	43.5	26.2	22.9
Neodiprion larva		4.1	3.6	3.6	3.4	1.4	.4	--	--	--	--	.5	--	.9
Diptera pupa		--	--	.3	--	--	--	--	--	--	--	--	--	.0
Asilidae adult		--	--	--	--	--	2.7	9.1	5.4	5.3	4.3	.7	1.4	3.0
Tipulidae adult		--	--	--	--	--	.9	--	--	--	--	--	--	.0
Coleoptera larva		--	--	--	--	.3	--	--	--	--	--	--	--	.0
Coleoptera adult		--	--	.3	--	--	--	.3	--	--	--	--	.4	.1
Dichelonyx adult		1.8	10.7	4.9	6.1	5.3	4.5	2.6	3.9	1.1	2.1	--	1.1	3.1
Buprestidae adult		--	--	--	--	--	.4	--	--	--	--	--	--	.0
Araneae unknown		4.1	3.6	2.6	5.4	1.6	4.9	1.7	.3	.5	.7	1.1	.5	1.8
Xysticus sp.		--	--	--	.7	--	--	--	--	--	--	--	--	.0
Philodromus sp.		--	3.6	1.6	.3	.1	--	.3	.1	--	--	--	.2	.3
Pelloctanes sp.		5.9	7.1	3.6	3.7	1.5	.9	3.8	2.2	.7	.7	.2	.4	1.9
Theridiidae unknown		--	--	--	--	.1	--	--	--	--	--	--	--	.0
Araniella displicata		1.2	--	1.3	.7	.7	--	.3	.6	.2	1.4	.2	.5	.5
Unknown		--	--	1.0	1.0	2.5	1.3	1.0	1.6	.7	2.1	1.1	.7	1.3
TOTAL food items	169		28	304	296	732	224	585	688	566	141	444	560	4737

3. Cocoon Predation Study. There were several types of predation observed on the cocoons (Table XI). Carpenter ants were observed on a number of occasions. This type of predation was distinctive as ants tear away pieces of the cocoon as they feed on the pupa. The end result is a ragged, torn and empty cocoon (Fig. 7). On two occasions pentatomids (Podisus sp.) were seen feeding on the pupae. They push their sucking mouth parts through the cocoon into the pupa. However, there is no overt evidence of this type of predation.

The most common type of predation which we are attributing to birds was not witnessed. In these cases the end of the DFTM cocoon was enlarged and the cocoon was empty (Fig. 8). Avian predation of egg masses was assumed when the entire egg mass was cleaned from the cocoon or when there were only a few eggs remaining. In some cases the masses appeared to be slashed as if by a bird's beak.

Ant predation was greatest on the bole by a substantial margin. Ants were observed frequently moving up and down the boles of white fir. This activity has been assumed to be due to the presence of aphids in the tree. Camponotus sp. have been observed tending Cinara sp. aphids on numerous occasions. Ant predation on the branches may have been incidental and associated with aphid presence.

The unknown predation category (Table XI) was probably avian predation also. In these cases the cocoon was empty and the opening at the end was smaller than in Figure 8 or the cocoon was opened laterally (Fig. 9). The total of the avian and unknown predation on the branches, regardless of the number and distribution of cocoons, was more than twice that on the bole.

There were 14 blocks in each area and five variables were tested: ant predation; bird predation; unknown predation (possibly bird); bird + unknown; bird + ant + unknown. For the analysis we used the transformation: arcsine ($\sqrt{}$percent of cocoons not preyed upon). BMD programs were used to perform a two-way analysis of variance (randomized block design). The only significant interactions were between areas and blocks (for bird predation and bird + unknown predation both at α = .05) and areas and treatments (for ant + bird + unknown predation at α = .05).

Results indicate that areas were significant for bird predation but not ant predation. This could have been because of nest box density which may translate to chickadee populations. Treatments are significantly different and this is no doubt due to the differences between ant and bird predation on the bole and the branches.

TABLE XI. Fate of Douglas-fir Tussock Moth Cocoons Placed on White Fir Branches and Boles at Different Densities

Area	Total Cocoons	Ant Pred.	Bird Pred.	Unk.[a] Pred.	No. Egg Masses	Egg Mass Pred.	Parasites	Unk.[b] Mortality	Males Emerged	Females Emerged[c]
Ten Cocoons per Tree--Spread Out										
Yellowjacket Sp.	140	37	56	14	9	2	16	1	7	0
Tom's Creek	150	27	47	7	20	2	17	3	26	3
Hilton Spike	140	31	59	6	7	0	18	3	16	0
Total	430	95	162	27	36	4	51	7	49	3
%		22.1	37.7	6.3	8.4	0.9	11.9	1.6	11.4	0.7
%		Total Pred. 67.0					Total Para. 11.9	Total Mort. 80.5		
Ten Cocoons per Tree--Grouped on a Single Branch										
Yellowjacket Sp.	140	40	79	4	0	0	8	3	6	0
Tom's Creek	150	43	48	9	13	1	8	6	23	0
Hilton Spike	140	31	50	6	14	2	14	7	13	5
Total	430	114	177	19	27	3	30	16	42	5
%		26.5	42.2	4.4	6:3	1.4	7.0	3.7	9.8	1.2
%		Total Pred. 72.8					Total Para. 7.0	Total Mort. 83.5		
Five Cocoons per Tree--Grouped on Bole										
Yellowjacket Sp.	70	44	16	2	0	0	4	1	3	0
Tom's Creek	75	49	9	0	5	2	1	0	11	0
Hilton Spike	70	50	13	0	2	0	1	1	2	1
Total	215	143	38	2	7	2	6	2	16	1
%		66.5	17.7	0.9	3.3	0.9	2.8	0.9	7.4	0.5
%		Total Pred. 86.0					Total Para. 2.8	Total Mort. 89.8		

TABLE XI. Continued.

One Cocoon per Tree--On a Single Branch Tip

Area	Total Cocoons	Ant Pred.	Bird Pred.	Unk.[a] Pred.	No. Egg Masses	Egg Mass Pred.	Parasites	Unk.[b] Mortality	Males Emerged	Females Emerged[c]
Yellowjacket Sp.	14	3	6	3	0	0	0	2	0	0
Tom's Creek	15	4	2	0	3	0	4	0	2	0
Hilton Spike	14	0	5	2	1	0	4	0	2	0
Total	43	7	13	5	4	0	8	2	4	0
%		16.3	30.2	11.6	9.3	0	18.6	4.6	9.3	0
%		Total Pred. 58.1		Total Para. 18.6		Total Mort. 81.4				

[a] Unknown predation.

[b] Pupae collapsed, possible virus.

[c] Females emerged, no egg masses present.

FIGURE 7.   A Douglas-fir Tussock Moth Cocoon That Has Been
Torn Apart by Ants.

In another study, Lühl and Watzek (1975) compared forested
areas in Europe with and without nest boxes to study avian predation on two insects closely related to the Douglas-fir tussock
moth--the nun and gypsy moths. In addition, they manipulated
the initial density of the insects artificially. They found considerable decreases in larval and pupal populations of both
insects in those areas with nesting boxes. As in our study Lühl
and Watzek assumed that birds were responsible for the dis-
appearance of the insects, and did not make any direct observations of the actual predation act. The results of our study were
not definitive. The data suggested that bird density as indicated by mountain chickadee nest box density may be important.

FIGURE 8.   Douglas-fir tussock moth cocoons opened along the
             side and the contents removed, presumably by birds.

ACKNOWLEDGEMENTS

     We thank Ms. Nancy X. Norick and Mr. David L. Rowney for
statistical assistance, Drs. Charles J. Henney and Lawrence J.
Blus, U. S. Department of Interior, Fish and Wildlife Service,
Corvallis, Oregon, for critical review of the manuscript, and
Dr. Charles David, formerly Post-graduate Research Entomologist,
Department of Entomological Sciences, University of California,
Berkeley for identification of Camponotus modoc Wheeler.

FIGURE 9.   Douglas-fir tussock moth cocoons opened along the
            side and the contents removed, presumably by
            birds.

REFERENCES

Bruns, H.   (1960).   Bird Study 7, 193.
Campbell, B.   (1968).   Forestry 41, 27.
Copper, W., Ohmart, C., and Dahlsten. D.   (1978).   Western Birds
    9, 41.
Dunn, E.   (1977).   J. Anim. Ecology 46, 633.
Franz, J.   (1961).   Ann. Rev. Entomol. 6, 183.
Furniss, R., and Carolin, V.   (1977).   USDA, Forest Service,
    Misc. Publ. No. 1339.
Haftorn, S.   (1954).   Det Kgl Norske Videnskabers Selskabs
    Skrifter 1953(4), 1.
Haftorn, S.   (1974).   Ornis Scandinavica 5, 145.
Kluyver, H.   (1951).   Ardea 39, 1.
Lack, D.   (1954).   "The Natural Regulation of Animal Numbers."
    Clarendon Press, Oxford, England.

Lack, D.   (1955).   Ardea 43, 49.

Lack, D.   (1966).   "Population Studies of Birds."   Clarendon
      Press, Oxford, England.

Laskey, A.   (1946).   Wilson Bull. 58, 217.

Lühl, V., and Watzek, G.   (1977).   Allg. Forst – u.J. – Ztg.,
      147.   Jg., 6/7, 113.

Ortiz de la Torre, F.   (1970).   Boletin del Servicio de Plagas
      Forestales 13, 57.

Perrins, C.   (1965).   J. Anim. Ecology 34, 601.

Royama, T.   (1959).   Tori 15, 172.

Staude, J.   (1968).   In Allg. Forstzeitschr. 23, 775.

Tichý, V.   (1963).   Rev. Appl. Entomol. 52, 208.

Tinbergen, L.   (1946).   Ardea 34, 1.

Van Balen, J.   (1967).   Ardea 55, 1.

Welty, J. C.   (1975).   "The Life of Birds."   2nd Edition, W. B.
      Saunders Co., Philadelphia.

SEASONAL POPULATIONS OF INSECTIVOROUS BIRDS IN A MATURE
BOTTOMLAND HARDWOOD FOREST IN SOUTH LOUISIANA

James G. Dickson

USDA Forest Service
Southern Forest Experiment Station
Nacogdoches, Texas[1]

When insectivorous birds were censused monthly in a mature
bottomland hardwood forest, populations were low and made up a
small proportion of total birds in winter. During the breeding
season, however, insectivorous bird populations were high and
dominated the total population. Population peaks occurred
earliest (March through May) for ground gleaners, slightly later
(April and May) for foliage and bark gleaners, and even later
(May through July) for air feeders.

I. INTRODUCTION

Though population data are available for some species of
insectivorous birds (IB) and bird communities in some habitats
during part of the year (Shugart and James, 1973; Conner and
Adkisson, 1975), information is scant on year-round populations,
especially in mature forests. This study was designed to
monitor monthly IB population fluctuations in a mature bottomland
hardwood forest in Louisiana.

---

[1]In cooperation with the School of Forestry, Stephen F.
Austin State University, Nacogdoches, Texas.

II. STUDY AREA

The study was conducted on the Thistlethwaite Wildlife
Management Area near Opelousas, Louisiana. The area is an old
floodplain of the Mississippi and Red Rivers and is classified as
hardwood bottom (Braun, 1950:293). Woody vegetation was dense,
28.2 m^2/ha basal area of trees > 2.5 cm diameter. Water oak
(Quercus nigra) and sweetgum (Liquidambar styraciflua) were the
2 most abundant tree species. The understory was dominated by
ironwood (Carpinus caroliniana), cane (Arundinaria gigantea),
and dwarf palmetto (Sabal minor).

III. METHODS

Birds were censused along a 1.6 km transect. Strips of
varying width were used for different species with different
effective detection distances (EDD). An EDD was determined for
each species from the distribution of detection points perpendic-
ular to the transect center line. It was assumed that all species
were detected effectively to a minimum of 18.9 m laterally from
the transect center line. Beyond this point, the EDD for each
species was the distance beyond which bird density dropped below
75% of bird density nearer the transect center line. Thus, EDD
generally corresponded closely to the inflection point that Emlen
(1971) used in determining the level of maximum detection for the
various species.

Each month 4 to 8 counts of birds on the transect were made,
beginning in January 1973 and ending January 1974. The transect
was traversed in alternate directions within 3.5 hours after
sunrise. Censusing was not conducted during moderate or heavy
rain or during high winds (> 18 mph).

Birds were classified according to food habits from the work
of Martin et al. (1951), substantiated, in part, by observations
on the study area. Birds were classified as primarily insectivorous during the months when Martin et al. (1951) reported that
over half of their diet was insects. The IB community was
divided into 4 primary feeding guilds: foliage gleaners, bark
gleaners, ground gleaners, and air feeders (Table I).

IV.   RESULTS AND DISCUSSION

Of the 66 species of birds censused, 45 (68%) were primarily
insectivorous at some season. The most abundant foliage gleaners
were wood warblers (Parulidae). Vireos (Vireonidae), old world
warblers (Sylviidae), and tits (Paridae) were also common. All
bark gleaners were woodpeckers (Picidae). Most of the ground
gleaners were thrushes (Turdidae), and most of the air feeders
were flycatchers (Tyrannidae).

A.   Seasonal Populations

Distinct seasonal differences in IB populations were evident.
During winter (Jan., Feb., Dec.) only 8 to 10 insectivorous
species were present, (Table II) with only 82 to 132 insectivorous birds per km^2 (Table III). The proportion of insectivorous
birds to total birds was also low during winter. Only 27 to 29%
of the bird species were predominantly insectivorous, and only
6 to 8% of the population was insectivorous. This was lower than
the proportion of insectivorous birds in mostly grass and brush
stands in south Texas (Emlen, 1972).
As new leaf growth occurred in spring insectivorous birds
increased, but total birds decreased. In March IB species
increased to 15, and they made up 45% of total bird species.
Insectivorous bird density increased to 256/km^2, 26% of total
bird density.
Insectivorous bird populations continued to increase in late
spring and early summer. Number of species was highest in April
(27) and May (24), but the proportion of IB species peaked later,
in June (83%) and July (86%). Insectivorous bird density
followed a similar pattern. Peak densities occurred April
through June (424-340/km^2) and peak IB proportions of total bird
density occurred in May (97%) and June (99%), after the last of
the winter residents, which were mostly not insectivorous, had
departed.
From mid-summer to winter both number of IB species and IB
density declined. In July there were 18 IB species comprising
an estimated population of 230/km^2. This was 86% of the total
species and population. By December number of IB species had
dwindled to 10 (29% of total) and density had declined to
132/km^2 (6% of total).
Insectivorous birds apparently responded to seasonal changes
in temperature and foliage density, and insect density changes
that accompanied these changes. During winter little substrate
existed for the few foliage gleaners in the deciduous forest.
Much of the energy of the ecosystem went into mast, mainly of

TABLE I. Composition of Insectivorous Bird Guilds

Feeding guild	Common name	Scientific name
Foliage	Yellow-billed Cuckoo	Coccyzus americanus
	Ruby-throated Hummingbird	Archilochus colubris
	Carolina Chickadee	Parus carolinensis
	Tufted Titmouse	Parus bicolor
	Carolina Wren	Thryothorus ludovicianus
	Brown Thrasher	Toxostoma rufum
	Blue-gray Gnatcatcher	Polioptila caerulea
	Golden-crowned Kinglet	Regulus satrapa
	Ruby-crowned Kinglet	Regulus calendula
	White-eyed Vireo	Vireo griseus
	Yellow-throated Vireo	Vireo flavifrons
	Red-eyed Vireo	Vireo olivaceus
	Prothonotary Warbler	Protonotaria citrea
	Swainson's Warbler	Limnothlypis swainsonii
	Worm-eating Warbler	Helmitheros vermivorus
	Orange-crowned Warbler	Vermivora celata
	Parula Warbler	Parula americana
	Yellow-rumped Warbler	Dendroica coronata
	Kentucky Warbler	Oporornis formosus
	Hooded Warbler	Wilsonia citrina
	American Redstart	Septophaga ruticilla
	Cardinal	Cardinalis cardinalis
	Summer Tanager	Piranga rubra
Bark	Pileated Woodpecker	Dryocopus pileatus
	Red-bellied Woodpecker	Melanerpes carolinus
	Red-headed Woodpecker	Melanerpes erythrocephalus
	Hairy Woodpecker	Picoides villosus
	Downy Woodpecker	Picoides pubescens
Air	Great Crested Flycatcher	Myiarchus crinitus
	Eastern Phoebe	Sayornis phoebe
	Acadian Flycatcher	Empidonax virescens
	Eastern Wood Pewee	Contopus virens
	Rough-winged Swallow	Stelgidopteryx ruficollis
Ground	Common Flicker	Colaptes auratus
	Mockingbird	Mimus polyglottos
	Gray Catbird	Dumetella carolinensis
	American Robin	Turdus migratorius
	Wood Thrush	Hylocichla mustelina
	Hermit Thrush	Catharus guttatus
	Swainson's Thrush	Catharus ustulatus
	Gray-cheeked Thrush	Catharus minimus

TABLE I.   Composition of Insectivorous Bird Guilds (Cont'd)

Feeding guild	Common name	Scientific name
	Eastern Bluebird	Sialia sialis
	Starling	Sturnus vulgaris
	Northern Waterthrush	Seirus noveboracensis
	Rufous-sided Towhee	Pipilo erythrophthalmus

oaks. Mast matured in fall and was available to birds in late
fall and winter. Bird populations greater than 1000/km^2 were
evident in winter but most birds ate mast then. Insectivorous
bird populations increased in spring as temperatures increased
and new foliage grew, and decreased in autumn as leaves hardened,
foliage quantity declined, and temperatures decreased.

B.   Feeding Guilds

All 4 guilds exhibited seasonal population peaks, but these
occurred at different times for different guilds. Populations
of ground gleaning birds peaked earliest. The number of species
of ground feeding insectivorous birds was highest March through
May, and density was highest in March and November. There were
fewer birds on the ground during spring than in winter (Dickson
and Noble, 1978), but a greater percentage of these were insectivorous. There was a shift of food habits to mainly insects
for common flickers, American robins, and hermit thrushes (Martin
et al. 1951). As ground temperatures increased during spring,
ground insects probably became active and available to birds.
Also insect larvae may have been abundant on the ground in early
spring.
Bark-gleaning birds were at peak abundance in April and May,
slightly later than ground gleaners. Although number of species
of bark gleaners was relatively stable, 1 to 3 each month,
density increased in April and May because red-headed and red-
bellied woodpeckers became primarily insectivorous. Bark gleaner
density declined after April as red-headed woodpeckers departed
the dense woods for more open nesting habitat.
Foliage-gleaning birds were more abundant than the other 3
guilds combined. Number of species and density were highest in
April and somewhat less in May. Wood warblers, vireos, and
other foliage gleaners, that were mostly summer residents, were
abundant (Dickson, 1978), probably because of the insect larvae

TABLE II. Monthly Number of Insectivorous Bird Species

	Jan	Feb	Mar	Apr	May	Jun	Jul	Aug	Sep	Oct	Nov	Dec
Total species	30	29	33	36	32	24	21	26	25	21	26	35
Insectivorous species												
Foliage	5	5	9	18	14	11	11	13	10	4	6	6
Bark	2	2	1	3	3	2	3	2	2	2	3	2
Ground	0	1	5	4	4	3	2	2	0	1	1	1
Air	1	0	0	2	3	4	2	2	1	0	0	1
Total	8	8	15	27	24	20	18	19	13	7	9	10
Insectivorous species as % of total species	27	28	45	75	75	83	86	73	52	33	35	29

TABLE III. Monthly Insectivorous Birds Per Km^2

	Jan	Feb	Mar	Apr	May	Jun	Jul	Aug	Sep	Oct	Nov	Dec
Total birds	1412	1234	974	744	386	344	266	294	292	348	680	2036
Insectivorous birds												
Foliage	100	62	212	366	302	280	180	196	182	94	78	122
Bark	12	14	10	44	24	16	16	8	8	4	14	6
Ground	0	6	34	12	18	8	4	6	0	4	4	0
Air	2	0	0	2	32	36	30	12	4	0	0	4
Total	114	82	256	424	376	340	230	222	194	102	134	132
Insectivorous birds as % of total birds	8	7	26	57	97	99	86	76	66	29	20	6

267

associated with the succulent new leaves in spring. Densities
of foliage gleaners decreased from April to a seasonal low in
October and November (< 100/km^2). A few foliage gleaners were
present throughout the winter.

Birds that fed mainly on flying insects were most numerous
May through July, later than the other guilds. Number of species
was highest in May (3) and June (4), and density was over 29
per km^2 in each of the 3 months. Flying insects were probably
more abundant in the forest during this period than earlier
because enough time had elapsed for them to develop into
maturity. After August, air-feeding bird populations plummeted
to less than 5 per km^2 each month.

ACKNOWLEDGMENTS

Data for this paper was taken from Dickson (1974), a
doctoral dissertation submitted to Louisiana State University,
Baton Rouge. The field work was financed by the L.S.U. School
of Forestry and Wildlife Management and the Agricultural
Experiment Station. Thanks go to Drs. Jack D. McCullough and
Robert R. Fleet for reviewing an earlier draft of the manuscript,
and to Dr. Richard N. Conner for suggestions in data analysis.

REFERENCES

Braun, E. (1950). "Deciduous Forests of Eastern North America."
    Blakiston Co., Philadelphia.
Conner, R., and Adkisson, C. (1975). J. For. 73, 781.
Dickson, J. (1974). "Seasonal Populations and Vertical
    Distribution of Birds in a South Central Louisiana
    Bottomland Hardwood Forest. Ph.D. Thesis. La. State Univ.,
    Baton Rouge.
Dickson, J. (1978). J. Wildl. Manage. 42, 875.
Dickson, J., and Noble, R. (1978). Wilson Bull. 90, 19
Emlen, J. (1971). Auk 88, 323.
Emlen, J. (1972). Ecology 53, 317
Martin, A., Zim, H., and Nelson, A. (1951). "American Wildlife
    and Plants: a Guide to Wildlife Food Habits." Dover Pub.
    Inc., New York.
Shugart, H., and James, D. (1973). Auk 90, 62.

IMPACT OF WOODPECKER PREDATION ON
OVER-WINTERING WITHIN-TREE POPULATIONS
OF THE SOUTHERN PINE BEETLE *(Dendroctonus frontalis)*

James C. Kroll[1]
Robert R. Fleet

School of Forestry
Stephen F. Austin State University
Nacogdoches, Texas

Impact of woodpecker predation on within-tree populations of
southern pine beetles *(Dendroctonus frontalis)* was studied during
the period November 1975 to February 1976. Exclusion studies
indicated that woodpeckers had a significant impact on pupa and
brood adult life stages of southern pine beetles (SPB), especially
at mid-bole where densities are greatest. Woodpecker foraging
activity paralleled increases in insect predator and parasite
densities.
   Downy, Hairy, and Pileated woodpeckers appeared to exploit
a food resource patch in which pine beetles were the most abundant
prey item. Birds foraged optimally on later stages of beetle
brood which were larger and more easily attainable than earlier
life stages.

I. INTRODUCTION

   Birds, as biological control agents of forest insects have a
substantial impact on many prey populations (Otvos, 1965).
Woodpeckers in particular have been observed to concentrate
feeding in bark beetle infestations (Hopkins, 1909; Baldwin, 1968;
Shook and Baldwin, 1970; Overgaard, 1970; Moore, 1972), and
respond numerically to the concurrent increases in food supply
(Koplin, 1969).

[1] Supported by a grant from U.S.D.A.-C.S.R.S. through the
Expanded Southern Pine Beetle Research and Applications Program.

269

The southern pine beetle, *Dendroctonus frontalis* Zimm., is a major pest of southern pine forests. In Texas alone, between 1958 and 1969, SPB destroyed 170,747 million board ft. of sawlogs and 201,704 cords of pulpwood valued at over $5.5 million (Coulson *et al.*, 1972). Previous control practices (e.g., cut-and-leave, insecticide treatment) have proved to be unsuccessful; SPB populations reached epidemic levels in 1976. Recently, severity and cost of damage by this insect pest have resulted in a South-wide research and applications program (=Expanded Southern Pine Beetle Research and Applications Program or ESPBRAP) in an attempt to alleviate these problems.

Studies were initiated during ESPBRAP to determine roles of invertebrate and vertebrate predators in SPB population dynamics. Woodpeckers were the only vertebrate predators studied. Findings of the woodpecker predation research group and reports in the literature of woodpecker-bark beetle interactions strongly suggest that woodpeckers are the greatest single mortality agent operating in the SPB-timber complex. Our paper presents information on impact of woodpecker predation on overwintering within-tree populations of SPB.

II. METHODS

Research was conducted in the forests of east Texas where pine, either pure or in mixture with hardwoods, predominates. The most widespread forest type is loblolly-shortleaf pine followed by oak-pine, oak-hickory, etc. (Sternizke, 1967). The area has been a center of SPB activity since 1957 (Coulson *et al.*, 1972). The SPB infested site selected for our study was a mixed pine-hardwood stand 3.3 km NW of Broaddus (St. Augustine Co.), TX. The stand was initially attacked by SPB during July 1975. The infestation had grown to 9 ha (=1000 + trees) by March 1976.

The study area was considered as mixed pine-hardwood, with basal areas of 19.3 ± 0.9 m^2 for pines and 11.7 ± 0.7 m^2 for hardwoods. Pines (loblolly and shortleaf) comprised 57.1% of overstory composition; tree species diversity was 1.98 (Equitability = 0.797).

In addition, we selected a mixed pine-hardwood stand (Nacogdoches Co., TX) comprising 45 ha and with no history of SPB infestation, to serve as a control. Pine basal area was 14.1 $\pm$ 0.6 m^2, while that for hardwood species was 10.3 ± 0.5 m^2. The stand was 59.1 percent pine (loblolly and shortleaf). The species diversity for the uninfested area was 2.24. Comparisons (e.g., basal area, diameters, etc.) between infested and uninfested areas suggested no significant difference, except for pine basal area (P < 0.01).

A census of numbers and species of woodpeckers present in each study area was conducted each month during the period

November 1975 – February 1976. Census lines 50 m apart were
established in each study area; lines were walked slowly,
stopping at 50 m intervals to listen for sounds of woodpeckers
(i.e., calls and/or tapping). If sounds were heard or if birds
were seen, we walked no more than 25 m at a right angle away
from the survey line to investigate. Location of each woodpecker
was recorded on a map of the area, along with sex, relative age
(adult, juvenile), species of tree used, location on tree, and
stage of SPB development, if present, within the tree. Transect
lines in the infestation were walked 3 times daily (dawn, mid-day,
evening) for 3 days per month. Control area transect lines were
walked twice daily (dawn, evening) for 3 days per month.

Twenty trees were selected for study; the main criterion
being that they had recently come under SPB attack. Ten trees
(4 shortleaf and 6 loblolly) served as treatments, while an equal
number (2 shortleaf and 8 loblolly) were used as controls.
Treatment trees were caged with 3 ft. (0.92m) wide strips of
$\frac{1}{4}$-in. hardware cloth (held 10 cm from the bark surface by wooden
blocks) at three standardized heights. Standardized heights
were obtained by dividing infestation height (i.e., distance from
bottom to top of infested tree area) by three.

Treated and control trees were compared at the late instar
larva (=X-ray No.2 of Coulson *et al.*,1976) SPB developmental stage.

Within-tree populations of SPB were determined by
radiographic analysis of bark discs (DeMars, 1963; Berryman, 1964;

FIGURE 1. Circular bark samples (100 cm^2) removed from
beetle infested trees.

Ragenovich and Coster, 1974; Coulson *et al.*, 1976). Circular
bark samples (100 cm^2) were removed from the north and south
sides of trees (Fig. 1) at three standardized heights. Samples
were placed in individual plastic bags and brought to the
laboratory and X-rayed. Developmental stages of SPB, and
predators and parasitoides were counted from radiographs (Fig. 2).
Estimates for within-tree populations of SPB life stages and
associates were obtained using the topological model (multiple
level sampling scheme) of Coulson, *et al.*, (1976).

It was also necessary to determine mortality of SPB and
associates in bark chips and flakes dislodged by woodpeckers
during feeding activity. Bark chips from 1 m^2 beneath ten
unscreened trees of late instar infestation stage were taken to
the laboratory and radiographed to determine densities of SPB
at various life stages of development. Bark phloem area was
derived by tracing radiographs with a planimeter. Bark chips
were then returned to the field within 24 h and caged with 18
mesh screening (common window screen) coated with Stick-Em^R.
(Stick-Em^R is a painted-on adhesive designed to capture insects
on contact.) Adult beetles and associates were counted upon
emergence. Environmental conditions (maximum and minimum
temperatures and relative humidity) were recorded daily
throughout these tests.

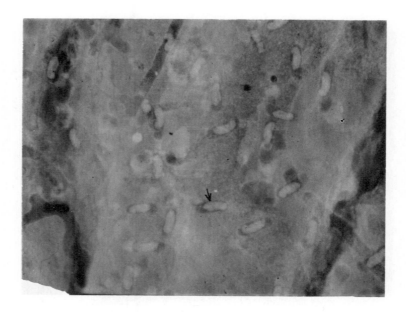

FIGURE 2.   Radiograph (x2) of SPB infested bark.   Arrow
indicates pupa.

Table I. Total Sightings for Woodpeckers Frequenting the Southern Pine Beetle Infested Study Area. Composition of Woodpecker Assemblage by Number and % for Infested & Control Areas. Numbers of sightings are not directly comparable since they are based on different number of censuses.

| | Woodpecker species | | | | | | |
| | | | | | Yellow-bellied | | |
Month	Downy	Pileated	Hairy	Red-bellied	Sapsucker	Flicker	Unknown
November	16	7	5	13	16		4
December	17	11	5	9	15	1	5
January	9	15	4	14		9	1
February	12	11	2	4	11	2	1
Total sightings	54	44	16	40	42	12	11
% of total woodpecker observations in infested area	24.6%	20.1%	7.3%	18.3%	19.2%	5.5%	5.0%
% of total woodpecker observations in uninfested area	11.2%	14.6%	5.2%	33.2%	21.6%	10.1%	4.1%
Total sightings	30	39	14	89	58	27	11

III.   RESULTS

A. Woodpecker Population Densities

Six woodpecker species were observed in both SPB infested and uninfested study areas during November 1975 - February 1976 (Table I). The Downy Woodpecker was the most commonly observed species in the infested area, followed by Pileated Woodpeckers. Yellow-bellied Sapsuckers, Red-bellied Woodpeckers, Hairy Woodpeckers and Common Flickers, respectively. Downy, Pileated and Hairy woodpeckers are the primary predators of SPB and insect associates (Kroll and Fleet, unpubl. data).

The Red-bellied Woodpecker was the most commonly observed species in the uninfested area, followed by Yellow-bellied Sapsuckers, Pileated Woodpeckers, Downy Woodpeckers, Common Flickers and Hairy Woodpeckers.

Size of infestation increased from 3.25 ha in November to 6.25 ha in February, while densities of primary woodpecker predators on SPB generally decreased during the period (Tables II & III). Since population densities did not increase with

Table II.   Comparisons of Estimated[a] Densities of Woodpeckers Observed in Southern Pine Beetle Infested and Uninfested Study Areas.

	Nov.	Dec.	Jan.	Feb.
Size of infestation (ha)	3.25	3.75	5.12	6.25
Infested area				
Density all species[b]	4.31	4.53	3.51	2.56
Density primary predators[c]	2.46	2.93	1.76	1.44
Size of control area (ha)	45.0	45.0	45.0	45.0
Uninfested area				
Density all species	0.36	0.27	0.73	0.62
Density primary predators	0.13	0.09	0.29	0.22

a/ The greatest number of each woodpecker species recorded during any one of the censuses was used as the minimum possible number of woodpeckers using the area.

b/ Downy, Hairy, Pileated, Redbellied, Common Flicker and Yellow-bellied Sapsucker.

c/ Downy, Pileated, and Hairy woodpeckers.

increasing infestation size, there was probably little
recruitment of birds into the area. Population densities of
primary predators in the infested study area were 6 (January) to
33 (December) times greater than in the uninfested area. The
three species of primary predators accounted for 52% of
woodpecker sightings in the infested area whereas in the control
area they accounted for only 31% of all sightings (Table I).

B. Woodpecker Foraging

We also examined types of trees used as foraging substrates
by woodpeckers within the SPB infested study area. Downy, Hairy,
and Pileated woodpeckers preferred beetle infested or killed
trees while Red-bellied Woodpeckers, Yellow-bellied Sapsuckers,
and Common Flickers fed extensively in hardwoods and uninfested
pines (Table IV).

C. Woodpecker Predation on SPB

Within-tree populations of SPB are typically distributed over
the entire bole with greatest densities of all life stages
occurring below mid-bole (Fig. 3). Initial attack by parent
adult SPB is usually at 30-50% of infested bole height. Mass
attack then proceeds above and below this region. Hence, brood
development is earlier at mid-bole than above and below point of
initial attack. Woodpeckers initially attacked SPB brood near
mid-bole, expanding foraging area as brood matured (Kroll and
Fleet, unpubl. data).

Table III.   Estimated Densities[a] of Six Woodpecker Species
             Frequenting the Southern Pine Beetle Infested Study
             Area During November - February.

Species	Estimated woodpecker density (birds per ha)			
	Nov.	Dec.	Jan.	Feb.
Downy woodpecker	1.23	1.33	0.58	0.80
Hairy woodpecker	0.62	0.53	0.39	0.16
Pileated woodpecker	0.62	1.07	0.78	0.48
Red-bellied woodpecker	1.54	0.80	0.58	0.32
Common flicker	0.31	---	0.58	0.16
Yellow-bellied sapsucker	---	0.80	0.58	0.64
Size of infestation (ha)	3.25	3.75	5.12	6.25

a/ based on minimum number of woodpecker using areas.

Table IV. Types of Trees Used by Woodpeckers Within the Southern Pine Beetle Infested Study Area.

Woodpecker species	Hardwood	Uninfested pines	Numbers (Percent) of observations Infested pine stage			
			Green	Brown	Snag	Total
Downy	8(19.0)		8(19.0)	21(50.0)	5(11.9)	34(81.0)
Pileated	3(8.8)	4(11.8)	9(26.5)	17(50.0)	1(2.9)	27(79.4)
Hairy		1(7.7)	4(30.8)	5(38.5)	3(23.0)	12(92.3)
Yellow-bellied sapsucker	27(65.8)	5(12.2)	6(14.6)	3(7.3)		9(22.0)
Red-bellied	14(46.7)	10(33.3)		4(13.3)	2(6.7)	6(20.0)
Common flicker	8(72.7)	1(9.1)			2(18.2)	2(18.2)

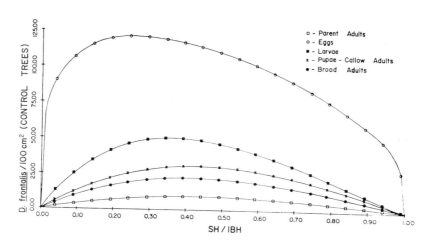

FIGURE 3.   Distribution of life stages of SPB over standardized bole height.   From Coulson *et al.* (1975).

Examination of SPB population data from screened (predation excluded) and unscreened (predation included) trees at the late instar larva stage suggested that woodpeckers have a substantial impact on SPB (Table V).

Analysis of variance (with covariates tree and sample height) indicated significant (P < 0.05) reductions of SPB pupae and brood adults in unscreened trees.   Since the above analysis of variance indicated a significant (P < 0.05) effect due to sample height, we compared estimates of SPB brood densities for lower, mid- and upper bole areas within screened and unscreened trees (Table VI).   Reductions in pupae and brood adults were significant (P < 0.05) only at mid-bole.

Woodpecker foraging appeared to increase slightly average within-tree densities of insect predators and parasites.   We observed general increases in insect predator and parasite densities in unscreened trees; insect predator densities were significantly (P < 0.05) increased in lower and upper bole areas (Table V & VI).

Reductions observed in population densities of SPB life stages as a result of woodpecker activity represent more than consumption of beetles.   In the process of foraging, woodpeckers dislodge large numbers of bark flakes which fall to the ground. Bark debris contained SPB broods with an average density (all life stages) of $1.23 \pm 0.24/dm^2$ phloem area.   Of this number, only 0.25 ± 0.10 (20.3%) survived to emergence.   Bark debris SPB population densities are presented in Table VII.

Table V.   Comparison of Mean Within-tree Population Densities of
           Southern Pine Beetles in Screened (=Predation Excluded)
           and Unscreened (=Predation Included) Trees.

Life stage or Insect associate	Mean within-tree density (per dm^2)	
	Screened trees	Unscreened trees
Larvae	23.9 ± 4.2	22.4 ± 3.0
Pupae	15.3 ± 3.7	$1.0 \pm 0.3*$
Brood adults	14.2 ± 2.0	$11.6 \pm 2.2*$
Insect predators	0.2 ± 0.2	0.6 ± 0.2
Insect parasites	0.2 ± 0.1	0.3 ± 0.1

* Significant difference at the 0.05 level

Bark debris population densities were considerably lower than
those observed for within-tree populations (cf., Tables VI and
VII). Comparisons between screened and unscreened tree
populations plus bark debris data suggest that the majority of
SPB removed from trees during the study were actually consumed by
woodpeckers. Average densities of insect predators and
parasitoides in bark debris were 0.01 ± 0.00 and 0.02 ± 0.01,
respectively. Low bark debris densities of mortality agents are
similar to within-tree population densities.

Woodpeckers were also responsible for indirect mortality of
SPB. Flaking of bark by woodpeckers exposed numbers of brood to
the external environment. These beetles either desiccated or
were consumed by other predators. We observed Brown Creepers
(*Certhia familiarus*) and Black-and-White Warblers (*Mniotilta
varia*) feeding on SPB brood. In addition, we observed that
fungi often invaded galleries of SPB via openings created by
woodpecker foraging. Unidentified fungi attacked and killed SPB
larvae and pupae within trees and in bark debris.

IV.   DISCUSSION

Southern pine beetle infestations represent one type of food
resource "patch" in southern pine forests. Certain woodpeckers
have apparently adapted foraging strategies which maximize
utilization of such patches. Garton (1979) noted that in order
to forage optimally a bird must locate the most profitable patch.
Two strategies are then available: 1) the bird can devote all of
its foraging time to a single patch type, or 2) the bird spends

Table VI. Comparison of Population Densities of Southern Pine Beetle and Insect Associates, by Standardized Tree Height, for Screened and Unscreened Trees.

Standardized height	SPB life stage or associate	Mean density (per dm^2 ± S.E.)	
		Screened tree	Unscreened tree
Lower bole	Larvae	36.1 ± 9.9	28.9 ± 6.2
	Pupae	13.8 ± 4.7	1.1 ± 0.6
	Brood adults	9.8 ± 2.5	11.4 ± 4.3
	Predators	0.0 ± 0.0	0.9 ± 0.2*
	Parasites	0.1 ± 0.1	0.2 ± 0.1
Mid-bole	Larvae	16.1 ± 5.5	11.8 ± 1.5
	Pupae	24.6 ± 9.2	0.8 ± 0.4*
	Brood adults	16.6 ± 2.9	14.1 ± 4.2*
	Predators	0.5 ± 0.4	0.4 ± 0.3
	Parasites	0.1 ± 0.1	0.3 ± 0.2
Upper bole	Larvae	19.5 ± 3.9	24.8 ± 4.3
	Pupae	7.6 ± 2.1	1.0 ± 0.5
	Brood adults	16.3 ± 4.5	9.6 ± 3.1
	Predators	0.0 ± 0.0	0.4 ± 0.2*
	Parasites	0.4 ± 0.2	0.4 ± 0.2

* Significantly different at the 0.05 level.

Table VII.   Average Densities of SPB Larvae, Pupae and Brood
             Adults in Bark Debris Dislodged by Woodpeckers from
             Unscreened Trees.

Life stage	Average density (per dm² ± S.E.)	Percent of unscreened within-tree population
Larvae	0.71 ± 0.26	3.16
Pupae	0.39 ± 0.18	39.00
Brood adults	0.04 ± 0.01	0.03

most of its time foraging in the best patch. Since we observed
woodpeckers feeding in areas uninfested by SPB, adjacent to our
study area we must assume that the latter strategy is in
operation.

Woodpeckers foraging within SPB infestations are apparently
able to forage on optimal life stages of SPB and use optimum
foraging areas within trees. Pine beetles typically deposit
their eggs in the cambial layer. Early instar larvae are
extremely small and feed solely within cambial tissues of the
tree. Since SPB eggs and early larvae lie deep beneath the bark
and are extremely small in size, we would assume that
profitability in feeding on such life stages would be extremely
low. Examination of data from the screening study support this
assumption (cf., Tables V & VI). Later life stages of SPB
burrowed into the outer bark where they were more easily
excavated by woodpeckers; woodpecker induced mortality was
greater for pupae and brood adults than for larvae.

Woodpeckers preferred to forage at or near mid-bole. Beetle
populations are more dense (Fig. 3) in this region, while bark
thicknesses are intermediate ($\bar{x} = 10.63 \pm 2.15$ mm) to those
observed at bottom ($\bar{x} = 18.3 \pm 4.18$ mm) and top of infested bole
($\bar{x} = 4.42 \pm 1.59$ mm).

During our study woodpeckers had a substantial impact on
within-tree populations of SPB. Although woodpeckers consumed
large quantities of beetles at various life stages, we feel that
the most important contribution to SPB population control
occurred among pupae and emerging brood adult beetles. Emerging
beetles will either attack adjacent trees or disperse to form
new infestations; hence, woodpeckers significantly (Table V)
reduced the potentially reproductive portion of the population.

Woodpeckers are certainly not the only natural mortality
agent operating in the SPB-pine-timber complex, however, these
birds do affect beetle populations greater than any other known
agent. Consideration for woodpeckers in land management
decisions would be beneficial to an integrated pest management
program.

REFERENCES

Baldwin, P. (1968). USDA For. Serv. Res. Note RM-105. Rocky Mt.
    Forest and Range Exp. Sta., Ft. Collins, Colo.
Berryman, A. (1964). *Can. Entomol. 96*, 883.
Coulson, R., Foltz, J., Mayyasi, A., and Hain, F. (1975). *J.
    Econ. Entomol. 68*, 671.
Coulson, R., Payne, T., Coster, J., and Houseweart, M. (1972).
    Tex. For. Serv. Publ. 108, 1.
Coulson, R., Pulley, R., Foltz, J., and Martin, W. (1976). Tex.
    Agric. Exp. Sta. MP-1267.
DeMars, C., Jr. (1963). *Can. Entomol. 95*, 1112.
Garton, E. (1979). *In* Proc. Symp. "Role of Insectivorous Birds
    in Forest Ecosystem" (J. Dickson, R. Conner, R. Fleet,
    J. Jackson and J. Kroll eds.), Academic Press, New York.
Hopkins, A. (1909). *In* "Contributions toward a Monograph of the
    Scolytid Beetles." U.S. Bur. Entomol. Tech. Ser. 17, 1.
Koplin, J. (1969). *Condor 74*, 436.
Moore, G. (1972). *Environ. Entomol. 1*, 58.
Otvos, I. (1965). *Can. Entomol. 97*, 1184.
Overgaard, N. (1970). *J. Econ. Entomol. 63*, 1016.
Ragenovich, I., and Coster, J. (1974). *J. Econ. Entomol. 67*, 763.
Shook, R., and Baldwin, P. (1970). *Can. Entomol. 102*, 1345.
Sternizke, H. (1967). USDA For. Serv., So. For. Exp. Sta. Res.
    Bull. SO-10.

THE PILEATED WOODPECKER IN FORESTS
OF THE NORTHERN ROCKY MOUNTAINS

B. Riley McClelland

School of Forestry
University of Montana
Missoula, Montana

A study of nesting habitat of the Pileated Woodpecker
(Dryocopus pileatus) was conducted in northwestern
Montana during 1974 through 1978. Fifty-four trees
with active Pileated Woodpecker nests were located.
Most nest cavities were in large western larch (Larix
occidentalis) snags with broken tops. Mean measurements at nest trees were: dbh = 75 cm (29.5 in), tree
height = 28 m (92 ft), nest hole height = 15.2 m (50
ft), and basal area of surrounding forest = 25 m^2/ha
(109 ft^2/acre). In the northern Rocky Mountains,
forests with an old-growth component of western larch,
ponderosa pine (Pinus ponderosa), or black cottonwood
(Populus trichocarpa) seem to be essential for long-
term support of Pileated Woodpeckers. The roles of
the Pileated Woodpecker include: a) pathfinder for
non-excavating hole nesters, b) predator on insects,
c) and key species in forest management plans.

I.   INTRODUCTION

The Pileated Woodpecker (Dryocopus pileatus) is the largest
woodpecker found in the Rocky Mountains (Fig. 1). It may be the
largest surviving woodpecker in North America, due to the apparent extinction of the Imperial Woodpecker (Campephilus imperialis)
from Mexico and the endangered (perhaps extinct) status of the
Ivory-billed Woodpecker (Campephilus principalis) in its limited
range in a few southcentral and southeastern states and Cuba.
The loss of the Imperial Woodpecker has been attributed to direct

FIGURE 1. Pileated Woodpeckers are about 40 centimeters long. In the northern Rocky mountains their future is dependent upon the perpetuation of forests with a component of old-growth.

killing (Tanner, 1964). According to Tanner (1942) and Dennis
(1967) the Ivory-bill has been most affected by loss of suitable
habitat. The Pileated is in jeopardy in the western portion of
its range, where a planned elimination of old-growth forest is
underway on most commercial forests.

Although the Pileated Woodpecker is widely distributed in
many forests of the eastern United States, in the far West south
of Canada it is confined to the Pacific Coast states of Washington, Oregon, and northern California, and to northern Idaho and
northwestern Montana in the Rocky Mountains (Bent, 1939; Skaar,
1975) (Fig. 2). The Pileated is absent from, and apparently
never has been known to occupy, the central and southern Rocky
Mountains. Its absence in these regions is probably related to
the scarcity of highly productive forests with numerous large
trees and snags (Bock and Lepthien, 1975).

Saunders (1921) called the Pileated "a common permanent
resident of the more heavily timbered mountains of northwestern
Montana, west of the continental divide." However, at the time
of his publication he had no report of an observed active nest.
Weydemeyer and Weydemeyer (1928) considered the Pileated to be
common in northwestern Montana; by 1975 it was considered rare
(Weydemeyer, 1975). During the 50 years that Winton Weydemeyer
has studied birds in northwestern Montana, he has located thousands of active nests of many species. But he has only 8 Pileated
nests recorded on his life list (pers. comm.). This illustrates
the difficulty of locating Pileated nests and the relative scarcity of the species. The Pileated is a "denizen of extensive
forests" (Bent, 1939) and nest trees are usually surrounded by
forest with an old-growth component.

In Montana, the range of the Pileated Woodpecker is similar
to the range of western larch (Larix occidentalis). Since about
1950, western larch and Douglas-fir (Pseudotsuga menziesii) have
been subjected to heavy cutting. They lead all other species in
annual volume cut in the northern Rocky Mountains (Schmidt et al.,
1977). Nearly one-half of the area where western larch was a
major component now has been cut (Schmidt et al., 1977). The
1974 final environmental impact statement prepared for the Flathead National Forest Timber Harvesting Plan called for conversion
of all mature sawtimber in commercial forests, within 50 years.
Harvesting rotations thereafter would approximate 100 years.
Some private companies with large forest holdings in the Northern
Rockies plan to complete conversion in an even shorter time.
Using 100 year rotations, few western larch would exceed 16 inches
dbh (Schmidt et al., 1977) and few stands would provide suitable
nesting and feeding sites for Pileated Woodpeckers. The habit-
able Rocky Mountain range of the Pileated thus is diminishing as
old-growth stands are cut. With this concern, a study to determine the nesting habitat requirements and roles of the Pileated
Woodpecker in western larch/Douglas-fir forests was undertaken in
1974. This was part of a broader research project concerning
nesting habitat requirements of all hole-nesting bird species in

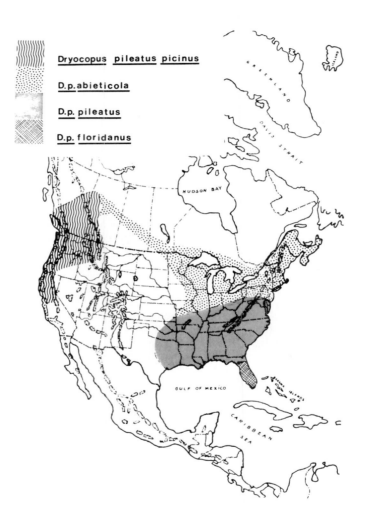

FIGURE 2.   The ranges of the four subspecies of Pileated Woodpeckers are shown on this map.   Note that the species is absent
from the central and southern Rocky Mountains (after Ridgway and
Friedmann (1914), Bent (1939), and Hoyt (1948).

the study area.

II.   STUDY AREA

The study focused on the Coram Experimental Forest (CEF), a 3,000 hectare portion of the Flathead National Forest, in north-western Montana.   Elevations range from 1,065 m to 1,920 m above sea level.   Annual precipitation averages about 90 cm.   Approximately 42% of the original available board feet of timber have been cut.   The CEF is classified as a biosphere reserve, part of the United Nations Educational, Scientific and Cultural Organization's (UNESCO) international system of reserves with primary objectives of conservation of genetic diversity, environmental research, and education (Franklin, 1977).   Other selected sites on the Flathead and Lolo National Forests, and Glacier National Park (also a biosphere reserve) were included in the study.

Study emphasis was on western larch/Douglas-fir forests because of their importance to timber production and their role as habitat for hole-nesting birds.

III.   METHODS

Forest stands of various ages were searched for active nest sites of Pileated Woodpeckers.   Nests were located by using auditory and visual cues.   Pileateds often drum or call in the vicinity of the nest tree during the excavation period and when male and female exchange incubation duties (Kilham, 1959).   Actual excavating sounds often can be heard a kilometer or more away when the wind is calm and there are no background noises (Fig. 3).   Although the adults become less vocal after eggs have hatched, the young become increasingly vocal as they grow and, for a week or so prior to fledging, give "high calls" while hanging out of the cavity opening.   Only those trees in which incubation or feeding young were confirmed are included in the data as active nest trees.

Locating Pileated nests can be very time-consuming.   Compounding the difficulty is the common occurrence of abandonment of nest excavations.   Most woodpeckers, especially Pileateds, have a well-deserved reputation of starting a nest hole, then abandoning it and moving to a new site (Bent, 1939; Hoyt, 1957; Kneitz, 1961).   One pair on the CEF excavated in a larch snag until the excavating Pileated could nearly disappear inside the cavity.   At this point the site was abandoned and the pair moved about 600 meters to a live larch in which they completed a nest cavity and successfully fledged two young.

Pileateds also roost in tree cavities at night and during stormy weather.   Roosts were located by listening for "cuk"

FIGURE 3. A Pileated Woodpecker throws chips from a nest cavity in a black cottonwood snag. With few exceptions new nest cavities are excavated each spring.

vocalizations given by Pileateds as they approached roost trees (Kilham, 1974). Roost trees were considered those in which a Pileated was observed to fully enter a cavity other than an active nest.

At each nest and roost tree a series of measurements related to the nest tree, the site, and the nest hole was made. Measurements included: tree species, tree status, dbh, tree height, decay evidence, fire evidence, percent bark, slope, aspect, elevation, distance to water, basal area of surrounding forest, nest-hole orientation and nest-hole height.

IV.   RESULTS AND DISCUSSION

During the 4 year study period, 54 active nest trees (58 active nests) and 28 active roost trees (32 active roost holes) of Pileated Woodpeckers were located. Nest and roost tree species are shown in Table I. The preference for western larch is not a result of that species simply being more abundant. On CEF sample plots, Douglas-fir outnumbered western larch 3.6 to 1 (live trees) and 1.1 to 1 (snags). The largest snags on the plots were Douglas-fir, but there was no significant difference in mean dbh of snags of the two species. Not a single Douglas-fir nest (Pileated) was found. The preference for western larch nest trees is highly significant (X^2 test, $p < .01$). Similar preference is shown in Oregon, where Bull and Meslow (1977) located 13 Pileated nests, all in western larch or ponderosa pine.

The avoidance of Douglas-fir may be related to decay characteristics in that species. Four years after death, the sapwood in Douglas-fir is essentially destroyed by decay (Wright and Harvey, 1967). In western larch, sapwood decay typically lags behind heartwood decay, leaving a decaying (easily-excavated) interior surrounded by a shell of relatively intact sapwood for many years.

I did not culture wood chips from the nest cavity vicinity as did Conner et al. (1976), so identification of fungi is uncertain. However, visible evidence of decay (e.g., white "pockets" of decay) was present in chips from the heartwood excavated by Pileateds at most nest sites. Shigo and Kilham (1968) and Kilham (1971) described similar nest site characteristics in paper birch (Betula papyrifera) and aspen used by Yellow-bellied Sapsuckers (Sphyrapicus varius). Conner et al. (1975, 1976) described decay characteristics of Pileated Woodpecker nest trees in forests of Virginia.

In riparian sites, stands of old-growth black cottonwood often support a Pileated Woodpecker pair. Where aspen grows sufficiently large, it also is a suitable nest tree species. Grand fir is used in some areas, but it does not appear to be preferred.

Nest trees were classified by "status" categories based on whether the tree was live or dead (a snag) and the condition of

TABLE 1. Species of Trees Used for Nesting and Roosting

	Number of Nest Trees	Percent of Total	Number of Roost Trees	Percent of Total
Western larch (Larix occidentalis)	29	54	15	53
Ponderosa pine (Pinus ponderosa)	12	22	3	11
Black cottonwood (Populus trichocarpa)	8	15	10	36
Aspen (Populus tremuloides)	4	7	0	0
Grand fir (Abies grandis)	1	2	0	0
TOTALS	54	100	28	100

the top (Table II).

Based on these data, broken-top snags are preferred as nest sites. In comparison with forest composition by tree status, the use of broken tops is statistically highly significant (X^2 test, $p < .01$). There were nearly twice as many intact-top snags as broken-top snags (all species) on the sample plots. Intact-top western larch snags outnumbered broken-tops 1.3 to 1. Conner et al. (1976) reported nest trees used by Pileated Woodpeckers in Virginia often had broken tops resulting from breakage at a nest cavity in the tree. Although I did not examine nest tree tops by climbing the trees, I believe that in western larch, broken-tops usually precede use by Pileated Woodpeckers. Evidence of top-breakage at the nest site was found in only one instance. Western larch are often subjected to top breakage by windstorms, lightning, snow or ice. Boyce (1948), Hepting (1971), and Partridge and Miller (1974) describe a broken-top as a common point of entrance for spores of fungi, particularly Laricifomes laracis, and Phellinus pini. Fire scars, branch stubs, and wounds also provide avenues of entrance. Conks were present on 10 nest trees.

Characteristics of nest trees reflect the relationship with old-growth forests in northwestern Montana (Table III). Favored nesting habitat is dense forest with a component of old-growth larch or ponderosa pine. Nest trees are typically large dbh.

Nest trees found by Bull (1977) in Oregon appear to have similar characteristics. Nest trees measured by Conner et al. (1975) in Virginia are considerably smaller. The smaller dbh of nest trees in that area may account for the more common occurrence of top-breakage at nest sites reported by Conner et al. (1976).

There was fire evidence at the base of most larch nest and roost trees. Large western larch have very thick bark on the lower portion of the trunk and often are able to survive fire. Western larch is a seral species, usually occupying sites where fire has played a major role in the past. In the northern Rocky Mountains the ecology of the Pileated Woodpecker is thus closely tied to forest structures which have been significantly influenced by fire.

Roost tree characteristics were similar to those of nest trees, probably because most roost cavities originally were nest cavities. Mean measurements of 28 roost trees were: dbh = 77 cm, roost tree height = 28 m, roost hole height = 14 m, and basal area of surrounding forest = 31 m^2/ha.

The nest tree search image concept, described by Kilham (1970, 1971, 1974) can be useful in summarizing characteristics of nest trees used by Pileated Woodpeckers and as a guideline for forest managers. The Pileated nest tree search image in northern Rocky Mountain forests is perceived as: a broken-top snag (western larch, ponderosa pine, or black cottonwood) at least 60 cm dbh, taller than 18 m (usually much taller), with

TABLE II. Number of Trees in Each Status Category

Status Description	Number of nest trees	Percent of Total	Number of Roost trees	Percent of Total
snag				
intact top	7	13	5	18
broken top	35	65	15	54
live				
broken top	4	7	6	21
intact dead top	2	4	1	3.5
intact live top	6	11	1	3.5
Totals	54	100	28	100

TABLE III. Mean Measurements of Pileated Woodpecker Nest Trees[a]

	Study area and reference		
Type of Measurement	Northwestern Montana (this study)	Northeastern Oregon (Bull and Meslow, 1977)	Southwestern Virginia (Conner et al., 1975)
Number of nest trees	54	13	18
DBH of nest tree (cm)	74.9 (39–119)	76.2 (58–99)	54.6 (33–91)
Height of nest tree (m)	28.0 (12.2–47.2)	21.0 (11.9–36.9)	20.3 (7.3–36.6)
Height of nest hole (m)	15.2 (5.5–29.9)	13.1 (7.0–18.9)	13.6 (5.5–19.2)
Basal area (m^2/ha)	25.1		31.5

[a]Ranges are shown parenthetically.

293

heartwood substantially affected by decay, within a forest with
an old-growth component and a basal area of at least 23 m^2/ha.

V.   THE ROLES OF THE PILEATED WOODPECKER

A.   The Pileated Woodpecker as a Pathfinder

Kneitz (1961) described two European Woodpeckers -- the
Variegated (Picoides major) and the Black (Dryocopus martius) --
as "pathfinders" for secondary hole nesters that cannot accom-
plish their own excavating.   The Pileated Woodpecker is a pathfinder in the northern Rocky Mountains.   In only one of 58 nests
in the study area was a Pileated pair observed to reuse a previous year's nest cavity.   With rare exceptions then, a new nest
cavity is excavated each spring, making previous nest holes
available for Pileated roosts or for nest or roost sites for
other species.   Nest holes made by Pileated Woodpeckers are subse-
quently used by cavity-nesters such as the Wood Duck (Aix sponsa),
Bufflehead (Bucephala albeola), Common Merganser (Mergus merganser), Hooded Merganser (Lophodytes cucullatus), Common Goldeneye
(Bucephala clangula), Barrow's Goldeneye (Bucephala islandica),
American Kestrel (Falco sparverius), Saw-whet Owl (Aegolius
acadicus), Pygmy Owl (Glaucidium gnoma), Boreal Owl (Aegolius
acadicus), Screech Owl (Otus asio) (Fig. 4), Common Flicker
(Colaptes auratus), red squirrel (Tamiasciurus hudsonicus), flying squirrel (Glaucomys sabrinus), and pine marten (Martes
americana).   Smaller hole nesters (e.g., bluebirds (Sialia sp.),
nuthatches (Sitta sp.) and chickadees (Parus sp.) prefer the
smaller nest holes provided by the other major pathfinder in the
study area, the Yellow-bellied Sapsucker.   Brown Creepers (Certhia
familiaris) and Hairy Woodpeckers (Picoides villosus) were observed roosting in Pileated feeding excavations.   The Pileated
Woodpecker thus creates nesting and roosting "paths" for many
birds and small mammals in a forest ecosystem.

B.   The Role of the Pileated as a Predator on Insects

The insectivorous habits of the Pileated Woodpecker are well-
documented in the literature.   Beal (1911) reported 2,600 ants
from a single Pileated stomach.   Bent (1939), Hoyt (1957), and
Conner and Crawford (1974), commented on the high proportion of
carpenter ants (Camponotus sp.) in its diet.   Downing (1940)
collected droppings in Iowa and found them composed of nearly
100% carpenter ant remains.   Martin et al. (1951) listed ants
as comprising over half of the animal diet of the Pileated.
Some impression of the biomass consumed by the species may be in-
ferred from a study by Rumsey (1968).   A Pileated Woodpecker kept

FIGURE 4.   An adult Screech Owl peers from its nest.   The
cavity was excavated by a pair of Pileated Woodpeckers the
preceding year.

in captivity consumed 100 grams of mealworms per day. Tanner (1942) found that Pileateds spend 77% of their feeding time digging (excavating for ants or other within-wood insects) and 23% scaling (removing bark and feeding on insects below the bark).

Pileated Woodpeckers in northwestern Montana are resident on feeding territories throughout the year and display close pair bonds year-round as they do in eastern areas studied by Kilham (1959, 1976). They are heavily dependent on carpenter ants and wood-boring insects. During summer months, Pileateds spend much of their feeding time on logs, low stumps, and on the lower portions of snags and trees. Based on examination of many feeding sites and the remains of droppings, in most cases, they had fed upon carpenter ants.

In the northern Rocky Mountains, logs and low stumps on most Pileated territories are snow-covered for at least half the year. Snags and tall stumps receive a proportionally greater share of feeding activity in that season. In many snags, deep cracks caused by lightning, frost, or drying are often used as sites for egg laying or refuge by insects. Especially during winter months, Pileateds enlarge many of these cracks in search of insects. Douglas-fir snags and stubs are often used as feeding sites in contrast to no use as nest sites (by the Pileated). Western larch, western red cedar (Thuja plicata), and lodgepole pine (Pinus contorta) are also favorite feeding trees. Bull and Meslow (1977) found Douglas-fir to be a preferred feeding tree species in Oregon.

Pileated Woodpeckers fed in recently logged areas (selection or shelterwood cuts -- rarely in small clearcuts) if a substantial number of logs or snags were available for feeding. They were observed flying over open fields, clearcuts, lakes, and even across busy highways in order to reach suitable feeding areas. However, they seldom fed within broad open areas and more rarely nested in such sites. Although the availability of suitable nest trees is an important management consideration and perhaps the limiting factor in intensively cut areas, it is more probable that the availability of a sufficient supply of the winter food is the general limiting factor -- the "bottleneck" for Pileated populations in the study area.

C. Managing the Pileated Woodpecker in the Northern Rocky Mountains

In the northern Rockies, the population status of this species appears to be directly related to the presence of old-growth stands that include western larch, ponderosa pine, or black cottonwood. Such stands contain decaying trees which provide suitable nest trees and feeding sites. Where extensive areas have been cut and no old growth remains, a Pileated pair usually cannot be sustained.

How much forest does a pair of Pileated Woodpeckers need to maintain themselves indefinitely? Pairs fed throughout areas ranging from approximately 200 to well over 400 hectares. Some of the larger feeding territories included interspersions of clearcut, agricultural, or "developed" areas, requiring more total land to comprise the necessary acreage of feeding area. Tanner (1942) estimated territory size of 40 hectares in dense forest of Louisiana. Bull and Meslow (1977) described territories ranging from 130-243 hectares in Oregon.

Minimum territory size for Pileated Woodpeckers is somewhat flexible because of numerous factors, particularly food availability. The number and productivity of carpenter ant colonies and the abundance of wood-boring insects probably is significant in determining how much forest is needed to sustain Pileateds. Based on the 54 territories in which active nest trees were located, it appears that a minimum of 200 hectares of suitable feeding area usually is needed in Northern Rocky Mountain areas. Thus, in a 400 hectare planning unit, half would need special attention to perpetuation of suitable feeding sites, e.g., leaving snags and logs and perhaps going back to high stumping to favor carpenter ant colonies. Twenty to 40 hectares with a substantial component of old-growth larch or ponderosa pine should receive special care for long-term nesting and stable feeding habitat. Along streams, corridors (at least 90 m wide) with large specimens of black cottonwood, particularly snags that have lost their bark, provide suitable nesting sites as long as numerous feeding trees and logs are also available.

On forest lands in national parks and wilderness areas where natural diversity is perpetuated, the Pileated probably will continue to find old-growth stands with essential heartwood decay present. On commercial forest lands in the northern Rocky Mountains, the elimination of old growth would mean the elimination of the Pileated Woodpecker and other species with similar requirements. Where forest managers have as one of their objectives provision of suitable habitat for Pileated Woodpeckers and associated species, management plans must recognize the need for an old-growth component.

D. The Role of the Pileated Woodpecker as an Indicator Species

"The mere presence or absence of a given species or group of species in a particular environment can be used to define normal or baseline environmental conditions and to determine the degree to which communities have been affected by ... influences such as pollution or man-made habitat alteration"

Ehrenfeld (1976)

In relation to suitable habitat, the Pileated Woodpecker has the narrowest ecological amplitude of all the hole nesters found in the study area. It is a steno species or ecological indicator (Graul et al., 1976: Adams and Barrett, 1976). Pileateds need forest with an old-growth component to enable long-term success, as do many other hole-nesting species. But other species are more adaptable, in varying degrees, to other habitats including those modified by man's activities. Thus, in the study area the Pileated Woodpecker can be the key to retaining a complete community of hole-nesting birds. The Pileated cannot be accomodated simply by leaving snags in areas otherwise clearcut. Very carefully conceived and implemented forest management plans will be needed to perpetuate this species. If managers succeed in perpetuating Pileated on forest planning units of about 400 hectares, they can feel reasonably confident that provisions of suitable habitat for the other hole-nesters will require comparatively simple measures, e.g., leaving snags and future replacements for use by Lewis' Woodpeckers (Melanerpes lewis) or Mountain Bluebirds (Sialia curricoides) in heavily cut areas.

Managers may find it useful to consider the Pileated Woodpecker as an indicator species, a key to the health of communities of hole-nesting birds in those northern Rocky Mountain areas in which it is a resident.

ACKNOWLEDGEMENTS

This study received financial support from the U.S. Forest Service Intermountain Forest and Range Experiment Station, and the School of Forestry and Montana Cooperative Wildlife Research Unit at the University of Montana, Missoula. Manuscript review and suggestions were provided by R. Conner, J.Dickson, W. Fischer S. Frissell, L. Pengelly, and P. Wright. Assistance from C. and G. Halvorson, P. McClelland, R. Trembath, and J. Whitney is gratefully acknowledged.

REFERENCES

Adams, D. and Barrett, G. (1976). Am. Mid. Nat. 96, 1979.
Beal, F. (1911). "Food of the woodpeckers of the United States." U.S.D.A. Biol. Survey Bull. 37.
Bent, A. (1939). U.S. Nat. Mus. Bull. 1974.
Bock, C. and Lepthien, L. (1975) Wilson Bull. 87, 355.
Boyce, J. (1948) "Forest pathology." McGraw-Hill, Inc., New York.
Bull, E. and Meslow, E. (1977). J. For. 75, 335.
Conner, R. and Crawford, H. (1974) J. For. 72, 564.

Conner, R., Hooper, R., Crawford, H., and Mosby, H. (1975) J. Wildl. Manage. 39, 144.

Conner, R., Miller, D., and Adkisson, C. (1976). Wilson Bull. 88, 575.

Dennis, J. (1967) Audubon 69, 38.

Downing, G. (1940). IA Bird Life 10, 43.

Ehrenfeld, D. (1976). Am. Sci. 64, 649.

Franklin, J. (1977). Science 195, 262.

Graul, W., Torres, J., and Denny, R. (1976). Wildl. Soc. Bull. 4, 79.

Hepting, G. (1971). "Diseases of forest and shade trees of the U.S." U.S. For. Serv. Agr. Handbook 386.

Hoyt, S. (1957). Ecology 38, 246.

Kilham, L. (1959). Condor 61, 377.

Kilham, L. (1970). Auk 87, 544.

Kilham, L. (1971). Wilson Bull. 83, 159.

Kilham, L. (1974). Wilson Bull. 86, 407.

Kilham, L. (1974). Auk 91, 634.

Kilham, L. (1976). Auk 93, 15.

Kneitz, G. (1961). Waldhygiene 4, 80.

Martin, A., Zim, H. and Nelson, A. (1951). "American wildlife and plants: a guide to wildlife food habits." McGraw-Hill, Inc., New York.

Partridge, A. and Miller, D. (1974). "Major wood decays in the inland northwest." For., Wildl., and Rge. Exp. Sta. Publication, Univ. ID., Moscow.

Ridgway, R. and Friedmann, H. (1914). "Birds of North and Middle America". Part VI. U.S. Natl. Mus. Bull. 50, Washington, D.C.

Rumsey, R. (1968). Bird-banding 34, 313.

Saunders, A. (1921). "A distributional list of birds in Montana." Pacific Coast Avifauna No. 14, Cooper Ornithological Society.

Schmidt, W., Shearer, R., and Roe, A. (1976). "Ecology and silviculture of western larch forests." U.S. For. Serv. Tech. Bull. 1520.

Shigo, A. and Kilham, L. (1968). "Sapsuckers and Fomes igniarius var. populinus. U.S. For. Serv. Res. Note NE-84.

Skaar, P. (1975). "Montana bird distribution." P.D. Skaar, Bozeman, MT.

Tanner, J. (1942). "The Ivory-billed Woodpecker." Nat'l Audubon Soc. Res. Rept. No. 1. New York.

Tanner, J. (1964). Auk 81, 74.

Weydemeyer, W. (1975). Condor 77, 281.

Weydemeyer, W., and Weydemeyer, D. (1928). Condor 30, 339.

Wright, K. and Harvey, G. (1967). "The deterioration of beetle-killed Douglas-fir in western Oregon and Washington." U.S. For. Serv. Res. Pap. PNW-50.

POPULATION DEMOGRAPHIES,
SPACING, AND FORAGING BEHAVIORS
OF WHITE-BREASTED AND PIGMY NUTHATCHES
IN PONDEROSA PINE HABITAT

Shaun M. McEllin

Department of Science and Mathematics
New Mexico Highlands University
Las Vegas, New Mexico

Comparison of various life-history features of White-breasted and Pigmy nuthatches demonstrated strikingly different patterns of population demography, spacing behavior, and foraging ecology. White-breasted nuthatches exhibited an *exclusivity strategy* in which isolation is of fundamental importance. Territoriality functioned to exclude conspecifics and in association with specialized nesting conditions resulted in low population densities. White-breasted nuthatches maintained a territory throughout the year; consequently, the population showed little yearly variability. Territorial behavior functioned in the dispersal and exclusion of young from the territory soon after fledging. This species demonstrated additional isolation from conspecifics through sexual differences in foraging habitat exploitation patterns. Through differential utilization of heights within trees and tree structures while foraging, the pair exploited the same habitat without intersexual interference.

Pigmy nuthatches exhibited an *inclusivity strategy* in which intraspecific tolerance is a prime characteristic. Territorial behavior was confined to the immediate vicinity of the nest tree. Nest site availability and less specialized nesting conditions in association with limited territorial behavior resulted in high population densities. Outside the breeding season Pigmy nuthatches formed gregarious monospecific flocks which resulted in comparatively high yearly population variability. The temporal period of population decline indicated that individuals dispersed in the spring prior to the onset of breeding which coincided

301

with the disruption of the winter foraging flocks. The foraging behavior of this species exhibited individual similarity in habitat utilization patterns.

    Comparisons of foraging behavior between species
revealed many dissimilarities. White-breasted nuthatches
foraged on coarse-featured habitat aspects. This species
exhibited foraging specialization in habitat features which
was complemented by feeding behavior and food item generalization. Pigmy nuthatches foraged on fine-featured habitat
characters. In foraging in these areas this species exhibited generalization in habitat features which was complemented by feeding behavior and food item specialization
during the reproductive period with less specialization
during the nonreproductive period. These patterns suggested
that White-breasted nuthatches foraged for evenly distributed food items and Pigmy nuthatches foraged for patchy food
items. Both species demonstrated behavioral stereotypy
among diverse ecological characters.

I.   INTRODUCTION

    North-temperate, forest-dwelling insectivorous birds demon-
strate a continuum of space utilization patterns with territoriality and flocking behaviors occupying the extremes. Some
species maintain a permanent territory excluding conspecifics
other than the mate and obtaining requisite resources entirely
with the territory (Nice, 1941; Hinde, 1956; Lawrence, 1967).
Other species, although exhibiting varying degrees of territorial behavior during the breeding season, form gregarious
winter foraging flocks with conspecifics and, at times, other
species (Brown, 1963; Carrick, 1963; Morse, 1970; Austin and
Smith, 1972). These contrasting patterns of foraging sociality
and spatial dispersion of individuals are often interpreted in
terms of distributions and abundances of principal food resources
(Brown, 1964, 1969; Crook, 1965; Schoener, 1971).

A.   Permanent Territoriality

    Species that maintain a permanent territory typically exploit
food resources that are evenly distributed but not locally abundant (Brown, 1964, 1969; Morse, 1971a; Schoener, 1971). Generally, these species forage for food items that are not readily
available to other species. Among permanent residents of

north-temperate forests woodpeckers, nuthatches, and creepers, specialized trunk foragers constituting a trunk foraging guild, typically exhibit this strategy (Lawrence, 1967; Ligon, 1968; Jackson, 1970; Willson, 1970). By excluding conspecifics other than the mate these species exploit an apparently evenly distributed suite of food resources without interference from conspecifics (Brown, 1964, 1969; Kilham, 1965; Ligon, 1968).

Investigations of foraging behavior of avian species, notably woodpeckers, exhibiting year-long territoriality in association with trunk foraging adaptations have demonstrated sexually dimorphic resource exploitation patterns (Kilham, 1965, 1970; Selander, 1965, 1966; Koplin, 1967; Ligon, 1968; Jackson, 1970; Willson, 1970; Kisiel, 1972; Wallace, 1974; Austin, 1976). The sexually divergent foraging patterns are often, but not necessarily, associated with morphological differences in body dimensions, quite often, bill dimensions (Schoener, 1965; Willson, 1969; Williamson, 1971; Morse, 1971b; Wallace, 1974). Both morphological and behavioral differences resulting in sexual foraging dimorphism are of ecological significance because they represent adaptations which decrease intraspecific contention for resources by permitting the sexes to occupy different "subniches" (sensu Selander, 1966) or adaptive subzones. The functional mechanisms that have been described which result in intersexual resource partitioning are essentially the same as the differential use of habitat or behavioral characters which isolate genetically or ecologically (guild) similar species (MacArthur, 1958; Root, 1967; Stallcup, 1968; Cody, 1974). Principally, species or sexes partition resources by segregation along vertical and horizontal habitat dimensions (MacArthur, 1958; Stallcup, 1968; Morse, 1970), differential use of various foraging substrates, such as trunks (Stallcup, 1968; Willson, 1970), behavioral differences in prey acquisition techniques (MacArthur, 1958; Root, 1967; Jackson, 1970), and differential exploitation of the various plant species (Betts, 1955; Willson, 1970; Edington and Edington, 1972). Recently, Schoener (1971, 1974) and Cody (1974) have extensively reviewed the various patterns of resource partitioning in birds and other animal groups.

Differences in foraging behavior between the sexes can result in increased foraging efficiency for the territorial mated-pair. Through spatial or behavioral isolation mechanisms, the sexes can reduce altercations between one another at foraging sites (Ligon, 1968). This reduction in aggressive interactions or displacements at feeding sites may also be important in the continued maintenance of the pair-bond (Kilham, 1965; Wallace, 1974). Within the territorial context, isolation of the sexes reduces the probability of one sex foraging at a location previously exploited by the other sex and provides a mechanism allowing greater predictability in resource exploitation

patterns. This particular facet of foraging ecology, whether
individuals (or groups) forage randomly, systematically, or in a
stratified-random system within the geographical foraging space
and whether temporal patterns of space utilization exist, has not
been thoroughly analyzed. The distribution and temporal patterns
of abundance of the food supply are undoubtedly important factors
directing geographical and temporal patterns of space utilization.

The increased incidence of sexual dimorphism among island
populations has suggested that reduced ecological pressure was an
important factor facilitating sexually divergent resource exploitation patterns by allowing expansion into areas along a resource
gradient typically occupied by other species' populations in more
densely packed faunas (Selander, 1966; MacArthur and Wilson,
1967; Morse, 1977). However, more detailed analyses are needed
to ascertain whether the increased incidence of foraging sexual
dimorphism exhibited by island populations resulted from reduced
interspecific ecological pressures, increased intraspecific ecological pressures, or both. Wallace (1974) and Morse (1977) have
demonstrated foraging dimorphism between the sexes in island
faunas under conditions of reduced interspecific ecological
pressures (reduced species packing). The observed expansion
along niche dimensions and intersexual segregation within the
expanded dimensions corresponded with theoretical predictions
concerning habitat utilization developed by MacArthur and Pianka
(1966). However, the study of Morse (1977) has also revealed
that, in reference to the mainland, territory size decreased and
population density increased under island conditions. Consequently, the possibility exists that increased intraspecific
pressures are influential in the development of sexually
divergent resource exploitation patterns.

Avian territory size fluctuates in response to both intra-
and interspecific ecological circumstances. Typically, territories fluctuate in size but do so within definite upper and
lower size limits (Wilson, 1975). A number of environmental
factors have been shown to influence territory size (Schoener,
1968; Verner, 1975). In reference to intra- and interspecific
population pressures, territory size varied inversely with
increasing population density (Holmes, 1970) and varied directly
with increasing numbers of competing species (Yeaton and Cody,
1974). These two differing forms of territory size accommoda-
tions to intra- and interspecific pressures have demonstrated
that territories can be adjusted in area to reflect the prevail-
ing ecological conditions. The high incidence of sexual
dimorphism among permanently territorial species, in light of the
nature of territorial alterations to both intra- and interspecific pressures, indirectly supports the hypothesis that sexual
foraging dimorphism may be a response to both increased intraspecific pressures (limiting territory size) and decreased

interspecific pressures (allowing species expansion of resource
utilization patterns within the geographically limited space).
    In general, sexual dimorphism can be viewed as an increase
in overall niche breadth in association with increased specialization by the sexes with reference to within habitat dimensions
(Selander, 1966). The reduction of interspecific ecological
pressures under conditions of uniform resource availability
theoretically has lead toward expansion along niche dimensions
(MacArthur and Pianka, 1966; MacArthur, 1970) and has been demonstrated in island avifaunas (Selander, 1966; Morse, 1977), while
the sexual divergence in foraging ecology may be the result of
increased intraspecific pressures brought about by high population density. The increased specialization resulting in dissimilarity between the sexes in certain aspects of foraging ecology
may be compensated through a broader utilization spectrum within
another aspect of foraging. This concept of complementarity,
dissimilarity along one dimension associated with similarity
along another, has been commonly observed in comparative studies
of avian foraging ecology (Cody, 1968; Hespenheide, 1971; Snow
and Snow, 1972).

B.   Foraging Flocks

    At the other extreme of space utilization patterns some
species form gregarious flocks that are maintained throughout the
majority of the nonreproductive period. Within a flock individuals are found to forage in close spatial association with one
another. The adaptive significance of this foraging strategy has
been related to the abundance and distribution of the food supply
(Rand, 1954; Crook, 1965; Brown and Orians, 1970). When the
distribution of the food supply is uneven, patchy, but abundant
within a patch, congregation could afford the advantage of more
thorough coverage and more rapid exploitation of the food
supplies (Brown, 1964, 1969; Horn, 1968).
    Krebs et al. (1972, 1974) have recently investigated some
aspects of flocking function and have obtained data indicating
that this strategy assists individual flock members within monospecific flocks in obtaining and possibly communicating information on the location or nature of food resources. When food
items occurred in localized patches, but were abundant, social
learning facilitated flock members in securing food items which
resulted in individuals altering specific searching behaviors
(Krebs et al., 1972). Presently, detailed behavioral analyses
designed to elucidate whether information is communicated among
flock members and, if so, the mechanisms by which information is
transferred between individuals, have not been reported.
However, Ward and Zahavi (1973), studying large monospecific
flocks, have hypothesized and presented some supportive evidence

indicating that communal roosts function as centers of information transfer concerning the location of food resources.

Other possible advantages of flocking elucidated by Cody
(1971) suggest that: (a) the flocking association reduces the
probability that an individual will forage at a specific location
recently depleted of food resources; (b) through the flock-
foraging strategy the habitat can be more thoroughly covered and
exploited; and (c) in cases of renewable resources, flocking
provides a means of co-ordination of temporal return schedules.

An intensive study of foraging flocks by Morse (1970) has
demonstrated that within mixed-species flocks the component
species exhibited more specialized foraging behavior than when
foraging individually.   Similar modifications of foraging
behaviors by species when foraging as components of mixed-species
flocks have been observed by Austin and Smith (1972).   Morse
(1970) observed that the component species of a mixed-species
flock were those similar in overall foraging behavior.   These
data suggest that mixed-species foraging flocks are, in essence,
socially-structured guilds which, in addition to partitioning
resources, offer a format providing advantages to both dominant
and subordinate component species.

The precise nature of flocking associations shows considerable variation along the continuum of space utilization patterns.
Considering only single-species flocks, both flock size and
foraging area exhibit vast differences among species.   Studies of
chickadees, in particular, have demonstrated typical flock sizes
of six to seven individuals foraging in an area of approximately
ten hectares (Dixon, 1963, 1965; Hartzler, 1970; Glase, 1973;
Smith, 1976).   The Black-capped chickadee's (*Parus atricapillus*)
winter space utilization pattern consisted of a small flock that
defended a group-territory with a dominance hierarchy established
among members of the flock (Glase, 1973).   Other flocking
species, notably herbivores and carrion-feeders, exhibit considerable variation in the size of the foraging group, size of the
foraging area, extent of group defense of space, and nature of
intraspecific aggressive behavior (Carrick, 1963; Brown, 1972;
Ward, 1965).   In some cases flock size may reach thousands with
the flock exploiting vast geographical areas (Ward, 1965).
Flock size and foraging area appear to be related to the nature
and abundance of the principal food resources utilized by the
flock.   Large flocks are common among granivorous and herbivorous species, such as Red-winged blackbirds (*Agelaius
phoeniceus*) (Neff and Meanley, 1957; Ward, 1965), while smaller
flocks are associated with an insectivorous diet (Bock, 1969;
Morse, 1970; Austin and Smith, 1972).   The extent of aggressive
interactions among individuals within flocks also exhibits
considerable variation.   Morse (1970) has demonstrated that
aggressive interactions among species in mixed-species flocks
result in a dominant-subordinant hierarchial system.   In

contrast, some monospecific flocks exhibit highly gregarious
social structure with considerable individual spatial tolerance
(Ward, 1965).

Brown (1974) interpreted the diversity of avian spacing
systems in reference to territoriality and food supply. The
varying degrees of flocking behavior appear closely associated
with differing patterns of breeding territoriality and the
nature and distribution of food resources. Territorial flocking
species, such as the Black-capped chickadee, have individual
mated-pair, multi-purpose territories (Type A) during the reproductive period and form small, group-territorial winter foraging
flocks. It is interesting that within this system Glase (1973)
has observed differences in foraging behavior between the sexes
of Black-capped chickadees. This pattern of flocking behavior
may be considered a link between a permanently territorial
system of a mated-pair and larger, more gregarious flocking
systems. The fact that sexually dimorphic foraging behaviors
were observed within this species in association with territoriality, adds additional evidence supporting the close interrelationship between these two phenomena.

Horn (1968) in a study of a colonial nesting species,
Brewer's blackbird (*Euphagus cyanocephalus*), has demonstrated
that the colonial nesting habit was closely associated with food
resources that were spatially and temporally variable. There
appeared to be a close interrelationship among the degree of
breeding dispersion, extent of flocking behavior, and the nature
of the food supply.

Avian space utilization strategies show considerable variation in a number of ecological parameters. The nature of this
variation appears to exhibit interrelationships among breeding
and nonbreeding dispersion, territoriality, social behavior,
foraging behavior and food supplies. Analysis of these aspects
of nuthatch (Order: Passeriformes; Family: Sittidae) life
histories offers a means to compare how closely related species
with contrasting patterns of breeding territoriality and winter
social structure differ in resource utilization patterns and
other ecological factors.

II.   STUDY AREA AND METHODS

The study site was approximately 72 km west of Fort Collins,
Larimer County, Colorado (40° 37' N, 105° 31' W). The site
enclosed an area of 173 ha with the elevation ranging between
2380 and 2500 m. The vegetation of the study site is best
described as a Ponderosa pine (*Pinus ponderosa*) forest and
parkland consisting of a relatively uniform stand of Ponderosa

pine trees interspersed with Quaking aspen (*Populus tremuloides*) along stream bottoms and in the more mesic depressions.

The study site was censused for nuthatches every two weeks for 112 weeks (1974-1975), using a technique developed by Emlen (1971). During the reproductive periods (April through June) nesting locations of nuthatches were recorded.

The investigation of the resource utilization patterns and foraging behaviors of nuthatches included: (a) recording the exact location of a bird while foraging including such aspects as tree height, foraging height, tree structures used while foraging, and distance of foraging from the trunk and (b) categorization of prey-capture techniques.

III. RESULTS AND DISCUSSION

A. Population Demography, Spacing Behavior, and Nest Site Location

1. Population Demography--Spacing Behavior. Both White-breasted nuthatches (*Sitta carolinensis*) and Pigmy nuthatches (*Sitta pygmaea*) are permanent residents of Ponderosa pine (*Pinus ponderosa*) forests of the Transition Zone in north-central Colorado (Bailey and Niedrach, 1965). Population estimates of these species revealed strikingly different demographic patterns (Fig. 1). In association with population censusing and foraging behavior observations, nuthatch intra- and interspecific spatial associations were recorded. A summary of these associations on a monthly basis for combined data for both field seasons is presented in Table I. These data elucidate the contrasting patterns of spatial associations between the two species.

White-breasted nuthatches had a significantly lower (Student's t-test and Wilcoxon test; $p < 0.001$) population density than Pigmy nuthatches. Mean yearly population densities for White-breasted and Pigmy nuthatches within the Ponderosa pine study site were 9.2 birds/100 ha and 38.8 birds/100 ha, respectively. Both species maintained relatively constant population densities between the two years of the study and no significant differences (Student's t-test and Wilcoxon test; $p > 0.05$) were observed within a species in comparisons of mean monthly populations between the two years. Stallcup (1968) in an earlier study of these species in similar habitat also observed a larger Pigmy nuthatch population density and constant between-year population patterns. White-breasted nuthatches maintained a mated-pair association throughout the entire year and were observed in close spatial association with conspecifics other than the mate only during a short time period immediately after the young fledged.

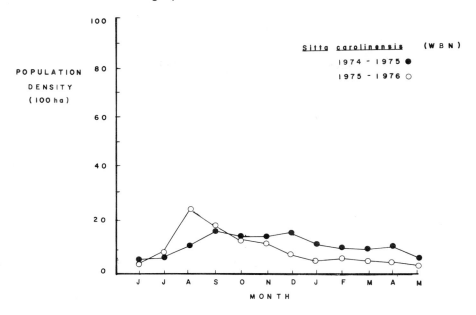

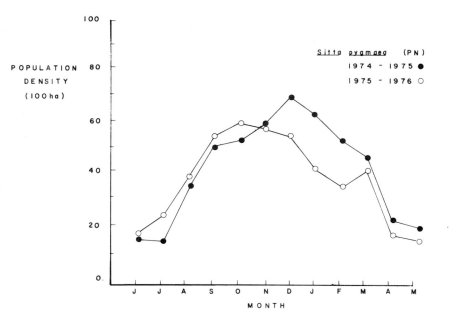

FIGURE 1.   Mean monthly population density estimates for
White-breasted Nuthatches (Sitta carolinensis) and Pigmy
Nuthatches (Sitta pygmaea) for the two field seasons (1974-1975;
1975-1976) in Ponderosa pine habitat.

TABLE I. Mean Monthly Variations in Intra- and Interspecific
Spatial Associations of Nuthatches in Ponderosa Pine Habitat.
The values in parentheses are the number of discrete observa-
tions from which the monthly means were derived.

	January			February	
	WBN[a]	PN[b]		WBN	PN
WBN	1.9(17)	1.7(14)	WBN	2.1(23)	1.5(11)
PN	–	14.3(21)	PN	–	11.3(27)

	March			April	
	WBN	PN		WBN	PN
WBN	2.1(23)	1.8(41)	WBN	2.3(30)	1.1(31)
PN	–	8.2(42)	PN	–	6.1(37)

	May			June	
	WBN	PN		WBN	PN
WBN	2.0(31)	0.7(19)	WBN	2.0(35)	1.2(27)
PN	–	2.7(43)	PN	–	2.5(48)

	July			August	
	WBN	PN		WBN	PN
WBN	6.7(41)	1.5(32)	WBN	4.5(30)	0.7(27)
PN	–	8.9(64)	PN	–	10.7(49)

	September			October	
	WBN	PN		WBN	PN
WBN	2.7(46)	8.3(39)	WBN	2.7(43)	0.3(38)
PN	–	12.3(53)	PN	–	12.3(63)

	November			December	
	WBN	PN		WBN	PN
WBN	2.3(31)	0.5(27)	WBN	2.1(36)	1.4(17)
PN	–	15.7(43)	PN	–	16.5(45)

[a] White-breasted Nuthatch (*Sitta carolinensis*)
[b] Pigmy Nuthatch (*Sitta pygmaea*)

Bent (1948) has presented information showing that mated-pairs of
this species maintain territories throughout the year.

In contrast, Pigmy nuthatches had a significantly higher population density than White-breasted nuthatches. Analysis of
intraspecific spatial associations demonstrated that individuals
of this species were found in close proximity to conspecifics
(greater than a pair) throughout the majority of the year.
Norris (1958), Stallcup (1968), and Bock (1969) observed similar
flocking behavior for this species. The yearly pattern of intraspecific Pigmy nuthatch spatial associations showed that flock
size increased in the fall, reached a peak in December, and
declined rapidly in April. Mean flock size during the nonreproductive period (September through March) was thirteen birds per
flock. In addition to differences in population densities and
flocking behavior, the species exhibited differences in the temporal pattern of population fluctuation with particular reference
to the periods of population decline. Both species' populations
increased due to recruitment at the end of the breeding season.
However, the White-breasted nuthatch population maintained a
nearly constant density because the period of population decline
occurred soon after the young fledged. The mean yearly population density for this species was just slightly higher than the
number of birds known to breed in the area. This temporal
pattern strongly indicated that young birds dispersed soon after
fledging due to aggressive interactions with the parents resulting in territorial exclusion. In comparison, the Pigmy nuthatch
population density increased to levels well above the number of
known breeding birds. This population remained at above-breeding
levels throughout the post-breeding period until early spring, at
which time the population rapidly declined. The higher overwinter population density coincided with winter flocking
behavior, while the period of population decline corresponded
with the advent of the spring breeding season and disruption of
winter foraging flocks.

2. <u>Nest Sites</u>. Both White-breasted and Pigmy nuthatches
nest in cavities; however, nest characteristics differ markedly
between the species (Bent, 1948; McEllin unpublished data).
White-breasted nuthatches principally nested in natural cavities
within live Ponderosa pine trees, while Pigmy nuthatches also
made use of pre-existing cavities but within dead Ponderosa
pines. Both breeding population densities and nest site characteristics indicated a greater nest site availability for Pigmy
nuthatches. Fig. 2 illustrates the distribution of both White-
breasted and Pigmy nuthatch nests during the two breeding
seasons. Pigmy nuthatches defended only the nest tree (Type C)
territory, and pairs were observed nesting in close proximity to
each other. White-breasted nuthatches defended a large (Type A)

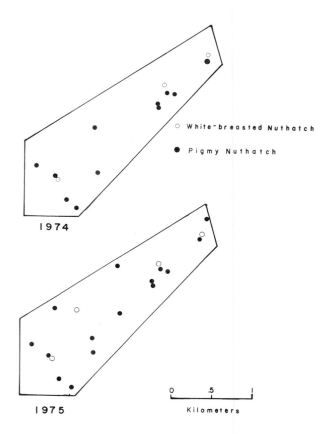

FIGURE 2. Nest tree locations occupied by White-breasted (o) and Pigmy (•) nuthatches during the 1974 and 1975 breeding seasons.

territory and there were considerably greater distances between nest sites.

Population studies of cavity nesting passerines have demonstrated that breeding population densities can be limited by nest site availability, and, by increasing the number of available nest sites, a significant increase in breeding density can be obtained (Lack, 1966; Krebs, 1970). Analysis of nesting characteristics of White-breasted nuthatches suggested that this technique must be subject to constraints set by territory size and function. Increasing nest site availability for White-breasted

nuthatches could have little or no impact on breeding population
levels because territoriality could exclude others from breeding
if the new nest site was within a pre-existing territory.

    3. Discussion. The population demographies exhibited by
these species suggest two different space utilization strategies.
For White-breasted nuthatches territorial exclusion appeared to
limit the number of breeding birds and overall population levels.
This aspect of territoriality has been discussed by Brown (1964,
1969) and Verner (1975) and detailed investigations indicated
that territoriality can be a means of population regulation
(Krebs, 1971; Manuwal, 1974). Since Pigmy nuthatches defend only
the nest site and do not exclude conspecifics from regions other
than the nesting tree, nest site availability coupled with small
territories were principal factors resulting in the higher population levels exhibited by this species. Increasing the number
of available nest sites for this species could result in
increased population densities because this species only excludes
conspecifics from the nest sites. Increasing nest site availability for White-breasted nuthatches may not result in increased
breeding populations because of territorial exclusion of conspecifics. Consequently, maximizing (or increasing) population
levels of White-breasted nuthatches (or species exhibiting
similar strategies) may be more a function of optimization of the
spatial distribution of nest sites. In addition, the different
patterns of population decline suggested different temporal
periods of population dispersal. The data indicated that White-
breasted nuthatches dispersed in late summer. Since it appeared
advantageous for this species to obtain a permanent territory and
the possibility existed that in the process of habitat selection
a suitable nest or roosting site was an important criterion,
availability of artificial nest sites during the period of
dispersal may be an important factor for consideration.
Detailed analyses of optimal spacing of artificial nests and
optimal periods of availability may have important management
implications.

B. Resource Utilization Patterns

    1. Vertical Dimension. Data on heights at which the two
species of nuthatches foraged within Ponderosa pine trees are
summarized in Table II for both reproductive (April--August) and
non-reproductive (September--March) periods of both field
seasons. The species exhibited significant differences (Student's t-test; $p < 0.01$) in vertical foraging height with White-
breasted nuthatches foraging lower in trees than Pigmy nuthatches. In addition, analysis of foraging height by sex for
White-breasted nuthatches also revealed significant differences

TABLE II, Data Summary of Nuthatch Vertical Foraging Location in Ponderosa Pine Trees for Reproductive (April--August) and Nonreproductive (September--March) Seasons. The data presented are sample size (N), Mean ($\bar{x}$), and standard error of the mean (SE).

	FORAGING LOCATION					
	Foraging Tree Height (m)			Foraging Height (m)		
	N	$\bar{x}$	SE	N	$\bar{x}$	SE
REPRODUCTIVE						
Sitta carolinensis (WBN)						
male	107	16.02	1.23	107	4.41	0.53
female	87	18.76	1.39	87	9.66	0.62
Sitta pygmaea (PN)	187	14.31	1.01	184	9.51	0.51
NONREPRODUCTIVE						
Sitta carolinensis (WBN)						
male	170	14.74	1.41	167	2.80	0.41
female	107	12.56	1.36	99	7.70	0.59
Sitta pygmaea (PN)	309	13.02	1.11	306	10.40	0.56

(Student's t-test; $p < 0.01$) with males foraging lower than females. The interspecific and intrasexual differences in foraging height were displayed during both reproductive and nonreproductive periods. No statistically significant differences (Student's t-test; $p > 0.05$) were observed in foraging height of nuthatches between season, but White-breasted nuthatches had lower mean foraging heights and Pigmy nuthatches had higher mean foraging heights during the nonreproductive period. The differential foraging heights of the sexes of White-breasted nuthatches resulted in females having a mean foraging height similar to Pigmy nuthatches. However, comparisons of foraging height in reference to foraging tree height (foraging height/tree height) between the sexes of White-breasted nuthatches and between the sexes and Pigmy nuthatches exhibited significant differences (Chi-square contingency test; $p < 0.01$). These data demonstrated that within a tree the birds foraged at different vertical levels (Fig. 3). The resultant pattern of spatial

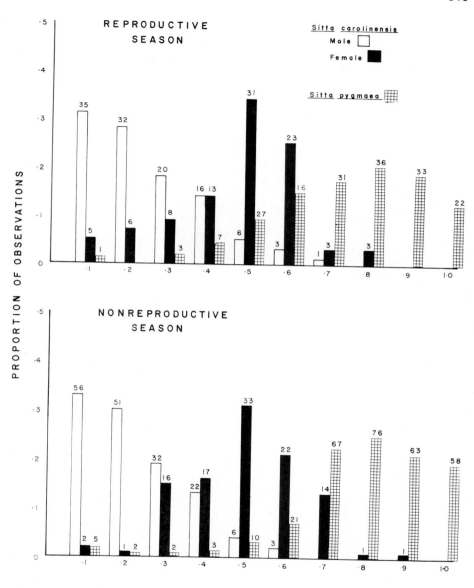

FIGURE 3.   The frequency of nuthatch utilization of vertical
tree height in reference to total tree height during the reproductive season and nonreproductive season.   The number above each
bar in the histogram is the total number of observations for each
species or sex within each height category.

utilization of trees in the vertical dimension is that male
White-breasted nuthatches forage lowest in the trees, females
forage above the males but below Pigmy nuthatches, and Pigmy nuthatches forage in uppermost portions of a tree. Similar foraging
patterns were observed by Norris (1958), Stallcup (1968), and
Austin and Smith (1972). Comparisons of the vertical range of
tree utilization between seasons showed that both sexes of White-
breasted nuthatches were more specialized; i.e., used a significantly (Chi-square contingency test; p < 0.01) smaller proportion
of vertical height during the nonreproductive period.   In
contrast, Pigmy nuthatches used a greater proportion of the
vertical height during the nonreproductive period.

    2. Horizontal Dimension. Nuthatch exploitation patterns of
the horizontal dimension within a tree also demonstrated significant (Student's t-test; p < 0.05) separation between the
species. These data are summarized in Table III for both reproductive and nonreproductive periods. Both sexes of White-
breasted nuthatches confined foraging activities to the central
portions of trees and made only limited use of trees in the

TABLE III.   Data Summary on Nuthatch Horizontal Foraging
    Location in Ponderosa Pine Trees for Reproductive (April--
    August) and Nonreproductive (September--March) Seasons. The
    data presented are sample size (N), mean ($\bar{x}$), and standard
    error of the mean (SE).

SEASON/SPECIES	FORAGING DISTANCE LATERAL TO TRUNK (M)		
	N	$\bar{x}$	SE
REPRODUCTIVE			
Sitta carolinensis (WBN)			
male	21	0.25	0.30
female	34	0.75	0.51
Sitta pygmaea (PN)	41	5.78	1.07
NONREPRODUCTIVE			
Sitta carolinensis (WBN)			
male	26	0.12	0.29
female	33	0.95	0.60
Sitta pygmaea (PN)	53	7.27	1.11

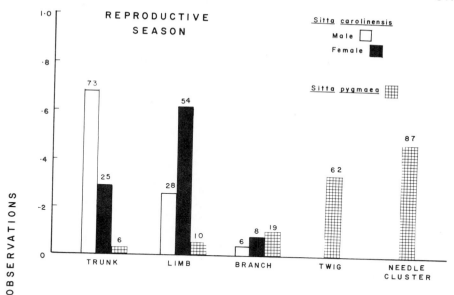

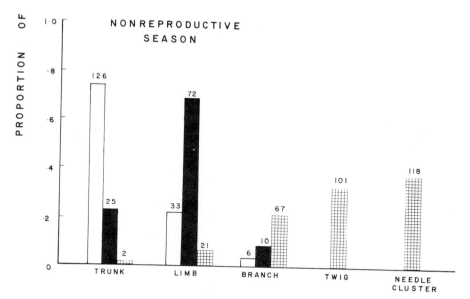

FIGURE 4. The frequency of nuthatch utilization of the five principal foraging substrates (tree trunks, limbs, branches, twigs, and needle clusters) during the reproductive season (April–August) and nonreproductive season (September–March). The number above each bar in the histogram is the total number of observations.

horizontal dimension. No differences (Student's t-test;
p > 0.05) were detected between the sexes of White-breasted nuthatches, although females had a greater mean foraging distance
lateral to the trunk. Pigmy nuthatches made extensive use of the
horizontal dimension having a yearly mean lateral foraging
distance of 6.62 meters from the trunk. Norris (1958) and
Stallcup (1968) reported similar patterns of space utilization in
this dimension.

   3. Tree Structures. Associated with nuthatch use of
vertical and horizontal tree dimensions, nuthatch foraging behavior was analyzed in reference to the tree structures from which
nuthatches were observed to forage and obtain prey items. A
summary of nuthatch utilization of tree structures for both
reproductive and nonreproductive periods is presented in Fig. 4.
Significant differences (Chi-square contingency test; p < 0.001)
were observed between the sexes with male White-breasted nuthatches principally foraging from trunks and females foraging
principally from limbs. Both sexes of White-breasted nuthatches
and Pigmy nuthatches exhibited significant differences (Chi-
square contingency test: p < 0.01) in the frequency of use of
foraging substrates, between seasonal periods. Both sexes of
White-breasted nuthatches increased the frequency of utilization
of their principal foraging substrate; that is, they became more
specialized. In contrast, Pigmy nuthatches exhibited a more
equitable utilization pattern during the nonreproductive period;
Pigmy nuthatches became more generalized during the nonreproductive period. Stallcup (1968) observed similar shifts in tree
structure utilization between seasonal periods for these species.

   4. Prey Acquisition Behaviors. Behaviors associated with
nuthatches obtaining food items were recorded and the results of
these observations are presented in Fig. 5. Significant differences (Chi-square contingency test; p < 0.001) occurred
between the species during both seasonal periods. Pigmy nuthatches obtained food items principally by probing; this technique consisted of a bird perching on a twig or branch and
searching the basal portions of Ponderosa pine needle clusters
(or pine cones). White-breasted nuthatches exhibited a more
generalized pattern of prey acquisition making almost equal use
of scaling, the chipping away of bark with the bill to expose
hidden food items from crevices in the bark by probing into these
areas with the bill.
   White-breasted nuthatches exhibited sexual differences in the
frequency of use of these techniques with males using scaling to
a greater frequency than females which made greater use of
peering-poking. However, this intersexual difference was only
significant (Chi-square contingency test; p < 0.01) during the
reproductive period. White-breasted nuthatches demonstrated

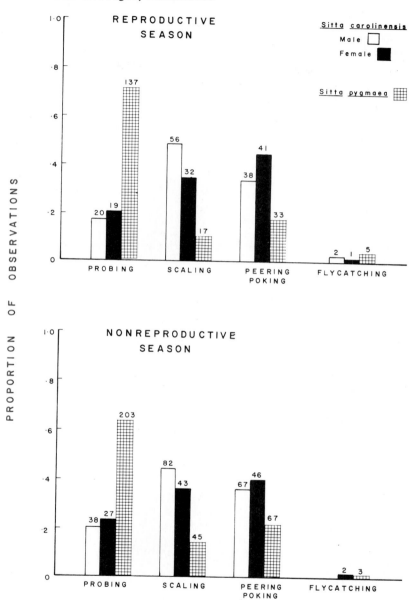

FIGURE 5. The frequency of nuthatch utilization of the four principal prey acquisition techniques (probing, scaling, peering-poking, and flycatching) during the reproductive season (April–August) and nonreproductive season (September–March). The number above each bar in the histogram is the total number of observations.

seasonal differences in the frequency of use of these techniques
which resulted in no differences (Chi-square contingency test;
$p > 0.05$) between the sexes during the nonreproductive period.
This seasonal shift resulted in the sexes making use of prey
acquisition techniques in a similar fashion during the nonreproductive period while exhibiting a more generalized prey regime.

   5.   Prey Items. Anderson (1976) compared dietary preferences
of nuthatches within a Ponderosa pine habitat in Oregon.   Major
food items of White-breasted nuthatches were wood-boring larvae
(Order: Coleoptera; Family: Buprestidae), while Pigmy nuthatches
fed on weevils (Order: Coleoptera; Family: Curciliionidae) and
leaf beetles (Order: Coleoptera; Family: Chrysomelidae).   The
larger-billed White-breasted nuthatch took a greater diversity of
food items and sizes than the smaller-billed Pigmy nuthatch.
Both species demonstrated seasonal changes in the diversity of
food items obtained.   The pattern of these seasonal changes
resulted in a greater diversity of food items incorporated in the
nonreproductive seasons diet.

   6.   Discussion. Analysis of resource utilization patterns of
White-breasted and Pigmy nuthatches within Ponderosa pine habitat
demonstrated ecological isolation between the species.   Ecological segregation was brought about through differential use of
vertical and horizontal tree dimensions, tree structures, prey
acquisition techniques and food items.   The utilization patterns
exhibited by the species demonstrated that the species exploited
different aspects of the habitat in different manners.   White-
breasted nuthatches were more specialized in vertical, horizontal
and tree structure dimensions than Pigmy nuthatches.   In comparison, Pigmy nuthatches exhibited greater specialization in food
items and prey acquisition techniques.   Both species demonstrated
complementarity between habitat-type and food-type dimensions;
specialization along one dimension was associated with generalization along another (Schoener, 1974).   Relationships between
both prey acquisition techniques and food items with bill lengths
for these species supported findings of Pulliam and Enders (1971)
that the larger-billed species (White-breasted nuthatch) obtained
a larger range of food items and employed a more diverse prey
acquisition repertoire than the smaller-billed species (Pigmy
nuthatch).
   White-breasted nuthatches exploited coarse-featured tree
structures, trunks and limbs.   In foraging from these structures,
White-breasted nuthatches displayed sexual differences in
habitat-type dimensions principally through differential use of
vertical tree heights and tree structures, while using similar
prey acquisition techniques.   The sexually dimorphic resource
exploitation pattern was maintained during both reproductive and
nonreproductive seasonal periods.   However, seasonal shifts in

resource exploitation occurred producing a reduction in overall foraging habitat space utilization (greater habitat specialization) during the nonreproductive period. This seasonal change resulted in both sexes foraging lower in trees during the non-reproductive period. Since both sexes foraged lower, they maintained significant intersexual foraging differences while making use of less overall foraging space during the nonreproductive season. In association with the downward shift in resource utilization, both sexes made more equitable use of prey acquisition techniques. The increased equitability in prey acquisition techniques corresponded to increased food item diversity observed by Anderson (1976) during the nonreproductive period. These results supported optimal diet theory of Emlen (1966, 1968) in that both food item diversity and diversity of prey acquisition techniques increased while habitat usage decreased during the nonreproductive period. The complementary pattern of decreased habitat usage and more diversified food-feeding technique usage indicated that this species foraged on evenly distributed food items.

The sexually dimorphic foraging behavior of White-breasted nuthatches showed close affiliations with territoriality and the coarse-featured aspects of the habitat used in foraging. Ligon (1968) observed similar associations in woodpeckers and suggested a causal relationship among these factors. Exclusion of conspecifics allowed the pair to forage without intraspecific interference, while trunk-limb foraging structure presented a foraging space with evenly-distributed food items that could be spatially subdivided. These data supported this conclusion in that the sexes exhibited vertical and tree-structure segregation in association with similarity in prey acquisition techniques. The sexes of White-breasted nuthatches maintained close spatial contact while foraging and were often observed foraging in the same tree but at different strata. This pattern allowed the species to maintain contact without interference in foraging location which, as Kilham (1965) suggested, may be important in pair-bond maintenance. Another advantage of this pattern is that it provides a format by which the pair could systematically exploit the geographical foraging space.

Pigmy nuthatches made use of fine-featured aspects of the Ponderosa pine habitat. The principal foraging substrates of this species were needle clusters and twigs, while vertical and horizontal foraging location measurements also showed that this species made use of peripheral portions of trees. For these parameters, this species could be viewed as a generalist; however, in both prey acquisition techniques and food items, Pigmy nuthatches exhibited greater specialization than White-breasted nuthatches.

Pigmy nuthatches exhibited seasonal differences in foraging behavior which resulted in a greater range of utilization of

habitat features and a wider range of utilization along food-
feeding dimensions during the nonreproductive period. This
pattern contrasted with the pattern exhibited by White-breasted
nuthatches and indicated that this species foraged on patchy
food items (Emlen, 1966, 1968; MacArthur and Pianka, 1966).

The increased range of habitat utilization and increased
range of food item-prey acquisition utilization occurred during
the temporal period when Pigmy nuthatches exhibited the flock-
foraging strategy. Pigmy nuthatches formed monospecific flocks
immediately after breeding; these flocks were primarily composed
of family units that remained together until the spring dispersal
period. Individuals within these flocks were highly gregarious
and socially tolerant of conspecifics. Flock integrity appeared
to be actively maintained by continued vocalizations. The
observed foraging patterns of individual Pigmy nuthatches within
the flocking context suggest that food items were unevenly distributed (increased habitat utilization) but locally abundant
(increased prey acquisition technique equitability and food item
diversity). Krebs *et al.* (1972) demonstrated that behavioral
mechanism can operate to facilitate individual flock members
finding food sources. These results are indirectly supported by
the flock-foraging behavior of Pigmy nuthatches. In addition,
cooperative breeding by nest-helpers (Norris, 1958; Skutch,
1961), breeding dispersion, and foraging dispersion indicated
that tolerance and cooperation are important behavioral characters of this species. The kinship relations of flock members
suggested that cooperation in finding food sources may be an
important aspect of the foraging strategy of Pigmy nuthatch
flocks. More detailed analysis of this possible feature of Pigmy
nuthatch (and other) flocks are needed.

The increased habitat exploitation pattern of monospecific
Pigmy nuthatch flocks contrasted with observations of Morse
(1970) and Austin and Smith (1972) on mixed-species flocks.
These differences (monospecific flock--greater species habitat
exploitation pattern; polyspecific flock--narrower species
habitat exploitation pattern) indicated separate, differing
strategies between these flock types. The narrower range of
habitat exploitation of species within a mixed-species flock
format produced reduced spatial overlap through specialization.
Since mixed-species flocks are typically composed of ecologically
similar species, they may offer a social format producing ecological isolation through aggressive interactions. The specialized
foraging habitat exploitation patterns exhibited by the component
species of the flocks indicated that cooperation between species
in finding food sources was not a strategy between species in a
mixed-species flock; rather, mixed-species flocks can be considered *socially-structured guilds* that function to maintain ecological habitat segregation. A possible advantage associated
with individual participation in a mixed-species flock could be

gained through reduced energy expended in maintaining spatial isolation. By individuals of different, but ecologically similar, species exploiting the geographically same habitat space at the same time, each could defend its own domain and reduce the probability of another individual having recently exploited the area. The strategic similarities existing between the spatial isolation of sexes through sexual foraging dimorphism and spatial isolation of species through aggression within mixed-species foraging flocks suggested that aggression may have played an important role in the evolution of ecological differences in foraging between sexes.

The habitat exploitation patterns demonstrated by White-breasted and Pigmy nuthatches can be viewed in reference to the nature and extent of intraspecific associations. White-breasted nuthatches maintained an *Exclusivity Strategy* throughout the entire year by territorial defense. Within the territory nut-hatches nested and obtained requisite resources without conspe-cific interference. The driving force of the strategy was iso-lation which was accentuated within the mated-pair's territory by sexually dimorphic foraging behavior. The nature of the princi-pal foraging substrates and evidence from dietary analysis indi-cated that the species exploited an evenly distributed food supply. Spatial isolation under these circumstances would function to sequester habitat space and resources for the indi-viduals.

The habitat exploitation pattern within monospecific Pigmy nuthatch flocks exhibited an *Inclusivity Strategy* in which tolerance and, possibly, cooperation resulted in group utiliza-tion of habitat space. The combination of increased habitat exploitation and increased diversity of feeding behavior and prey items suggested that monospecific flocking was a strategy to increase habitat coverage in species exploiting finely-structured habitat space in which food items were patchy in distribution. By exploiting food-finding abilities of others the habitat could be more thoroughly covered and exploited.

Both exclusivity and inclusivity strategies may exist within a mixed-species foraging flock. The flocking association presented individuals with a format which permitted interspecific segregation of optimal foraging space. Interspecific isolation occurred within a temporal-geographical setting which resulted in a synchronized floating territory. Since many mixed-species flocks have a numerically dominant species (Moynihan, 1962; Morse, 1970), this segment may demonstrate the inclusivity strat-egy through intraspecific tolerance. Bock (1969) observed higher levels of interspecific aggression than intraspecific aggression in mixed-species flocks containing Pigmy Nuthatches.

Both foraging strategies exhibited by nuthatches, exclusive and inclusive, offered the potential for a logistic approach to habitat exploitation on temporal and geographical bases. The

TABLE IV. Comparison of Life-History Features of Sympatric
White-breasted and Pigmy Nuthatches in Ponderosa Pine Habitat
Elucidating Interspecific Differences and Intraspecific
Similarities among Varied Life-History Phenomena.

	Species	
	Sitta carolinensis (White-breasted Nuthatch)	*Sitta pygmaea* (Pygmy Nuthatch)
1. Population Density		
a. Breeding	Low	High
b. Nonbreeding	Low	Higher
2. Population Spacing		
a. Breeding	Even	Less even
b. Nonbreeding	Even	Clumped
3. Nest Sites		
a. Location	Natural cavities Live trees	Natural cavities Dead trees
b. Availability	Few	Numerous
c. Nesting conditions	More specialized	Less specialized
4. Territorial Behavior		
a. Type	Type A	Type C
b. Duration	Permanent	Breeding only
5. Population Dispersal Period	After breeding/Late summer-early fall	Before breeding Spring
6. Factors Regulating Population Density		
a. Primary	Territoriality	Nest site availability
b. Secondary	Nest site availability (?)	Territoriality (?)
7. Foraging Habitat Utilization		
a. Major foraging substrate	Coarse-featured Trunks; limbs	Fine-featured Needle clusters; twigs
b. Tree dimensions		
i. Horizontal	Center	Peripheral
ii. Vertical	Low-medium	High
c. Seasonal pattern	Nonreproductive-increased specialization	Nonreproductive-increased generalization

TABLE IV   (continued)

	Species	
	Sitta carolinensis (White-breasted Nuthatch)	*Sitta pygmaea* (Pygmy Nuthatch)
d. Sexual difference	Yes/Height; tree structure	No
8. Food Resource Utilization		
a. Prey acquisition techniques	Diverse Generalized	Less diverse Specialized
b. Food items		
i. Insect group	Wood-boring larva	Weevils
ii. Diet	Diverse	Less diverse
iii. Location	Hidden	Exposed
c. Seasonal trends	Nonreproductive-increased generalization	Nonreproductive-increased generalization
9. Foraging Social Unit		
a. Reproductive	Individual	Individual
b. Nonreproductive	Mated-pair	Monospecific flock
10. Nature of Intraspecific Competition	Interference	Exploitation
11. Intraspecific Relations	Exclusion Isolation	Tolerance Cooperation
12. Overall Trends	Hatibat specialization Food-feeding generalization Intraspecific isolation Sexual differences (intraspecific isolation) Exclusivity	Habitat generalization Food-feeding specialization Intraspecific tolerance Similarity (food-feeding generalization) Inclusivity

basic foraging formats afforded the opportunity for definitive
temporal and geographical schedules of resource utilization.
Whether individuals (or groups) use logistic approaches in foraging and, if so, the nature of these approaches are important
ecological questions that remain to be answered.

A comparative summary of various life-history features of
White-breasted and Pigmy nuthatches describing numerous ecological differences between the species is presented in Table IV.
In contrast to interspecific differences, many intraspecific
similarities existed for a number of varied life-history features
revealing interrelationships (similar strategies) among diverse
ecological characters. These similarities suggested that the
ecological strategy of a species represents a best fit (on a
compromise of adaptations) to meet an array of environmental
variables. Consequently, interrelationships can be observed in
population density, dispersal period, territorial behavior,
availability of nest sites, flocking behavior, foraging behavior,
and prey item distribution. The behavioral stereotypy (Klopfer,
1967) exhibited within a species for these diverse ecological
traits strongly suggested a holistic ecological strategy.

Cautious analyses of ecological interrelationships and extent
of behavioral stereotypy are necessary requirements within any
forest management program for avian species. Artificial production of favorable conditions for one life-history aspect, such as
breeding, might result in unfavorable conditions for another; or
production of favorable conditions for one species may result in
unfavorable conditions for another species. Detailed analyses of
space utilization patterns within and between species under different forest management practices, in association with analyses
of the influence of management practices on species with varying
resource utilization strategies can provide important ecological
information applicable to forest management programs for non-game
species. In addition, analyses of species' life-history patterns
under differing environmental conditions could provide needed
information on whether behavioral characteristics can be adjusted
to accommodate different environmental conditions.

ACKNOWLEDGMENTS

This research was part of a dissertation submitted to the
Department of Zoology and Entomology of Colorado State University
in partial fulfillment of the requirements for the Ph.D. degree.
Sincere gratitude is expressed to Dr. Paul H. Baldwin, chairman
of my graduate committee, for his invaluable aid, direction, and
encouragement throughout this study.
I thank Drs. William Gern, Ronald Ryder, and Terrol F.
Winsor for critical advice on aspects of this work. I also want

to thank J. A. R. Bly, M. K. E. Kilkelly, and L. Werner for
assistance in field work.

REFERENCES

Anderson, S. (1976). *Northwest Science* *50*, 213.
Austin, G. (1976). *Condor* *78,* 317.
Austin, G., and Smith, E. (1972). *Condor* *74*, 17.
Bailey, A., and Niedrach, R. (1965). "The birds of Colorado,"
    Vol. 2. Denver Mus. Nat. Hist.
Bent, A. (1948). "Life histories of North American nuthatches,
    wrens, thrashers, and their allies." U. S. Natl. Mus.,
    Bull. 195.
Betts, M. (1955). *J. Anim. Ecol. 24*, 282.
Bock, C. (1969). *Ecology* *50*, 903.
Brown, J. (1963). *Condor* *65*, 460.
Brown, J. (1964). *Wilson Bull. 76,* 160.
Brown, J. (1969). *Wilson Bull. 81,* 293.
Brown, J. (1972).
Brown, J. (1974). *Amer. Zool. 14*, 61.
Brown, J., and Orians, G. (1970). *Annu. Rev. Ecol. Syst. 1*, 239.
Carrick, R. (1963). *Proc. 13th Inter. Ornithol. Congr.* 740.
Cody, M. (1968). *Amer. Natur. 102*, 107.
Cody, M. (1971). *Theoret. Popul. Biol. 2*, 142.
Cody, M. (1974). "Competition and the structure of bird commu-
    nities." Princeton Univ. Press, Princeton, N. J.
Crook, J. (1965). *Symp. Zool. Soc. Lond. 14*, 181.
Dixon, K. (1963). *Proc. 13th Intern. Ornithol. Congr.* 240.
Dixon, K. (1965). *Condor* *67*, 291.
Edington, J., and Edington, M. (1972). *J. Anim. Ecol. 41*, 331.
Emlen, J. M. (1966). *Amer. Natur. 100*, 611.
Emlen, J. M. (1968). *Amer. Natur. 102*, 385.
Emlen, J. T. (1971). *Auk* *88*, 323.
Glase, J. (1973). *Living Bird* *12*, 235.
Hartzler, J. (1970). *Wilson Bull. 82*, 427.
Hespenheide, H. (1971). *Ibis* *113*, 59.
Hinde, R. (1956). *Ibis* *98*, 340.
Holmes, R. (1970). *Symp. Brit. Ecol. Soc. 10*, 303.
Horn, H. (1968). *Ecology* *49*, 682.
Jackson, J. (1970). *Ecology* *51*, 318.
Kilham, L. (1965). *Wilson Bull. 77*, 134.
Kilham, L. (1970). *Auk* *87*, 544.
Kisiel, D. (1972). *Condor* *74*, 393.
Klopfer, P. (1967). *Wilson Bull. 79*, 290.
Koplin, J. (1967). "Predatory and energetic relations of wood-
    peckers to the Englemann spruce beetle." Colorado State Univ.
Krebs, J. (1970). *J. Zool.*, London *162*, 317.

Krebs, J. (1971). *Ecology* *52*, 2.
Krebs, J., MacRoberts, M., and Cullen, J. (1972). *Ibis* *114*, 507.
Krebs, J., Ryan, J., and Charnov, E. (1974). *Anim. Behav.* *22*, 953.
Lack, D. (1966). "Population studies of birds." Clarendon Press, Oxford.
Lawrence, L. (1967). "A comparative life-history study of four species of woodpeckers." Ornithol. Monogr. No. 5, American Ornithologists' Union.
Ligon, J. (1968). *Auk* *85*, 203.
MacArthur, R. (1958). *Ecology* *39*, 599.
MacArthur, R. (1970). *Theoret. Popul. Biol.* *1*, 1.
MacArthur, R., and Pianka, E. (1966). *Amer. Natur.* *100*, 603.
MacArthur, R., and Wilson, E. (1967). "The theory of island biogeography." Princeton Univ. Press, Princeton, N. J.
Manuwal, D. (1974). *Ecology* *55*, 1399.
Morse, D. (1970). *Ecol. Monogr.* *40*, 119.
Morse, D. (1971a). *Annu. Rev. Ecol. Syst.* *2*, 177.
Morse, D. (1971b). *Ecology* *52*, 216.
Morse, D. (1977). *Condor* *79*, 399.
Moynihan, M. (1962). "The organization and probable evolution of some mixed species flocks of neotropical birds." Smithson. Misc. Coll. 143.
Neff, J., and Meanley, B. (1957). "Blackbirds and the Arkansas rice crop." Ark. Agri. Expt. Sta. Bull. 584.
Nice, M. (1941). *Amer. Midl. Natur.* *26*, 441.
Norris, R. (1958). *Univ. Calif. Publ. Zool.* *56*, 119.
Pulliam, H., and Enders, F. (1971). *Ecology* *52*, 557.
Rand, A. (1954). *Fieldiana: Zool.* *36*, 1.
Root, R. (1967). *Ecol. Monogr.* *37*, 327.
Schoener, T. (1965). *Evolution* *19*, 189.
Schoener, T. (1971). *Annu. Rev. Ecol. Syst.* *2*, 369.
Schoener, T. (1974). *Science* *185*, 27.
Selander, R. (1965). *Wilson Bull.* *77*, 416.
Selander, R. (1966). *Condor* *68*, 113.
Skutch, A. (1961). *Condor* *63*, 198.
Smith, S. (1976). *Auk* *93*, 95.
Snow, B., and Snow, D. (1972). *J. Anim. Ecol.* *41*, 471.
Stallcup, P. (1968). *Ecology* *49*, 831.
Wallace, R. (1974). *Condor* *76*, 238.
Ward, P. (1965). *Ibis* *107*, 173.
Ward, P., and Zahavi, A. (1973). *Ibis* *115*, 517.
Williamson, P. (1971). *Ecol. Monogr.* *41*, 129.
Willson, M. (1969). *Amer. Natur.* *108*, 531.
Willson, M. (1970). *Condor* *72*, 169.
Wilson, E. (1975). "Sociobiology, the new systhesis." Harvard Univ. Press, Cambridge, Massachusetts.

Verner, J. (1975). *In* Proceedings Symposium Management of Forest
    and Range Habitats for Nongame Birds. U. S. For. Ser.,
    Tucson, Arizona.
Yeaton, R., and Cody, M. (1974). *Theoret. Popul. Biol.* 5, 42.

WINTER FEEDING NICHE PARTITIONMENT BY CAROLINA
CHICKADEES AND TUFTED TITMICE IN EAST TEXAS

R. Montague Whiting, Jr.

School of Forestry
Stephen F. Austin State University
Nacogdoches, Texas

Evidence presented indicate that Carolina Chickadees and
Tufted Titmice partitioned the winter feeding niche in mixed
hardwood-pine forests in East Texas. Chickadees chose taller
trees and spent longer foraging periods in a given crown area
than did titmice. Likewise, chickadees fed in the upper and
outer areas of the trees whereas titmice utilized lower and
interior areas. Each species seemed to prefer to feed in
different species of trees. These differences in feeding areas
indicate that chickadees were probably feeding on smaller insects
than were titmice.

I. INTRODUCTION

Carolina Chickadees (*Parus carolinensis*) and Tufted Titmice
(*Parus bicolor*) are two closely related species sympatric over
90% of their range (Peterson, 1947, Robbins *et al.*, 1966).
Although plumage color is different and titmice are slightly
larger and have longer (11.9 vs. 8.8 mm, Schoener, 1965) and
more wedge-shaped bills than chickadees, the two species live in
spatial and temporal coexistence. Primary foods for both species
are small insects and insect eggs hidden in the bark and among
the leaves of trees. Both species acrobatically search
throughout trees of various species for such food. Thus it
appears that Carolina Chickadees and Tufted Titmice occupy
similar feeding niches.

Gause (1934) hypothesized that "... as a result of
competition two similar species scarcely ever occupy similar
niches..." More recently, effects of prey availability on
competition have been argued (Weatherley, 1963; Weins, 1976).
However, it seems unlikely that the abundance of prey necessary
to restrict chickadee-titmouse competition exists in East Texas
during winter. If Gause's hypothesis is true, then there must be
some differences in the feeding niche occupied by the two species.
Herein I examine observations of winter feeding behavior of
both species to determine if chickadees and titmice partition
that niche, and if so, how.

Considerable work has been done to determine niche
differences in similar species of birds. Lack (1944) examined 38
closely related pairs of passerine species in England and was
able to determine differences between 32 of the pairs. He
thought that there were differences between the other six pairs,
but that he was unable to define them. Brewer (1963) researched
Black-capped Chickadees (*P. atricapillus*) and Carolina Chickadees
in the Midwest and found that the two species utilized available
tree species at different heights. Sturman (1968), working with
Black-capped Chickadees and Chestnut-backed Chickadees (*P.
rufescens*) during the breeding season in western Washington
state, found that they were exploiting different sources of food
in different tree species. In conifers, Chestnut-backed
Chickadees were feeding in the upper crowns whereas Black-capped
Chickadees foraged in lower parts of the crown. In hardwoods,
Black-capped Chickadees fed in the top of the canopy; Chestnut-
backed Chickadees seldom utilized hardwoods, but when they did,
they utilized the lower canopy parts. In Tennessee, Shugart
et al. (1975) used discriminant function analysis to show that
Carolina Chickadees preferred larger trees and increased
evergreenness than did Tufted Titmice during fall and winter.

Dixon (1954) found no difference in feeding heights of
Chestnut-backed Chickadees and Plain Titmice (*P. inornatus*) in
central California, rather feeding height reflected height of
the tree. Titmice utilized more tree species and fed on larger
arthropods than did chickadees; likewise they fed on the ground
whereas chickadees did not. Betts (1955) showed that there was
little similarity in diets of four species of tits in England.
She concluded that beaks of all four were adapted to size and
type of food taken, and thus all four species were able to feed
sympatrically with minimal competition.

There are at least ten different ways that two species of
birds can utilize the same food source within the same habitat
(Sturman, 1968). In this study, I attempted to evaluate four of
these ways; congeneric species foraging: 1) at different rates;
2) at different heights; 3) in different parts of the tree
crown; and, 4) in different species of trees.

II. METHODS

To quantify time spent by birds foraging and feeding in various parts of trees, each tree was divided into seven regions. Each region was considered as a proportion of the total tree volume. As shown in Fig. 1, regions one and two were the top one-third of the crown, regions three and four the middle one-third, regions five and six the lower one-third and region seven the trunk. Regions two, four and six represented the outer half of the canopy; one, three and five the inner half.

Field observation sheets contained such a sketch; time in seconds that a bird spent in each region was recorded. If the bird moved to another region within the tree, or another tree, the first timed observation ended and another started. Also recorded were bird species, tree species, tree height', number of individual birds, date, time, weather, place and forest type.

Data were collected during the winter of 1973 in a mature hardwood-pine forest in the Stephen F. Austin Experimental Forest located in East Texas. Generally birds were located by sound, then tracked using binoculars. Observations were made during both morning and afternoon periods as individual species were encountered.

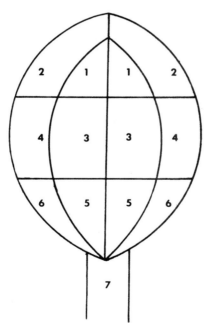

FIGURE 1. A schematic tree divided into foraging regions.

A vertical zonal index, as developed by Colquhoun and Morley (1943) was calculated for each species. To compute this index, the schematic tree was divided into vertical zones, then proportion of time a species spent in a zone was multiplied by a constant for that zone. Values for the four zones were summed by species and divided by 100, this value was the zonal index for that species. For this study, vertical zone one was tree regions one and two, zone two was regions three and four, zone three was regions five and six and zone four was region seven. Constants used were 87.5 for zone one, 62.5 for two, 37.5 for three and 12.5 for four.

Niche breadths of chickadees and titmice were evaluated using the Shannon equation (Shannon, 1948; Levins, 1968). Statistical analyses used as indicated, were t-tests, chi-square tests and tests of proportionality (Natrella, 1963). A probability of .05 or less was considered as significant throughout the study.

III.   RESULTS

Seventy-nine observations (i.e. number of seconds that one bird spent feeding in one region of one tree) totaling 2505 seconds were made on the two species. Time per observation was compared between species using a t-test. Observations of Carolina Chickadees were significantly longer than those of Tufted Titmice, indicating that titmice foraged in a given location for shorter time periods than chickadees (Table I).

TABLE I.   Summary of Observations by Species.

	Number Individuals	Number Observations	Number Seconds	Ave. Number Sec./Observ.
Carolina Chickadee	10	30	1420	47.3
Tufted Titmouse	25	49	1085	22.1
TOTAL	35	79	2505	

Heights that birds used were examined in two ways: 1) different height trees, and 2) different heights within a tree. Trees utilized by the two species ranged 3 to 30 m in heights. Carolina Chickadees used trees that averaged five m taller (t-test, P < .05) than trees used by Tufted Titmice (Table II). These results are similar to those of Brewer (1963) and Shugart et al. (1975), but contradictory to Dixon's (1954) findings.

Foraging heights within tree crowns were evaluated two ways. First the vertical zonal index for number of seconds and number of observations was computed for each species. Data showed that the chickadees' average feeding time zone was 80, in the top one-fourth of the tree (Table III). Titmice average feeding time zone was 54, in the next lower one-fourth of the tree. Zonal index for the number of observations, regardless of time, was 76 for chickadees and 62 for titmice. For each species feeding time zonal index and observation zonal index were in the same one-quarter of the tree. This indicated that, within their feeding range, Carolina Chickadees spend more time feeding in higher zones than in lower zones. The opposite held true for Tufted Titmice. Time value for chickadees was higher than the observation value; the relationship was reversed for titmice. Although differences between values was not great, this would indicate that chickadees moved upward upon entering a tree, and titmice downward. The higher index value for chickadees is in agreement with the findings of Dickson and Noble (1978) in Louisiana. However, these results are dissimilar to Dixon's (1954) California study on Chestnut-backed Chickadees and Plain Titmice.

TABLE II. Summary of Height of Trees Utilized.

	No. tree species visited	Ave. ht. of trees	No. tree regions vistited
Carolina Chickadee	8	18.7 m	5
Tufted Titmouse	11	13.7 m	7

TABLE III. Zonal Indices for the Two Bird Species.

	Carolina Chickadee	Tufted Titmice
Time	80	54
Observations	76	62

The second quantification of foraging height within tree
crowns was made by examining use patterns of the seven tree
regions by Carolina Chickadees and Tufted Titmice. Figure 2
illustrates the data using a bar graph. Chickadees spent more
time feeding in the upper regions of the trees than did titmice
(Table IV). Time spent by chickadees in each tree region was
compared with time that titmice spent in each tree region. Using
a test of proportionality, significant differences between
species were found for all seven regions. However, when number
of observations by regions were statistically compared,
differences were not significant. This suggests that birds were
moving into the tree regions randomly, but spending significantly
more time feeding in certain desirable regions than in other
less desirable regions.

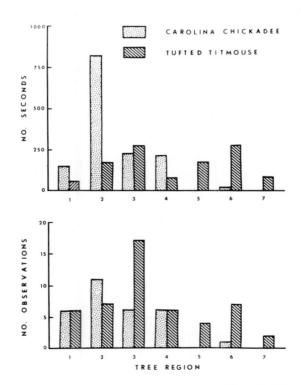

FIGURE 2.   Use of tree regions by Carolina Chickadees and
Tufted Titmice.

Differences in horizontal zonation between the two species were also compared (Table V). A test of proportionality revealed a significant difference between the two species when timed use of exterior and interior parts of the trees were compared. Chickadees used exterior portions of trees significantly more than titmice. Conversely, titmice used interparts of the crown significantly more than chickadees. Comparison of exterior and interior use of the crown by each species showed that there was no significant difference in use of exterior and interior parts of the trees by titmice. However, chickadees used exterior portions of the tree to a significantly greater degree than interior portions.

TABLE IV.   Feeding Activities by Tree Region.

| | Tree Region | | | | | | | Total | Breadth |
	1	2	3	4	5	6	7		
Chickadee									
No. sec.	150	830	225	210	0	5	0	1420	1.14
No. obs.	6	11	6	6	0	1	0	30	1.44
Titmouse									
No. sec.	55	170	270	75	165	270	80	1085	1.79
No. obs.	6	7	17	6	4	7	2	49	1.77
Total									
No. sec.	205	1000	495	285	165	275	80	2505	
No. obs.	12	18	23	12	4	8	2	79	

TABLE V.   Summary of Horizontal Stratification of the Trees by the Two Species of Birds.

| | Carolina Chickadees | | Tufted Titmice | |
	Time	Observations	Time	Observations
Exterior	1045	18	595	22
Interior	375	12	490	27
Breadth	.577	.673	.688	.68

The birds foraged in 12 species of trees that I could
identify, and two I could not (Table VI). Excepting sweetgum and
maple, there was a significant difference in the proportion of
time that each of the two bird species used each tree species.
Yet where both birds used the same species of trees, the number
of observations by tree species was not significantly different.
Again, this suggested that the birds were randomly choosing tree
species, but were remaining longer in preferred trees than others.
Titmice used white oak, hickories and sweetgum 54% of their
feeding time. Chickadees used elm, sweetgum, hickory, and pine
80% of their feeding time.
    Although not statistically tested, it is worth noting that
both birds seemed to prefer sweetgum and hickory. Also, that
there were several observations of chickadees on pines
("increasing evergreenness"), but none of titmice.

                            IV. DISCUSSION

    Carolina Chickadees and Tufted Titmice observed in this study
did not utilize identical winter feeding niches. Significant
differences in foraging rates and tree heights, regions and
species used by the birds occurred. Chickadees averaged longer
periods of time feeding per observation, and fed in taller trees
and higher and more exterior parts of tree crowns than did
titmice. Both species moved into tree species and regions
randomly, then spent significantly more time in certain areas.
This would indicate that each species fed where it had the best
success. Through evolution of bill size and foraging behavior,
this particular aspect of the feeding niche has been partitioned
by the two species.
    This partitionment probably resulted from only one real
division of the feeding niche; chickadees fed on smaller prey
than titmice. Generally size of insects that live in trees
decrease with increased height above the ground (J. E. Coster,
pers. comm.). Likewise smaller insects are usually found on
smaller twigs and branches in exterior crown parts. Logically
it would seem that smaller insects would be more difficult to
locate, thus chickadees would have to spend longer periods of
time searching a given area for that source of food.
    Brewer (1963) wrote that chickadees were not particularly
specialized in their foraging. Results of this study indicate
the opposite. Niche breadth values for Carolina Chickadees
were consistently smaller than those of Tufted Titmice (Table
VII). Reduced niche breadth would be associated with a higher
degree of specialization, indicating that chickadees were more
specialized than titmice. Differences in the species' realized
niches has probably reduced competition (no data to demonstrate
an actual reduction) in East Texas hardwood-pine habitats
during winter.

TABLE VI. Tree Species Used by Carolina Chickadees and Tufted Titmice.

| Tree Species | | Carolina Chickadees | | Tufted Titmice | |
| Common Name | Scientific Name | Total Number | | Total Number | |
		Seconds	Observations	Seconds	Observations
White oak	*Quercus alba* L.	140	4	245	8
Willow oak	*Quercus phellos* L.	0	0	50	4
Water oak	*Quercus nigra* L.	0	0	65	7
Red oak	*Quercus falcata* Michx.	15	2	70	2
Post oak	*Quercus stellata* Wang.	35	2	85	3
Hickory	*Carya* spp.	170	6	200	5
Sweetgum	*Liquidambar styraciflua* L.	170	6	155	10
Elm	*Ulmus* spp.	600	1	40	2
Maple	*Acer* spp.	10	1	10	2
Hackberry	*Celtis occidentalis* L.	100	4	10	2*
Pine	*Pinus* spp.	180	4	0	0
Blackgum	*Nyssa sylvatica* Marsh.	0	0	20	1
Unknown		0	0	135	3
Total		1420	30	1085	49
Breadth		1.728	2.038	2.153	2.279

TABLE VII. Breadth of Foraging Niche of Carolina Chickadees and
            Tufted Titmice.

	Tree Region		Horizontal Crown Parts		Tree Species	
	Sec.	Obs.	Sec.	Obs.	Sec.	Obs.
Carolina Chickadees	1.146	1.447	.577	.673	1.728	2.038
Tufted Titmice	1.797	1.773	.688	.688	2.153	2.279

ACKNOWLEDGMENTS

   Editorial comments and criticisms of Richard Conner, Susan
de Milliano and Robert Fleet are sincerely appreciated.

REFERENCES

Betts, M. (1955). *J. Anim. Ecol. 24*, 282.
Brewer, R. (1973). *Auk 80*, 9.
Colquhoun, M., and Morley, A. (1943). *J. Anim. Ecol. 12*, 75.
Dickson, J., and Noble, R. (1978). *Wilson Bull. 90*, 19.
Dixon, K. (1954). *Condor 56*, 113.
Gause, G. (1934). "The struggle for existence." Hafner and
    Sons, New York, N.Y. (Reprinted 1964).
Lack, D. (1944). *Ibis 86*, 260.
Levins, R. (1968). "Evolution in changing environments."
    Monogr. Pop. Biol. No. 2, Princeton Univ. Press, Princeton.
MacArthur, R., and Pianka, E. (1966). *Amer. Nat. 100*, 603.
Natrella, M. (1963). "Experimental statistics." Nat'l
    Bureau of Standards Handbook 91. U.S. G.P.O., Wash. D.C.
Peterson, R. (1947). "A field guide to the birds." Houghton
    Mifflin Co., Boston.
Robbins, C., Bruun, B., and Zim, H. (1966). "Birds of North
    America." Golden Press. New York.
Schoener, T. (1965). *Evolution 19*, 189.
Shannon, C. (1948). *Bell Sys. Tech. Journ. 27*, 379.
Shugart, H., Anderson, S., and Strand, R. (1975). *In*
    "Proceedings of the symposium on management of forest and
    range habitats for nongame birds." (D. Smith, Tech. Coord.)
    USDA. For. Ser., Gen. Tech. Rep. WO-1, p. 115.
Sturman, W. (1968). *Condor 70*, 309.
Weatherly, A. (1963). *Nature 197*, 14.
Weins, J. (1976). *Auk 93*, 396.

THE EFFECTS OF INSECTIVOROUS BIRD ACTIVITIES IN FOREST ECOSYSTEMS: AN EVALUATION

Imre S. Otvos

Newfoundland Forest Research Centre
Canadian Forestry Service
Department of Fisheries and Environment
St. John's, Newfoundland

Insectivorous birds play an important role in the forest ecosystem. They have a significant influence on the population dynamics of many forest insects. Birds act as direct mortality agents of insect pests, and they can also affect their prey indirectly through influencing insect parasites and predators of the prey, by spreading entomogenous pathogens, or in some cases by altering the microhabitat of the prey. Birds exert the greatest influence on insect populations at endemic levels; they suppress and delay population build-up to epidemic levels, and thus may increase the interval between insect outbreaks. Insectivorous birds may also accelerate the decline of an outbreak. They may also have an effect on the forest ecosystem by feeding on and by dispersing seeds of various forest trees and shrubs. Some birds may be involved in the spreading of wood rotting fungi and thus contribute to the nutrient recycling.

I. INTRODUCTION

From early, and often cursory observations it was generally concluded that certain species of birds consume large numbers of insects; therefore considerable economic importance was assigned to birds in the control of both agricultural and forest insects. The value of birds in controlling forest insects was recognized as early as 1335 when the civic authorities of Zürich, Switzerland issued an order for the protection of birds as enemies of insect pests (Boesenberg, 1970). This presumed economic significance of birds in insect control has been partly perpetrated without sufficient backing of scientific data even in recent times.

341

Insectivorous birds are usually opportunistic feeders and
were used in biological control attempts earlier than invertebrate predators or parasites. The first known successful
importation of a vertebrate predator occurred in 1762 when the
mynah bird (*Acridotheres tristis* L.) was introduced from India
to Mauritius to control the red locust (*Nomadacris septemtasc-
iata* (Serville) (Moutia and Mamet, 1946). The first known tran-
sportation of an insect predator occurred 112 years later, when
Coccinella unidecimpunctata L. was introduced along with other
enemies of aphids to New Zealand from England (Doutt, 1964).
Other early introductions of birds, include the importation of
the English or House Sparrow (*Passer domesticus* L.) and the
European starling (*Sturnus vulgaris* L.) from Europe to the
eastern United States. The sparrow was introduced in 1851 to
control shade tree pests in eastern U.S. cities (Barrows, 1889;
Forbush, 1921); the starling was introduced in 1872 but did not
become established until 1890 (Howard, 1959). Similar introductions of passerine birds also occurred in New Zealand in the
1860's (Wodzicki, 1950).
Opinions about the role played by birds in regulating
insect numbers differ widely. One extreme view is quoted in
Henderson (1927). "If it were not for birds, no human being
could live upon the earth, for the insects upon which birds live
would destroy all vegetation." McAtee (1925), reviewed the
American literature on the role of birds in insect control and
concluded that there were 109 cases of control and 88 cases of
local suppression of insects by birds. However, he acknowledged
that neither of these figures was exact because it was impossible to classify such expressions as "several times birds were
observed to clear up" certain infestations, or "... many fields
were kept clean." He believed that birds were capable of
controlling insects in local outbreaks, but that large outbreaks
were beyond the capacity of birds to control. On the other
hand, Ainslie (1930) felt that the role of birds in insect
control was over-rated and that their importance in reducing
insect outbreaks cannot be supported by facts. Both of these
are sweeping generalizations in my opinion; one point of view
can be just as erroneous as the other. Recent reviews present a
more balanced evaluation of the value of birds in biological
control (Bruns, 1960; Franz, 1961; Buckner, 1966; Murton, 1971;
McFarlane, 1976). This paper discusses the role of birds as
insectivorous predators in the forest ecosystem.

II. THE ROLE OF INSECTIVOROUS BIRDS IN THE FOREST ECOSYSTEM

Response of a predator to changes in prey density was
divided into two major components by Solomon (1949): functional
and numerical response. The former is primarily a behavioral
change by the individual predator reflected in its tendency to
attack more individuals of the prey species as prey density
increases. The latter is the change in the number of predators
as a result of changes in prey density; numerical changes of
bird populations can be caused by breeding and by invasion or
aggregation. These two responses have been further studied and
categorized by Holling (1959a, 1959b) who also demonstrated that
each of these categories can be affected by both predator and
prey density (Holling, 1961). Most avian predators exhibit both
numerical and functional responses to some forest insects, as
will be seen later, and the sum of these responses is the total
number of prey killed by the predators. The combined effects of
these numerical and functional responses to insect prey I define
as the "direct influence of birds on insects".

Birds can also influence insect populations indirectly by
altering the microenvironment of their prey thus making them
more susceptible to weather (Otvos, 1969; Moore, 1972), parasitism (Otvos, 1970) and possibly diseases, and by also spreading insect viruses (Franz et al., 1955; Entwistle et al., 1977a,
1977b). Birds influence their prey population directly or
indirectly and they also affect the forest ecosystem in other
ways such as spreading plant viruses, reducing number of tree
seeds as well as spreading them (Turček, 1954a, 1961), providing
sources of entry for wood decaying organisms which may reduce
the value of salvaged wood or help to return nutrients to the
soil by hastening decay.

Most insectivorous birds are facultative feeders utilizing
a multitude of available prey species opportunistically. Some
authors have examined the economic value of a single bird
species on a single prey while others have chosen to examine the
value of all or the most important predators on a single prey
species. It is probably easier to evaluate the role of a single
predator species in the control of its prey; however, forest
entomologists are more interested in the control value of all
predators on the pest species. It is a very difficult task to
obtain reliable and concrete data on percent prey reduction in
either case. Although estimates are subject to reservations,
they do give a valuable idea of the importance of bird predation
on insects. With these reservations in mind, a selected number
of forest pests and their population reduction by avian predators will be discussed. Defoliating insects will be treated

first, followed by a discussion of a few bark beetles and other
wood-boring insects. Most of the discussion is based on available published material together with some personal data from
unpublished manuscripts. Although it was not feasible to
complete a world-wide literature search, an attempt has been
made to present a cross-section of the different roles of avian
predators in the control of forest insects.

A. Direct Effect of Birds as Predators
of Forest Defoliators

   1. Spruce Budworm. *Choristoneura fumiferana* (Clemens) is
the most important defoliator of the boreal spruce-fir forest in
North America and has been extensively studied. Tothill (1923)
working in a declining budworm outbreak in New Brunswick estimated that birds consumed about 13% of the larvae. Kendeigh
(1947) showed both functional and numerical avian responses
during an outbreak of the spruce budworm in Ontario and estimated that only 4% of the epidemic population (which he estimated
at 376,000 individuals per acre or about 929 000 per hectare)
were consumed by birds. George and Mitchell (1948) calculated
that the degree of control by birds ranged from 3.5% to 7% in
budworm infestations of 1 235 000 to 2 471 000 per hectare.
Mitchell (1952) using stomach analysis determined that spruce
budworm made up about 40% of the diet of birds at outbreak
levels. A similar, four-year study in Newfoundland, revealed
that spruce budworm contributed between 7% and 46% of the diet
of the birds (Otvos, unpubl. manuscript). Dowden *et al.* (1953)
found that budworm populations between full-grown larval and
late pupal stages were reduced only 11% and 29% in areas where
bird populations were reduced by shooting but by 48% and 71% in
the control areas where birds were not molested.
   Morris *et al.* (1958) considered the effect of birds on
budworm populations during outbreak years negligible, despite
that several bird species increased numerically: "The budworm
increased 8000-fold, while the most responsive predator the bay-
breasted warbler increased only 12-fold."
   Birds exhibit both functional and numerical responses during
budworm outbreak. Several ground, tree-trunk and seed feeders
have been found to alter their feeding behavior and such species
as Ovenbirds (*Seiurus aurocapillus* (L.)), Juncos (*Junco* sp.),
woodpeckers (*Picidae*) and Crossbills (*Loxia* sp.) have begun to
take budworm larvae and pupae during the outbreaks (Zach and
Falls, 1975; Morris *et al.*, 1958; Mitchell, 1952; Kendeigh,
1947, Otvos, unpubl. manuscript). Several parulid warblers have
exhibited the greatest numerical response during budworm outbreaks; Bay-breasted Warbler (*Dendroica castanea* (Wilson))
population increased twelve-fold and the Blackburnian Warbler
(*Dendroica fusca* (Müller)) nine-fold - although some species
declined (Kendeigh, 1947; Hensley and Cape, 1951; Morris *et*

al., 1958; MacArthur, 1958). Some warblers are known to increase their clutch size during budworm outbreaks (MacArthur, 1958). Sanders (1970) determined bird densities during endemic budworm population levels in the same area where Kendeigh (1947) did his work during a budworm outbreak. The difference between the two census densities are almost entirely due to absence in the later census of warblers which are known to respond numerically to increasing budworm numbers.

Blais and Parks (1964) credit the control of a declining population of the spruce budworm to a large flock of Evening Grosbeaks (*Hesperiphona vespertina* (Cooper)) which invaded the remaining pockets of the infestation after it was treated with DDT. The Evening Grosbeak is known to have extended its summer breeding range and increase in numbers in outbreak areas (Shaub, 1956). This species has also extended its breeding range into Newfoundland recently.[1] However, whether this represents a response to outbreaks of another forest defoliator (Otvos *et al.*, 1971) or the spruce budworm, or both is difficult to judge at this time.

According to Royama (pers. comm.) the population increases and range expansions reported for some species during budworm outbreaks are probably coincidental. He believes that unless birds can be dependent on the budworm for food during the entire length of their breeding season they cannot respond to budworm abundance. This, in my opinion may only apply to some of the "local" species but not to invading ones.

The general concensus is that the role of birds in the control of a large scale outbreak of the spruce budworm is negligible. However, most agree that birds are important at endemic population levels, at the beginning or the declining phase of an outbreak, which may be the most important control periods.

2. Jack Pine Budworm. Mattson *et al.* (1968) in their study of vertebrate predation on the jack pine budworm (*Choristoneura pinus* Freeman) found that when larvae reached the 4th and 5th instars, many species of birds altered their feeding behavior to take advantage of this abundant food source. Predation continued from these late instars throughout the pupal and adult stage. During the second year of observation a great influx of Blackbirds (*Turdus merula* L.) and other species occurred at the smaller (240 hectare with a low budworm density of 24 700/hectare) of the two infestations studied and Mattson *et al.* (1968) estimated that predation on late instar larvae and pupae increased to 60%–65% in the second year from 40%–45% in the first

---

[1]Dr. L. Tuck, Memorial University of Newfoundland, St. John's, formerly with the Canadian Wildlife Service, Fisheries and Environment Canada.

year. This influx of non-resident birds and the subsequent
increase in prey mortality did not occur in the larger infestation (2 430 hectare with an outbreak population density of
247 100/hectare). From this the authors concluded that bird
predation has a greater potential value in small, restricted or
localized forests even when budworm population is moderately
high because of the influx of the non-resident birds.

     3. Other Defoliators. Buckner and Turnock (1965) demon-
strated that both numerical and functional responses also occur
among birds preying on the larch sawfly (*Pristiphora erichsonii*
(Hartig)). From their studies they concluded that birds are
significant factors in the population regulation of this sawfly
at low densities, and perhaps up to moderate infestation levels,
and there was preference by birds for adult sawflies. The birds
preferred sawfly adults perhaps due to the defensive or pro-
tective reactions exhibited by some species of tenthredinid
larvae against birds and parasites (Prop, 1960).
     Preying on the adult stage before egg laying is likely to
have a more pronounced effect on subsequent generations than
mortality factors at earlier life stages.
     Betts (1955) found various species of titmice (*Parus
coeruleus* L., *P. ater* L. and *P. major* L.) predation on winter
moth (*Operophtera brumata* (L.)) larvae exceptionally low, 0.3%-
2.6%. However, predation on emerging adults was about 10% and
stomach analysis showed that these were predominantly females,
thus about 20% of the females may have been consumed.
     Keve and Reichart (1960) reported from Hungary that two
sparrows, *Passer domesticus* L. and *P. montanus* L. might exert an
important role in the control of *Hyphantria cunea* (Drury) by
killing up to 98% of the adults. In Japan, Tree Sparrows
(*P. montanus*) and Grey Starling (*Sturnus cineraceus* L.) were
estimated to kill between 40% and 50% of *H. cunea* adults as
these emerged before dusk. As sexual activity of these moths
take place between dawn and sunrise (Hasegawa and Ito, 1967),
the mortality caused by birds is especially important in the
population dynamics of this insect.
     Some birds exhibit a preferential feeding on one of the
sexes of the adult insect and this may be potentially important
in the control of the insect. According to Readshaw (1965)
various species of birds in Australia exhibited a distinct
functional response to changes in the density of a stick insect,
Didymuria violescens (Leach) and had "a preference for the fat
juicy adult females."
     Predation by birds can be especially important when it takes
place over extended periods, such as winter, and when birds tend
to aggregate and move in flocks.

Sloan and Coppel (1968) and Coppel and Sloan (1971) found that 23.5% of the winter mortality of the larch casebearer (*Coleophora laricella* (Hübner.)) can be attributed to resident birds and an additional, though undetermined, mortality is caused in the spring. Preliminary studies on the introduced pine sawfly (*Diprion similis* (Hartig)) showed that resident overwintering birds, mainly the Black-capped Chickadee (*Parus atricapillus* L.) and the Downy Woodpecker (*Picoides* (= *Dendrocopos*) *pubescens* (L.)) may remove all but 5% of the overwintering sawfly cocoons above the snow line (Coppel and Sloan, 1971).

Holmes and Sturges (1975) reported an increase in the population of foliage gleaning species which was positively correlated with the outbreak phase of the saddled prominent caterpillar (*Heterocampa guttivitta* (Walker)); although lagging 1-2 years behind. A decline in the outbreak was followed by a decline in bird densities.

Gibb (1958, 1960, 1966) working with titmice (*Parus* spp.) and Goldcrests (*Regulus regulus* (L.)) found that tits regularly consumed the eucosmid moth larvae (*Ernarmonia* (= *Laspeyresia*) *conicolana* (Heyl.)). Intensity of predation varied with density of the larvae in the cone. There appeared to be a threshold of prey density (10 larvae per 50 cones)"... above which it became profitable for the birds to feed on them." Although tits may destroy more than 50% of the overwintering full grown larvae, according to Gibb (1960) it is not sufficient to regulate population of *Ernarmonia conicolana*.

Tinbergen (1960) conducted a detailed study on some members of *Paridae* yearly measuring their population, density of their main preys, and the frequency of prey species in the diet of birds rather than the proportion of the population consumed. He found that birds accept a new prey only gradually after its appearance, after birds encounter a prey often enough they develop a "specific search image" for it. Further evidence was presented by Mook *et al.* (1960) on the role of this "search image." Tinbergen (1960) has also shown that birds took fewer insects than expected when prey population was low, predation increased rapidly and exceeded expectation at moderate densities and then declined below expectation at high prey densities. Tinbergen's (1960) theory on "developing searching images" in my opinion is the same as "... a threshold in larval numbers above which it became profitable for the birds to feed on them" (Gibb, 1966, p. 52), or Royama's (1970, p. 657) concept of profitability which he defined "... as the amount of food the predator can collect for a given amount of hunting effort ..." Royama (1970) carries the theory one step further when he says that "... the profitability of a prey species is determined not only by the density but also the size of the prey and the method of hunting of the predator..."

Tinbergen (1949) working in a pine forest in Holland determined the percentages of the different insect species killed by tits (*Parus* spp.) alone. He found that the pine beauty moth (*Panolis flammea* Schiffermüller) was reduced by 24% and 34%, the barred red moth (*Ellopia prosapiaria* L.) by 23%, the sawfly (*Acantholyda nemoralis* Thomson) by 18%, the pine looper (*Bupalus piniarius* L.) by 10% and the fox-colored sawfly (*Diprion sertifer* Geoffroy) by 3%.

Gage *et al.* (1970) working on the blackheaded budworm (*Acleris variana* (Fernald)) in New Brunswick found that birds responded both functionally and numerically to increases in prey density. The authors estimated that birds consumed 3% to 14% of the blackheaded budworm population.

Clark (1964) working on a psyllid (*Cardiaspina albitextura* Taylor) in Australia, found that birds tended to consume an increasing proportion of late instar nymphs and adults (up to about 77%) with an increase in psyllid numbers at the lower end of the population density of the prey. However, birds were unable to prevent the outbreak when environmental factors favored population increase of the psyllid.

Even the "hairy" caterpillars which are generally avoided by birds have not completely escaped avian predation. Witler and Kulman (1972) reviewed the literature and concluded that 60 species of birds have been recorded to feed on tent caterpillars (*Malacosoma* spp.). Most birds tend to feed on small larvae rather than large larvae. Cuckoos appear to be adapted to feed on all sizes of caterpillars; they seem to follow tent caterpillar outbreaks and there is indication that the size of their brood appears to increase when food is abundant (Forbush,1927).

According to Royama (pers. comm.), because of the limited capacity of insectivorous birds for numerical response, the equilibrium densities of prey-bird interaction systems can only be stable "locally", as opposed to the equilibrium densities in host-parasite systems which can be stable over much larger areas. In other words, control by birds is only "local", and hence, the prey population can at times "escape" more easily to an outbreak level. The extent of the size of such local stability by bird predation depends on the life history of both predator(s) and prey concerned. For instance, if insects are exposed to bird predation during a relatively "short" period in which the insects are growing fast, such as with the spruce budworm during the birds' breeding season, the region of local stability, if any, must be very limited. On the other hand, if birds prey on insects that overwinter as mature larvae or pupae and are exposed to bird predation over a prolonged period, such as during winter, the region of stability could be quite extensive. It is, therefore, the latter situation in which bird predation may exert a considerable controlling influence.

There are several examples in the literature of birds "cleaning up" and possibly preventing small insect infestations from expansion. McAtee (1927) reported that 36 species of birds were known to feed on the forest tent caterpillar eggs and larvae and these birds often eliminated populations of this insect in localized areas. Starlings have been reported to terminate a small, about 0.1 ha (1/4 acre), sawfly (*Pristiphora abietina* Christ) infestation in Germany (Bruns, 1960). Lühl and Watzek (1973-74) gives an example of how an invasion of various species of birds eliminated a concurrent infestation of the nun moth (*Lymantria monacha* L.) and the gypsy moth (*Lymantria dispar* L.) covering about 3 ha. Davidson and Gimpel (1975) cite an example where a large flock of overwintering birds, mainly Robins (*Turdus migratorius* L.) settled in 1.6 ha (4 acre) area of *Ilex cornuta* Lindl. nursery, heavily infested with the Indian wax scale (*Ceroplastes ceriferus* Anderson), and cleaned the holly bushes of the wax scale. Mattson *et. al.* (1968) recorded a great influx of birds, primarily Blackbirds, into a 243 ha (600 acre) of jack pine budworm infested stand resulting in a considerable reduction of budworm population.

One of the reasons for this inability to influence (control) insect populations at outbreak densities is obviously related to the vast difference in the reproductive potential of insects and birds — insect populations can multiply several thousand-fold whereas birds can rarely multiply even twelve-fold. Other reasons may be that birds are opportunistic feeders (even under outbreak conditions the diet of a species is not composed entirely by the same insect, e.g. spruce budworm). Although Franz (1961) is of the opinion that the above disadvantages are partly compensated for by the ability of the birds to aggregate into infested areas, and by the fact that birds do specialize to a certain degree on abundant prey. Aggregation of birds at times can be impressive. Tichy and Kudler (1962) reported a 308% increase in bird densities, mainly as a result of invasion and aggregation to an outbreak of *Bupalus piniarius* in Czechoslovakia.

According to some authors densities of many, if not most, birds are limited by availability of food and they tend to support this by the food gathering behavior of birds, the variability of food-gathering anatomy, and territorialism (Lack, 1966). Others cite lack of nesting sites as a limiting factor and support their theory by the fact that bird densities can be increased several-fold by providing nesting boxes or other types of nesting material. Several types of nest and habitat manipulations have been used successfully over the years to protect and encourage birds to increase their density (Haartman, 1951). This was achieved even without any noticeable change in insect populations (i.e. increase of food supply) (Bruns, 1960) and populations of birds did not decline when the insect outbreak was controlled by insecticides (Henze and

Görnandt 1957 in Bruns, 1960).   Habitat manipulation for the
enhancement of bird populations is extensively practiced in
Europe, especially in Germany and Russia. Results of these works
show that a 5 to 20-fold increase in density of hole-nesting
birds can be achieved (Haartman, 1957; Bruns, 1960; Franz, 1961;
Gillmeister, 1963). Nesting boxes even attract woodpeckers
(Semenov, 1956). Pfeifer (1963) has demonstrated that numbers
of both hole-nesting and open-nesting birds can be increased by
providing artificial nest sites even over relatively large
areas. These suggest that, perhaps, the availability of nesting
sites, rather than the amount of food, acts as the ecological
limiting factor determining the number of nesting pairs of
birds. Haartman (1971) tried to reconcile these two opposing
views when he proposed the compensation principle or the principle of interchangeability which says that excellent breeding
opportunities may compensate to some degree for poor food source
or vice versa.

Reviewing three decades of bird protection (colonization by
artificial nesting boxes) in Germany Herberg (1960) concluded
that *Bupalus piniarius* was kept under control in the area with
nest boxes while in the control area (with no bird boxes)
chemical insecticides had to be used.

In Russia experiments were even conducted in the mass
transfer of the migratory Pied Flycatcher (*Ficedula* (= *Muscicapa*)
hypoleuca (Pallas)) at a young age with their parents. The
experiments were considered successful (Polivanov, 1956).

In North America only limited experiments were conducted
with nesting boxes to augment populations of insectivorous birds
(Dahlsten and Herman, 1965; Dahlsten, 1979). Because of the
extensive forest areas in North America, more consideration
needs to be given to the habitat requirements of insectivorous
birds, and manipulation of the habitat, in order to increase
bird density (Hardin and Evans, 1977; Bull, 1978; Edgerton and
Thomas, 1978).

B.   Direct Effect of Birds as Predators of Bark Beetles

Woodpeckers are probably the most studied avian predators
foraging on trunks and limbs of trees. It has been known for
some time that woodpeckers are capable of strong numerical
response, mainly by invasion, to infestations of bark beetles
and wood borers (Beal, 1911; Van Tyne, 1926). Their control
value has been studied to various degrees on several bark
beetles and wood borers. The more important ones will be
discussed by individual prey species.

1. Spruce Beetle. The spruce beetle, earlier known as the Engelmann spruce beetle (*Dendroctonus* (= *engelmanni* = *obesus*) *rufipennis* (Kirby)) and the relationships among its avian predators were extensively studied by Knight (1958), Koplin (1969, 1972), Koplin and Baldwin (1970), Baldwin (1960) and others. In the Rocky Mountain forests the principal predators of this bark beetle are: the Hairy (*Picoides villosus* (L.), Downy (*P. pubescens* (L.)), and the Northern three-toed woodpeckers (*P. tridactylus* (L.)). These predators exhibit both functional and numerical responses, and are known to aggregate during winter at local epidemics of the spruce beetle (Massey and Wygant, 1954; Knight, 1958). Numerical response (aggregation of woodpeckers in high prey density areas) is probably more important, than functional response especially in localized areas.

Koplin (1969) recorded a 50-fold increase in woodpecker density in response to insect prey buildup in fire-killed trees. In epidemic population of the spruce beetle up to 85-fold increase in woodpecker density has been noted over their density in subalpine forest with endemic beetle population (Koplin, 1969; Baldwin, 1960). At endemic spruce beetle population levels (which usually exist in windthrown trees or logging slash), woodpeckers consume between 20% and 29% of the beetles (Koplin and Baldwin, 1970). At moderate to high beetle population levels (which usually develop in standing trees) woodpeckers eat or otherwise destroy from 24% to 98% of the beetle population (McCambridge and Knight, 1972; Knight, 1958; Baldwin, 1960).

McCambridge and Knight (1972) noted that desiccation of the tree host causing larval mortality can be enhanced by feeding activity of woodpeckers.

Koplin (1972) used a deterministic model to predict predatory impact of woodpeckers on endemic, epidemic and "panepidemic" (large epidemic) populations of spruce beetle larvae with the corresponding densities of 2500, 395 400 and 1 581 400 larvae per hectare. The model predicted 20%, 85% and 51% reduction in the endemic, epidemic and panepidemic population levels. These predicted predations compare favorably with the larval mortality estimated from protected bark areas and those exposed to woodpecker predation. The latter gave 19%, 83% and 55% mortality at the three above mentioned prey densities.

Shook and Baldwin (1970) have demonstrated that stand density also influences the extent of predation; in open, semi-open and dense areas spruce beetle population was reduced by 71%, 83% and 52%, respectively by woodpeckers.

According to Stallcup (1963) eight species of avian predators other than woodpeckers reduced flying spruce beetle populations by about 10%. This mortality is normally referred to as "in-flight" mortality.

2. Southern Pine Beetle. Considerable research has recently been done on the southern pine beetle (*Dendroctonus frontalis* Zimmermann), and it is discussed in detail by Kroll and Fleet (1979).

Estimates of southern pine beetle mortality caused by woodpeckers vary considerably and this may be due to local variation in climate, outbreak size and history, and perhaps to the number of insect generations per year. According to Moore (1972) woodpeckers, mainly Downy Woodpeckers, were the most important predators of the southern pine beetle in North Carolina causing about 5% reduction on the average, and over 50% in individual trees. In Virginia the predatory value of woodpeckers seem to be higher as it is estimated that woodpeckers remove about 86% of the beetle population (Steirly, 1965). Moore (1972) suggested that beetle mortality caused by the "side effects" of the feeding activity of woodpeckers is greater than the reduction caused by the actual consumption of bark beetles. The removal of the outer bark causes the remaining bark to dry faster and allows cold weather and possibly disease to have a greater impact on the bark beetles than these factors would normally have. Moore (1972) measured the moisture content of the inner bark of infested trees "scaled" or woodpeckered, and in infested but intact trees, the moisture content was 33% and 60%, respectively. The influence of phloem moisture on brood development of bark beetles is getting increased attention (Webb and Franklin, 1978). The indirect effect of woodpecker predation on the prey will be discussed in more detail later.

3. Western Pine Beetle. Avian predation on the western pine beetle (*Dendroctonus brevicomis* LeConte) has been investigated in recent years (Otvos, 1965, 1969, 1970). The effects of "woodpeckering" (the process by which woodpeckers thoroughly puncture, flake, or drill the bark in search of food) on the western pine beetle include direct and indirect effects. The direct effect of woodpeckering is the reduction of *D. brevicomis* densities through actual consumption; indirect effects are increased mortality due to some modification of the beetle's environment, and an affect on densities of insect predators and parasites. Four species of woodpeckers, the Hairy, Downy, Pileated (*Dryocopus pileatus* (L.)), and White-headed Woodpeckers (*Picoides albolarvatus* (Cassin)) are the most important avian predators of the western pine beetle larvae at moderately high population levels in the central Sierra in California, and were estimated to cause 32% mortality (Otvos, 1965) during the early part of an outbreak.

In a later attempt to determine all of the avian predators of the western pine beetle, 11 species of birds from four families were observed to feed on some stage of the western pine beetle (Otvos, 1970) (Table I). Table I also shows that the same birds captured other bark beetles as well. The "in-flight"

TABLE I.　Avian Predators Known to Prey on Bark Beetles in California[a]

	Beetles												
Birds	*Dendroctonus brevicomis*	*Dendroctonus ponderosae*	*Dendroctonus valens*	*Ips confusus*	*Ips emarginatus*	*Ips latidens*	*Ips montanus*	*Orthotomicus sp.*	*Scolytus ventralis*	*Scolytus sp.*	*Gnathotrichus sp.*	*Trypodendron sp.*	*undetermined scolytid*
Tachycineta thalassina	x		x								x		x
Contopus sordidus													x
Nuttallornis borealis	x	x	x	x			x				x		x
Myiarchus cinerascens			x										
Sitta canadensis	x												
Certhia familiaris	x		x	x									x
Turdus migratorius			x										
Piranga ludoviciana			x										
Junco oreganus	x		x						x				x
Pheucticus melanocephalus	x												
Picoides villosus	x	x	x	x	x	x		x					x
Picoides pubescens	x	x		x	x	x			x	x			x
Picoides albolarvatus	x		x						x	x		x	x
Picoides tridactylus	x												

[a] Woodpeckers prey on both adults and larvae, while most other birds prey on adult beetles in flight, the Red-breasted Nuthatch and Brown Creeper have been observed to take bark beetle larvae from the tree after the bark has been "scaled" or "thinned" off by woodpeckers.

mortality of the western pine beetle, caused by six species of
birds (excluding *Picidae*) was estimated during 3 years, to range
between 8% and 26% (Otvos, 1969). There was a gradual increase
in woodpecker population in the infestation area, which was
probably caused at first by immigration; later the bark beetle-
killed trees increased availability of nesting or roosting
sites. Woodpecker populations are known to increase in response
to an increase in their food supply (Blackford, 1955; Yeager,
1955; Baldwin, 1960, 1969; Koplin, 1972). The higher population
densities may be due to aggregation, reproduction (increased
fecundity), or both. It may be on a small or large scale,
temporary or more permanent, depending on conditions.

In an insect outbreak which persists for a long period over
an extensive area (such as the spruce budworm outbreak in
eastern Canada from 1912-1920), when the dead or dying trees are
attacked by secondary wood-boring insects, the response of
woodpeckers is impressive. Van Tyne (1926) described one of
several heavy flights of Black-backed Three-toed Woodpeckers
(*Picoides articus* (Swainson)) in 1923-24 and theorized that
these birds were from an abnormally large population in Canada
which increased in the forests following the spruce budworm
outbreak.

In a western pine beetle infestation area when snags and old
beetle-killed trees (not housing western pine beetles) were
removed woodpecker populations declined considerably (Otvos,
1969). Using exclusion technique, linear regression was applied
and the coefficient estimated between percent bark area worked
over by woodpeckers and percent bark beetle mortality. Brood
density of woodpeckered bark was expressed as a percentage of
the brood density of the non-woodpeckered bark and was then
correlated with the percentage of bark area woodpeckered
(Fig. 1). The regression equation ($Y = 30.15 + 0.42X$) predicts
percent bark beetle mortality due to woodpeckers (Y) from the
bark area woodpeckered (X). This equation, I believe, gives a
conservative estimate of percent mortality due to woodpeckers.
When regression analysis was performed separately for the spring
or summer generations and overwintering generations, the resulting equations differed somewhat from each other, but the differences were not significant.

a. Indirect effects of woodpeckering. Woodpeckers in the
process of feeding on bark beetles flake, puncture or drill the
bark of infested trees. This alters the microhabitat of the
prey and can affect the survival of the remaining insects in the
trees. These indirect (side) effects of woodpecker predation
can be important and some of these effects are discussed below.

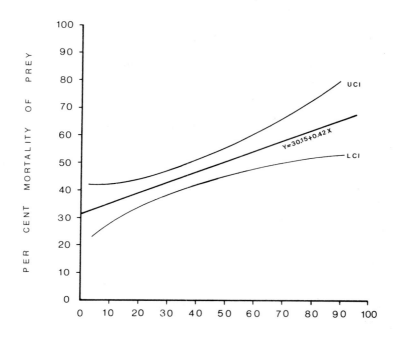

FIGURE 1. Regression line between percentage of mortality of western pine beetle (Y) and percent bark area woodpeckered (X) and confidence limits at the 95 percent level. UCI = upper confidence interval, LCI = lower confidence interval. (Redrawn from Otvos, 1970).

(i) Bark flaking – The western pine beetle differs from most other bark beetles in that the major part of its larval development takes place in the outer bark. The larvae and pupae of the western pine beetle occur at varying depths in the bark, and the portion of the bark removed to a more or less uniform thickness by woodpeckers contains a considerable quantity of brood (Figs. 2 and 3). There was no significant difference in the proportion of beetles removed in the summer (generation 1) and overwintering generation (generation 2). Approximately 58 percent of the larvae are located in the portion of the bark removed by woodpeckers. This estimate is somewhat lower than that obtained by regression analysis.

Not all larvae in the removed bark portion can be considered as consumed because undoubtedly some western pine beetle larvae as well as some of its predators and parasites remain in the bark chips flaked off by the birds. These insect inclusions in the bark flakes were few (Otvos, 1965) and practically none

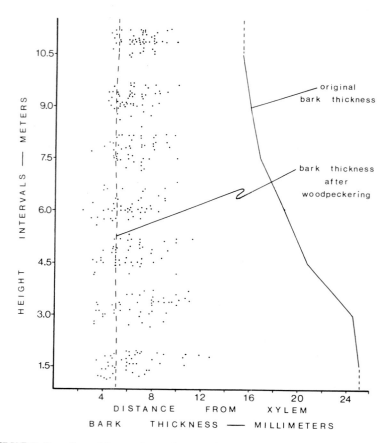

FIGURE 2.   Location of western pine beetle larvae and pupae
(see dots) in ponderosa pine bark in the summer generation. Bark
thickness measurements are averages (n = 18).

emerged in the field during the fall or second generation from
the bark flakes that spent the winter on the ground. Emergence
may occur during the summer or first generation when microclimatic conditions in the bark flakes may be more favorable for
insect development.

       (ii) Effect on parasites and predators - Foraging
activity of woodpeckers results in higher parasite densities in
woodpeckered bark areas (Fig.4). It is due to the feeding
activity of woodpeckers which places the remaining brood within
the ovipositor range of parasites with short ovipositors (Fig.
5). Otvos (1965) showed that parasitism increased in bark areas
where bark thickness was reduced by woodpeckers.

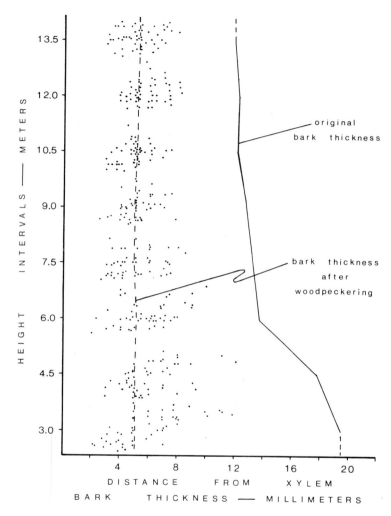

FIGURE 3. Location of western pine beetle larvae (see dots) in ponderosa pine bark in the overwintering generation. Bark thickness measurements are averages (n = 20).

The most important parasites of the western pine beetle are *Roptrocerus xylophagorum* (Ratzeburg) and *Cecidostiba burkei* Crawford (Dahlsten and Bushing, 1970). Both are small parasites with short ovipositors and *C. burkei* normally oviposit through the bark. Although *R. xylophagorum* usually enters the bark beetle gallery to oviposit (Györfi, 1952; Dahlsten and Bushing,

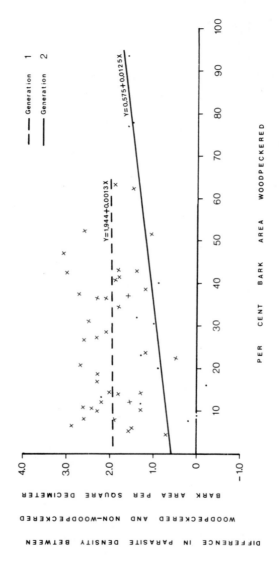

FIGURE 4. Regression line between percentage of bark area woodpeckered (X) and the difference between parasite densities per dm2 of the woodpeckered and non-woodpeckered bark (Y).

FIGURE 5.   Western pine beetle parasite, *Cecidostiba burkei*
ovipositing through woodpeckered bark.

1970) it is also known to oviposit through the bark, especially
when the bark is thin (Györfi, 1952). Perhaps this explains why
both *C. burkei* and *R. xylophagorum* occurred in consistently
greater numbers in the thinned (woodpeckered) than unthinned
bark.

A regression coefficient was estimated between percentage of
bark area woodpeckered and the difference in parasite density
between woodpeckered and non-woodpeckered bark (Fig. 4).
There was positive correlation between parasite density and the
bark area searched by woodpeckers for both the summer (generation
1) and overwintering generation (generation 2) indicating that
higher parasitism is associated with woodpecker activity (Otvos,
1969; Berryman *et al.*, 1970). At times there was a 3 to 10-fold
increase in percent parasitism (Otvos, 1965). However, this
increase in parasite density was significant only in the overwintering or second generation. Insect predator densities are
lowered in the woodpeckered bark (Otvos, 1969; Berryman *et al.*,
1970). It is probably caused by consumption of predators by
woodpeckers, the predators leaving the area of woodpecker
activity because they are "harassed," the prey is reduced,

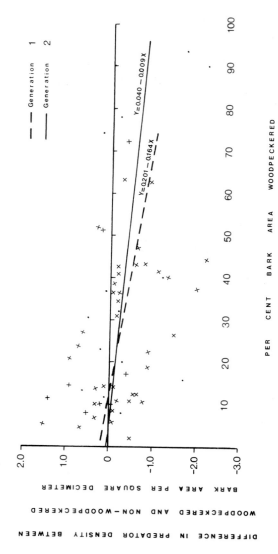

FIGURE 6. Regression line between percentage of bark area woodpeckered (X) and the difference between predator densities per dm² of the woodpeckered and non-woodpeckered bark (Y).

or by the altered microclimatic conditions in the remaining
bark. The principal insect predators of the western pine beetle
are highly mobile and follow the changing abundance of the prey
(Berryman, 1970).

Regression was performed for percentage of bark area woodpeckered and the difference in predator density between woodpeckered and non-woodpeckered bark (Fig. 6). There was a negative correlation between predator density and woodpecker
activity both in the summer and overwintering generation indicating that there is a decrease in predator numbers as woodpecker
activity increases. In the second or overwintering generation
the predators decreased slightly with increasing woodpecker
activity. In the first or summer generation, the initial
predator density is higher and the apparent impact of woodpeckers is greater. According to Berryman *et al.* (1970) however,
although negative correlation is observed in some cases between
woodpecker activity and insect predator density, mean predator
density is not reduced consistently by woodpecker activity.

(iii) Effect on microclimate – Reduction in bark thickness, due to the foraging activity of woodpeckers, causes marked
changes in the microenvironment of the bark beetle larvae remaining in the bark. The reduced bark thickness offers less
protection against the effect of changing weather. Temperature
and moisture conditions in woodpeckered bark are different from
those in non-woodpeckered bark.

Bark temperatures were measured at the approximate depth of
western pine beetle brood, at 1.5 m intervals along the infested
bole, during a continuous 35 hour period in late summer (Otvos,
1969). Four bark temperature readings were taken at each
height level – two on the north-northeast (NNE) side and two on
the south-southwest (SSW) side of the tree. One thermocouple
was placed in woodpeckered, the other in non-woodpeckered bark
at both sides of each height level. Air temperatures were
recorded by using a single unshielded thermocouple at each
height level. There were differences in bark temperature
between the NNE and the SSW side of the same tree and also
between woodpeckered and non-woodpeckered bark on the same side
of the tree.

Measurements of bark temperatures indicated a vertical
temperature gradient along the infested bole, the low being at
the base of the tree where bark is thicker. Bark temperatures
of woodpeckered bark responded faster and to a greater degree to
changes in air temperature than those of non-woodpeckered bark
both at the NNE and SSW aspects of the tree (Otvos, 1969).
Temperature of the woodpeckered bark on the NNE side can rise
higher than the ambient temperature on the SSE side at the same
height level. This is illustrated at a selected height in

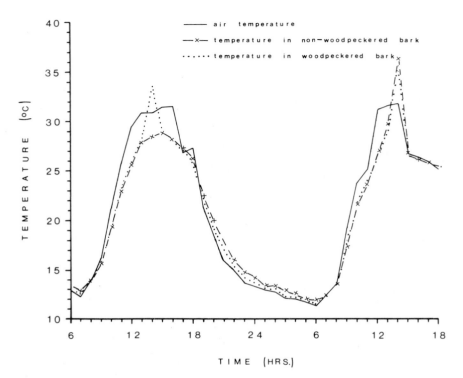

FIGURE 7. Air temperature on the SSW side and bark temperature on the NNE side of a bark beetle infested ponderosa pine
at the 19.5 m height, August 31-September 1, 1967.

There is considerable diurnal temperature fluctuation
in the bark. Bark temperature deviates most from ambient
temperature at night and in the early afternoon.

The greatest temperature difference recorded between ambient
temperatures and woodpeckered bark was 11.0° C at 19.5 m
(Fig. 8). Mortality due to lethal temperatures was not considered to be important in the central Sierra Nevada in California (Otvos, 1969). Under more rigorous winter conditions at higher elevations, such as in Colorado, or hotter summer conditions
like in the southern United States, bark beetle mortality due to
lethal temperatures may be important in heavily woodpeckered
trees, especially in trees with thinner bark than ponderosa pine
(*Pinus ponderosa* Lawson). In such trees desiccation during the
summer may be another important mortality factor.

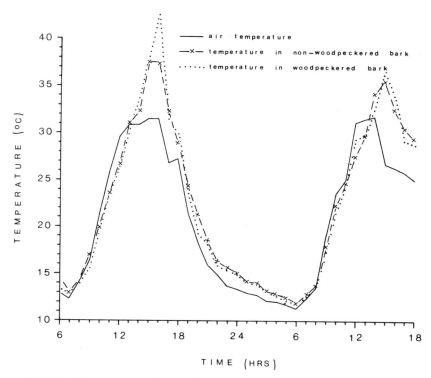

FIGURE 8. Air and bark temperatures on the SSW side of a
bark beetle infested ponderosa pine at the 19.5 m height,
August 31-September 1, 1967.

C. Direct Effect of Birds as Predators of Woodborers and Weevils

Solomon (1969) studying the effect of woodpecker predation
on various woodborers in hardwoods, found that woodpeckers
removed 32% of the white oak borer (*Goes tigrinus* (De Geer)).
Later he revised this predation at 65% (Solomon 1977). Solomon
(1969) also estimated that woodpeckers captured 39% of the
living beech borer (*G. pulverulentus* (Halderman)), and 13% and
65% respectively, of the poplar borer (*Saperda calcarata* Say)
studied at two locations. Working on a related poplar borer,
(*Saperda carcharias* L.) in Czechoslovakia Tichy (1963) on the
other hand found that although up to 98% of the diet of *Picoides
major* (L.) can be composed of *S. carcharias*, this woodpecker on
the average reduced, he estimated, the larval population of

S. carcharias by only about 12%. Studying the ash borer
(*Podosesia syringae* (Harris)) Solomon (1975) concluded that
woodpeckers are its most important natural enemy, causing 67%
to 81% mortality.

The relationship between woodpeckers and a weevil, *Pissodes
piniphilus* Herbst in Holland (Voûte, 1951) also suggests a
functional and perhaps a numerical response as well. At low
weevil population levels predation by woodpeckers is negligible.
Predation increases as weevil larvae become more numerous and
woodpeckers may remove 95% of the larvae and pupae, especially
during a period of food shortage (Voûte. 1951).

Finnegan (1958) working on the pine weevil, *Pissodes
approximatus* Hopkins, found that the Downy Woodpecker destroyed
up to 90% of the weevil population in the trunks of individual
trees; weevils in the stumps, however, were practically free
from bird predation.

D. Effect of Avian Predators on Parasitic Insects

Birds can affect parasitic insects: (a) by direct consumption
of adult parasites, or (b) indirectly by preferentially selecting (or avoiding) parasitized prey as food, or (c) altering prey
habitat to make it more susceptible to parasitism. The latter
has been illustrated earlier in the case of woodpeckers.

Most avian predators eat parasitic insects. However, the
relative frequency of parasites in the stomach contents of birds
is usually low in comparison to the frequency of other prey or
to the density with which parasites occur in the field. The
important question is whether the relative frequency of the
parasitized larvae in the diet of the birds is the same, less or
more than the proportion of the parasitized larvae in the field.
There is no universal agreement on this.

Buckner and Turnock (1965) reported that examination of 305
gizzards of 54 species of birds showed that 151 specimens representing 43 species had fed on the larch sawfly, and nine gizzards contained a total of 14 parasites. Two of these parasites
were the ichneumonid *Mesoleius tenthredinis* Morley and the other
12 were the tachinid *Bessa harveyi* (Townsend). The former was
considered to be rare, the latter common in the field.

Sloan and Coppel (1968) stated that birds do not discriminate against parasitized larch casebearer. In a later publication (Coppel and Sloan, 1971) however, they estimated that up
to 38% mortality of the parasites can be caused by predation.

Futura (1976) found during a study of low density gypsy moth
population in Japan that birds consumed most of the internal
parasites with the host before they (parasites) leave the host
to pupate.

Behavior or size of the parasitized prey may also influence
the fate of these larvae. Betts (1955) reported that parasitized larvae of the winter moth are sluggish and were frequently
observed on the ground where these larvae are conspicuous.
Tits, especially the Great Tit (*Parus major*) which frequents the
ground to a greater extent than the other species eat these
parasitized larvae. On the other hand, there was not any
detectable difference in the amount of parasitized and non-
parasitized budworm larvae consumed by birds in Newfoundland, in
spite of the fact that parasitized larvae are exposed to predation for longer time. Parasitized spruce budworm larvae
develop more slowly and are smaller than non-parasitized larvae.
Perhaps the longer exposure to predation was minimized by the
smaller size of the parasitized larva.

MacLellan (1958) reported that parasitized codling moth
(*Laspeyresia pomonella* (L.)) larvae are smaller in size than
non-parasitized larvae and only 3% of the codling moth larvae in
the diet of the woodpeckers were parasitized versus 14% of the
uncaptured larvae. This may indicate that woodpeckers may have
had difficulty in locating the parasitized larvae which were
smaller (MacLellan, 1971) or that parasitized codling moth
larvae were avoided by woodpeckers. There is mounting evidence
to support the latter theory that birds, like mammals (Holling,
1955), can differentiate between parasitized and non-parasitized
larvae. Korol'kova (1956) reported that birds prefer to prey on
non-parasitized larvae of the gypsy moth. In a caged experiment
Chipping Sparrows (*Spizella passerina* (Bechstein)) recognized
parasitized jack pine budworm larvae and pupae and avoided them
(Sloan and Simmons, 1973).

E.   Insectivorous Birds as Dispersal Agents
of Insect Pathogens

Birds have been demonstrated to assist in the spread of
entomopathogenic viruses by eating infected insects (Franz *et
al.*, 1955; Entwistle *et al.*, 1977a). Entwistle *et al.* (1977a)
working with the European spruce sawfly (*Gilpinia hercyniae*
(Hartig)) found that 44 of 49 bird droppings collected from
trees were infective: 89.8% of the test larvae contacted
nuclear polyhedrosis virus (NPV) and died from the disease. They
regarded birds as passive carriers of the virus and presented a
summary of records of survival of baculovirus infectivity
through the avian gut of several forest and agricultural insect
pests. In a subsequent study Entwistle *et al.* (1977b) found
that birds pass infective droppings during winter and other
months when no living stages of the host insect are present on
the trees. Their second study also suggested that during winter
the virus can be spread for at least 6 km from the source. The

question arises here how far migratory species could spread
these diseases. It is postulated that entomogenous fungal
resting spores may also be spread by avian predators. However,
Buse (1977) working in previously virus-free areas of European
spruce sawfly infestations found NPV virus only in some larvae
and none of the bird droppings he collected. Therefore, he
considers birds to be of minor importance in the dispersal of
virus diseases into virus-free areas.

Such woodland inhabiting insectivores as Starlings (Eccles
1939 in Murton, 1971) and sparrows (Broadbent 1965 in Murton,
1971) have been implicated in the transmission of some diseases
other than those infecting insects. Among the truly forest
inhabiting birds several species have been found to spread plant
or vertebrate diseases. Heald (1933) found almost one million
spores of the fungus *Endothia parasitica* (Murr.) P.J. & H.W.
And., causing the chestnut blight disease, on a single Downy
Woodpecker. Several species of European woodpeckers, possibly
titmice and blackbirds are involved in spreading the spores of
Monilinia laxa (Aderh. & Ruhl.) Honey [Perfect stage: *Sclerotinia
cinerea* auct.], a fungus causing brown rot in various pomiferous
and stone fruits (Turĉek, 1950). It is possible that future
work will implicate some of those cavity nesting insectivores
which preferentially excavate their nest in trees softened by
heart rot in the spread of the spores of wood rotting fungi.

A number of arboviruses seem to be dependent on a close
association between *Culex* mosquitoes and forest birds, including
passerines. For example, the transmission of western equine or
St. Louis encephalitis viruses depends on the association of
populations of *Culex tarsalis* Coquillett and avian hosts in
woodland ecosystems (Reeves, 1965).

F. Consumption and Distribution of Seeds in the Forest by Insectivorous Birds

Birds have potentially important roles in the forest ecosystem through consumption and dispersal of seeds. The value of
birds in the distribution of seeds in the forest is probably
underestimated. Some forest inhabiting insectivorous birds may
also feed on seeds of trees or shrubs, especially during adverse
weather conditions, or during food shortage (Tinbergen, 1960).
This feeding may be beneficial to the forest manager only if
birds disperse the seed of desirable species via feeding or
storing. Hagar (1960) considered forest birds as a "... hind-
rance to natural seeding ..." because they attack cones of
trees, especially around the edges of cutovers.

Turĉek (1961) discussed at length the species of seeds and berries consumed by different bird species and provided extensive tabular data. Some species of birds store excess food gathered in times of abundance for winter consumption, and usually not all the food stored is found. Among insectivorous birds Turĉek (1954a, 1961) considers woodpeckers (mainly *Picoides* spp.), thrushes (*Turdus* spp.), titmice and nuthatches (*Sitta* spp.), besides jays, important in seed dispersion. The jay, both palaeartic and nearctic forms are the "... most important and valuable planter among birds ..." (Turĉek, 1954a). This "planting" is an indirect result of the bird's food storing activity. Eurasian Jays (*Garrulus glandarius* L.) usually store acorns under leaves, moss, stones and debris. The extent of this food storing perhaps can best be illustrated with an observation made on a group of 30 to 40 Jays in England which collected an estimated 200,000 acorns and buried them about 1 km away (Chettleburg, 1952). Seed dispersal can occur over short or long distances. The dispersal of viable seeds in the digestive tracts of migrating birds has been recently documented (Proctor, 1968).

The Acorn Woodpecker (*Melanerpes formicivorous* (Swainson)) has an interesting food storing habit that deserves mention. This species riddles with holes the bark of various forest trees, as well as telegraph and telephone poles, fence posts and gables, if these are near a source of acorn supply, and stuff the holes with acorns. Up to 50,000 acorns may be stored in a single tree (Bent, 1939). Holes made in the bark of living trees do not appear to harm the tree as the holes do not reach the living layer of the bark.

G. The Role of Woodpeckers in Damaging Healthy Trees, Aiding in the Spread of Wood Rot and Damaging Telecommunication Poles

Turĉek (1960) reviewed damage done by birds, mainly woodpeckers, to power and communication lines, a damage that is widespread in several countries. He examined the distribution of damaged poles in relation to their distance from trees and found "over half of the damaged poles were in lines through woods in forested areas, or in the near vicinity of trees". Therefore, lack of opportunity to excavate "natural" nesting or roosting sites should not have been a contributing factor. In describing the type of damage Turĉek says that "... the mechanical properties of the poles are weakened, in extreme cases the poles break down; access to wood-decomposing fungi and to the weather is opened". The same statement equally applies to woodpeckers excavating nesting or roosting holes in green trees or branches. There are indications, however, that several

Imre S. Otvos

woodpeckers excavate their nest cavities in trees with heartwood
softened by fungal rots (Shigo and Kilham, 1968; Conner *et al.*,
1976). A shortage of suitable trees for roosting and nesting is
a possible reason for birds excavating holes in power lines
(Petterson 1951 and Brander 1956 in Turĉek, 1960). However,
this may not be entirely true because the Black Woodpecker
(*Dryocopus martius* (L.)) and the Great Spotted Woodpecker
(*Picoides major*) regularly excavate holes in sound and living
trees (Turĉek, 1960). Another theory, regarding excavation of
holes in telecommunication poles, is that the vibration and buzz
of the pole acts as an acoustic stimulus to attract woodpeckers
and trigger feeding behavior (Brander 1956 in Turĉek, 1960).
Turĉek (1960) discredits this by saying that the excavation
holes are mainly at the usual height for nesting and roosting
holes and the shape of the hole is round and not conical, the
latter being typical for "feeding" holes.
      Certain members of the woodpecker family which feed on
insects in dead and decaying wood have a beneficial effect on
the nutrient cycle in the forest ecosystem by speeding up decom-
position of the trees as an indirect result of their feeding
activity.
      Sapsuckers are known to drill holes in living trees and
drink the sap. These wounds initiate a process resulting in
discoloration or ring shakes (a condition where cracks form
between the annual rings) (Shigo, 1963), or may serve as a site
of fungal infection.
      Turĉek (1954b) in an earlier study noted that about half of
the species of woodpeckers in Europe and about a quarter of the
species in North America "ring" or drill trees in a sapsucker
fashion. According to him, in Europe, ringing occurs mainly in
the spring when the sap of trees begins to flow, and the purpose
of the drilling is sap-sucking. Turĉek (1954b) concluded that
most of the trees damaged in this way are abnormal, i.e. stunted,
damaged, "mixed-in" or "foreign to the community," and only a
few appear to be healthy.
      Repeated puncturing of healthy trees by sapsuckers or wood-
peckers will usually lead to callus forming around the wound
which sometimes may serve as site of infection from fungi or
attack from secondary insects.

                         III. CONCLUSIONS

      Estimation of the effectiveness of birds in forest insect
control is a complex task, especially when the effect of alternate
prey is to be considered as well. It would be highly desirable
and mutually beneficial for entomologists and ornithologists to
cooperate and investigate insect-bird problems jointly.

Insectivorous birds play an important role in the population dynamics of many forest insects, especially at low to moderate population levels. Birds can act as direct mortality agents or can affect their prey indirectly through influencing insect parasites and predators of the prey, or by altering the micro-environment of the prey.

The major role of birds is in the retardation or prevention of insect outbreaks and in suppression of insect pest populations before they reach outbreak levels. Although numerous species of birds respond both functionally and numerically to an increase in insect population levels birds alone cannot bring about the collapse of insect outbreaks of considerable size. The reproductive potential, feeding habits and territorial relationships of birds are such that these reduce their effectiveness as predators during outbreaks. Insect outbreaks can only be controlled by birds if the outbreaks are localized (small in size) and there is a considerable aggregation of resident birds or a large influx of migratory birds into the outbreak area. Insectivorous birds can be of considerable importance in accelerating the decline of an outbreak that has peaked.

Birds not only eat harmful insects but they also consume beneficial ones. However, recent evidence suggest that birds tend to avoid parasitized insects; thus avian predation may be complementary rather than competitive with parasitic insects. In evaluating the importance of insectivorous birds in insect control the various developmental phases of the insect on which predation takes place (i.e. egg, larva, pupa and adult) should be valued differently. The closer a mortality factor operates just prior to egg deposition of the prey the greater its value.

Birds can spread entomogenous pathogens of insect pests and thus contribute to the mortality of the prey indirectly as well. Some of the woodland birds can also be carriers of plant and vertebrate disease.

Insectivorous birds can have an affect on the forest eco-system by feeding on and dispersing seeds of various forest trees and shrubs. Woodpeckers may be involved in the spreading of wood rotting fungi by excavation of their food, nest or roosting cavities.

Although it has been demonstrated that the population of avian predators of forest insects can be increased by environmental modification leaving snags or stumps for cavity nesting species and by providing nest boxes, this type of manipulation on such a large scale as the boreal forest of North America is not practical.

ACKNOWLEDGMENTS

    I would like to express my appreciation to Drs. T. Royama,
Canadian Forestry Service, Fredericton, New Brunswick, L. Tuck,
Memorial University of Newfoundland, St. John's, Newfoundland
and M.F.D. Udvardy, California State University, Sacramento,
California for their helpful suggestions and critical reviews of
the manuscript. I would also like to thank Mr. D.S. Durling for
redrawing the figures and Mrs. D.J. Didham for typing the
manuscript.

REFERENCES

Ainslie, C. (1930). *Wilson Bull. 42*, 193.
Baldwin, P. (1960). *Proc. 12 Int. Ornith. Congr.,*
    Helsinik (1958), 71
Baldwin, P. (1969). *Proc. North Central Branch Ent. Soc.*
    Am. 23, 90.
Barrows, W. (1889). *USDA Div.* of *Economic Ornithology* and
    Mammalogy. Bull. No. 1, 1.
Beal, F. (1911). *USDA Biol. Surv. Bull. 37*, 1.
Bent, A. (1939). "Life Histories of North American Wood-
    peckers". *U.S. Nat'l. Mus. Bull. 174.* Washington, D.C.
Berryman, A. (1970). *In* "Studies on the Population Dynamics
    of the Western Pine Beetle, *Dendroctonus brevicomis* LeConte
    (Coleoptera: Scolytidae)" (R. Stark, and D. Dahlsten,
    eds.), p. 102. Univ. Calif., Div. Agric. Sci.
Berryman, A., Otvos, I., Dahlsten, D., and Stark, R. (1970).
    In "Studies on the Population Dynamics of the Western Pine
    Beetle, *Dendroctonus brevicomis* LeConte (Coleoptera:
    Scolytidae" (R. Stark, and D. Dahlsten, eds.), p. 174.
    Univ. Calif. Div. Agric. Sci.
Betts, M. (1955). *J. Anim. Ecol. 24*, 282.
Blackford, J. (1955). *Condor 57*, 28.
Blais, J., and Parks, G. (1964). *Can. J. Zool. 42*, 1017.
Boesenberg, K. (1970). *Deutsche Akademie der Landwirt-*
    Schaftswissenschaft (Tagungsberichte) *110*, 71.
Bruns, H. (1960). *Bird Study 7*, 193.
Buckner, C. (1966). *Ann. Rev. Ent. 11*, 449.
Buckner, C., and Turnock, W. (1965). *Ecol. 46*, 223.
Bull, E.L. (1978). *In* "Proc. of the Workshop on Non-game
    Bird Habitat Management in the Coniferous Forest of the
    Western United States" (R. DeGraaf, Tech. Coord.), p. 74.
    USDA For. Serv., Gen. Techn. Rpt. PNW-64.
Buse, A. (1977). *Ent. exp. appl. 22*, 191.
Chettleburg, M. (1952). *Brit. Birds 45*, 359.
Clark, L. (1964). *Austr. J. Zool. 12*, 349.

Coppel, M., and Sloan, N. (1971). *In* "Proc. Tall Timbers
    Conf. on Ecol. Anim. Contr. by Habitat Mgmt.", p. 259.
    Feb. 26-28, 1970. Tallahassee, Florida.
Conner, R., Miller, O., Jr., and Adkisson, C. (1976). *Wilson
    Bull. 88,* 575.
Dahlsten, D., and Herman, S. (1965). *Calif. Agr. 19,* 8.
Dahlsten, D., and Bushing, R. (1970). *In* "Studies on the
    Population Dynamics of the Western Pine Beetle, *Dendroctonus
    brevicomis* LeConte (Coleoptera: Scolytidae)" (R. Stark
    and D. Dahlsten, eds.), p. 113. Univ. Calif., Div. Agric.
    Sci.
Dahlsten, D. (1979). *In* Proc. Symp. "The Role of Insectivorous
    Birds in Forest Ecosystems" (J. Dickson, R. Conner, R.
    Fleet, J. Jackson, J. Kroll, eds.). Academic Press, New
    York.
Davidson, J., and Gimpel, W., Jr. (1975). *Proc. Ent. Soc.
    Wash. 77,* 165.
Doutt, R. (1964). *In* "Biological Control of Insect Pests
    and Weeds" (P. Debach, ed.), p. 21. Reinhold Publ. Corp.,
    N.Y.
Dowden, P., Jaynes, H., and Carolin, V. (1953). *J. Econ.
    Entomol. 46,* 307.
Edgerton, P., and Thomas, J. (1978). *In* "Proc. Workshop on
    Non-game Bird Habitat Management in the Coniferous Forests
    of the Western United States" (R. DeGraaf, Tech. Coord.),
    p. 56. *USDA For. Serv. Gen. Tech. Rpt. PNW-64.*
Entwistle, P., Adams, P., and Evans, H. (1977a). *J. Invert.
    Path. 29,* 354.
Entwistle, P., Adams, P., and Evans, H. (1977b). *J. Invert.
    Path. 30,* 15.
Finnegan, R. (1958). *Can. Entomol. 90,* 348.
Forbush, E. (1921). *Mass, Dept. Agric. Bull. 9,* 1.
Forbush, E. (1927). "Birds of Massachusetts and Other New
    England States". *Vol. 2.* Mass. Dept. Agr., Boston.
Franz, J. (1961). *Ann. Rev. Entomol. 6,* 183.
Franz, J., Kreig, A., and Langenbuch, R. (1955). *Viren. Z.
    Pfl. Ban. 62,* 721.
Futura, K. (1976). *Jap. Govt. For. Exp. Stn. Bull. 279.*
Gage, S., Miller, C., and Mook, L. (1970). *Can. J. Zool. 48,*
    359.
George, J., and Mitchell, R. (1948). *J. For. 46,* 454.
Gibb, J. (1958). *J. Anim. Ecol. 27,* 374.
Gibb, J. (1966). *Ibis 102,* 163.
Gibb, J. (1966). *J. Anim. Ecol. 35,* 43.
Gillmeister, H. (1963). *Allgem. Forstz. 18,* 470.

Györfi, J. (1952). *Orszagos Természettudományi Múzeum Évkönyve n.s. 2*, 113. Budapest.
Haartman, L. von. (1951). *Acta Zool. Fenn. 67*, 1.
Haartman, L. von. (1957). *Evolution 11*, 339.
Haartman, L. von. (1971). *In* "Avian Biology". Vol. 1 (D. Farner, J. King and K. Parkes, eds.), p. 381. Academic Press, New York and London.
Hagar, D. (1960). *Ecology 41*, 116.
Hardin, K. and Evans, K. (1977). *USDA For. Service, Gen. Tech. Rpt.* NC-30.
Hasegawa, H., and Ito, Y. (1967). *Appl. Entomol. Zool. 2*, 100.
Heald, F. (1933). "Manual of Plant Diseases". McGraw-Hill Book Co., New York & London.
Henderson, J. (1927). "The Practical Value of Birds". Mac-Millan Co., New York.
Hensley, M., and Cape, J. (1951). *Auk 68*, 483.
Herberg, H. (1960). *Arch. Forstwesen 9*, 1015.
Holling, C. (1955). *Can. J. Zool. 33*, 404.
Holling, C. (1959a). *Can. Entomol. 91*, 293.
Holling, C. (1959b). *Can. Entomol. 91*, 385.
Holling, C. (1961). *Ann. Rev. Entomol. 6*, 163.
Holmes, R., and Sturges, F. (1975). *J. Anim. Ecol. 44*, 175.
Howard, W. (1959). State of Calif., *Dept. Agric. Bull. 48*, 171.
Kendeigh, S. (1947). Ontario Dept. Lands & Forests, *Biol. Bull. 1*, 1.
Keve, A., and Reichart, G. (1960). *Falke 7*, 20.
Knight, F. (1958). *J. Econ. Entomol. 51*, 603.
Koplin, J. (1969). *Condor 71*, 436.
Koplin, J. (1972). *J. Wildl. Manage. 36*, 308.
Koplin, J., and Baldwin, P. (1970). *Amer. Midl. Natur. 83*, 510.
Korol'kova, C. (1956). *In* "Ways and Means of Using Birds in Combatting Noxious Insects" (L. Poznanin, ed.), p. 55. Ministry of Agriculture USSR, Moscow. Transl. from Russian, 1960. Israel Program for Scientific Translation, Jerusalem.
Kroll, J., and Fleet, R. (1979). *In* Proc. Symp. "The Role of Insectivorous Birds in Forest Ecosystems" (J. Dickson, R. Conner, R. Fleet, J. Jackson, J. Kroll, eds.). Academic Press, New York.
Lack, D. (1966). "Population Studies of Birds". Oxford. Clarendon.
Lühl, R., and Watzek, G. (1973-74). *Angew. Ornith. 4*, 95.
MacArthur, R. (1958). *Ecology 39*, 599.
MacLellan, C. (1958). *Can. Entomol. 90*, 18.
MacLellan, C. (1971). *In* "Proc. Tall Timbers Conf. on Ecol. Anim. Contr. by Habitat Mgmt". *2*, 273. Feb. 26-28, 1970 Tallahassee, Florida.
McAtee, W. (1925). *Ann. Rept. Smiths. Inst.*, p. 415.

McAtee, W. (1927). *Roosevelt Wildlife Bull. 4*, 1.
McCambridge, W., and Knight, F. (1972). *Ecology 53*, 830.
McFarlane, R. (1976). *The Biologist 58*, 123.
Massey, C., and Wygant, N. (1954). *USDA Circ. No. 944*, Washington, D.C.
Mattson, W., Knight, F., Allen, D., and Foltz, J. (1968). *J. Econ. Entomol. 61*, 229.
Mitchell, R. (1952). *J. For. 50*, 387.
Mook, J., Mook, L., and Heikens, H. (1960). *Arch. Neerl. Zool. 13*, 448.
Moore, G. (1972). *Environ. Entomol. 1*, 58.
Morris, R., Cheshire, W., Miller, C., and Mott, D. (1958). *Ecol. 39*, 487.
Moutia, A., and Mamet, R. (1946). *Bull. Ent. Research 36*, 439.
Murton, R. (1971). "Man and Birds". Collins, London.
Otvos, I. (1965). *Can. Entomol. 97*, 1184.
Otvos, I. (1969). "Vertebrate Predators of *Dendroctonus brevicomis* LeConte (Coleoptera: Scolytidae), with Special Reference to Aves" Ph.D. Thesis. University of California, Berkeley.
Otvos, I. (1970). *In* "Studies on the Population Dynamics of the Western Pine Beetle, *Dendroctonus brevicomis* LeConte (Coleoptera: Scolytidae)" (R. Stark and D. Dahlsten, eds.), p. 119. Univ. Calif., Div. Agric. Sci.
Otvos, I., Clark, R., and Clarke, L. (1971). *Envir. Canada, For. Serv., Inf. Rept.* N-X-68.
Pfeifer, S. (1963). *Proc. 13th. Int. Ornithol. Congr. 1962*, p. 754.
Polivanov, V. (1956). *In* "Ways and Means of Using Birds in Combatting Noxious Insects" (L. Poznanin, ed.), p. 111. Ministry of Agriculture USSR, Moscow. Transl. from Russian, 1960. Israel Program for Scientific Translation, Jerusalem.
Proctor, V. (1968). *Science 160*(3825), 321.
Prop, N. (1960). *Arch. Neerl. Zool. 13*, 380.
Readshaw, J. (1965). *Austr. J. Zool. 13*, 475.
Reeves, W. (1965). *Ann. Rev. Entomol. 10*, 25.
Royama, T. (1970). *J. Anim. Ecol. 39*, 619.
Sanders, C. (1970). *Amer. Midl. Nat. 84*, 131.
Semenov, S. (1956). *In* "Ways and Means of Using Birds in Combatting Noxious Insects" (L. Pozanin, ed.), p. 78. Ministry of Agriculture, USSR, Moscow. Transl. from Russian, 1960. Israel Program for Scientific Translation, Jerusalem.
Shaub, M. (1956). *Bird-Banding 27*, 157.
Shigo, A. (1963). *USDA For. Serv. Res. Pap.* NE-8.
Shigo, A., and Kilham, L. (1968). *USDA For. Res. Note* NE-84.
Shook, R., and Baldwin, P. (1970). *Can. Entomol. 102*, 1345.

Sloan, N., and Coppel, H. (1968). *J. Econ. Entomol. 61*, 1067.
Sloan, N., and Simmons, G. (1973). *Amer. Midl. Nat. 90*, 210.
Solomon, J. (1969). *Ann. Entomol. Soc. Am. 62*, 1214.
Solomon, J. (1975). *Ann. Entomol. Soc. Am. 68*, 325.
Solomon, J. (1977). *Ann. Entomol. Soc. Am. 70*, 57.
Solomon, M. (1949). *J. Anim. Ecol. 18*, 1.
Stallcup, P. (1963). "A Method for Investigating Avian Pred-
    ation on the Adult Black Hills Beetle". M.S. Thesis.
    Colorado State University, Fort Collins.
Steirly, C. (1965). *Raven 36*, 55.
Tichy, V. (1963). *Prace vyzkumnych ustavu lesnickych* CSSR
    26, 49.
Tichy, V., and Kudler, J. (1962). *Lesnictvi 35*, 151.
Tinbergen, L. (1949). *Ned. Bosb. Tijdschr 4*, 91.
Tinbergen, L. (1960). *Arch. Neerl. Zool. 13*, 265.
Tothill, J. (1923). *Acad. Entomol. Soc.* (1922), *8*, 172.
Turĉek, F. (1950). *Vĕr. Fĝgelvärld 9*, 210.
Turĉek, F. (1954a). *Aquila 55-8*, 51.
Turĉek, F. (1954b). *Ornis Fenn. 31*, 33.
Turĉek, F. (1960). *Bird-Banding 7*, 231.
Turĉek, F. (1961). "Ökologische Beziehungen der Vögel und
    Gehölze." Verlag der Slowakischen Akademie der Wissen-
    schaften, Bratislava.
Van Tyne, J. (1926). *Auk 43*, 469.
Voûte, A. (1951). *Z. angew. Entomol. 33*, 47.
Webb, W., and Franklin, R. (1978). *Envir. Entomol. 7*, 405.
Witler, J., and Kulman, H. (1972). Agr. Expt. Sta., Univ.
    Minnesota, *Tech. Bull. 289*, 1.
Wodzicki, K. (1950). *Bull. D.S.I.R. N.Z. 98*, 1.
Yeager, L. (1955). *Condor 57*, 148.
Zach, R., and Falls, J. (1975). *Can. J. Zool. 53*, 1669.

CONCLUDING REMARKS

Stanley H. Anderson

Migratory Bird and Habitat Research Laboratory
U.S. Fish and Wildlife Service
Laurel, Maryland

I. INTRODUCTION

The role that birds have in the ecosystem has been debated
for many years. Early naturalists often saw birds feed on
large numbers of insects. They felt that as insects were such
an important part of the birds' diet that bird populations
exerted a major controlling influence on insect populations.
Studies conducted in Europe and North America on the regulation of natural populations of animals indicate that birds
do take insects and exert some degree of population control,
but this degree is very poorly understood. For example,
Kendeigh (1947) estimated that birds destroyed 4.3% of the
spruce budworms in a heavy infestation in Ontario in 1945.
George and Mitchell (1948) calculated that the degree of control was from 3.5 to 7% in heavy infestations. When the budworms were found at levels of a half a million to a million
per acre, this meant that birds were taking 35,000 per acre.
Tothill (1923) indicated that birds consumed about 13% of the
larvae in a New Brunswick outbreak in 1918 and 39% of the
adults in a different habitat in British Columbia.
Other studies have indicated that birds take a considerably higher proportion of insect populations. Korol'kova (1956)
felt that as much as 75% of the hibernating beetles were destroyed in a Soviet forest. Bruns (1960) showed that about
20% of the Cabbage White Butterflies were killed by Tree Sparrows (Passer montanus). Keve and Reichart (1960) reported
that House Sparrows (Passer domesticus) in Hungary killed 98%
of the fall webworm moth. All these data serve to indicate
there is a great deal of variability in the role birds play in
controlling insect populations.

375

Data presented during this Symposium reenforces this conclusion. Robert Coulson and James Kroll indicated birds partially control barkbeetles. Donald Dahlsten showed the impact
of chickadees on the Douglas fir tussock moth. In the last
paper, Imre Otvos discussed the direct and indirect effects
birds have on insect populations.

To properly understand these effects, we must construct
life tables and determine what proportion of the insect population mortality is due to birds. We must distinguish the
effect birds have on increasing, declining, or stationary
insect populations.

II. CENSUS TECHNIQUES

A. Birds

One of the important problems facing those who census
birds in forest environments is that of detectability. William
Shields discusses this matter in his paper. He indicates that
trend data are easier to gather than absolute population size.
Several conspicuousness indices discussed in the literature
can be used. More important, the observer must be aware that
factors such as time of day, addressed by Thomas Smith, and
seasonal changes addressed by Richard Conner and James Dickson,
influence detectability of birds in the forest.

Robbins (1978) summarizes many of the bird census techniques tried in the forest ecosystem. He indicates the spot-
map technique is generally acknowledged as most dependable in
the breeding season. He also states that transect and point
counts, especially when well standardized and corrected for
bias, may be preferred if large areas must be sampled in a
short period of time.

B. Prey

Many papers have information on the insects found in the
stomachs of birds. Very few address the question of prey
availability. Robert Coulson discusses some of the problems
of sampling prey. Jerome Jackson shows how substrates affect
prey availability and how prey populations have different periods of activity on tree surfaces. Avian temporal and seasonal
activity patterns are in all likelihood influenced by changes
in prey availability and action.

Determination of the prey base has been very difficult.
Donald Dahlsten spoke of a large western data base at the
University of California, Berkeley. Data are thus being

accumulated on avian food availability in different habitats.
A better understanding of the role of birds in the forest
ecosystem will be developed when we have adequate knowledge
about the prey base. Thus, we have a challenge before us,
which as Robert Coulson indicates, requires the support of
statisticians and engineers.

III. AVIAN ACTIVITY PATTERNS

Many of the papers given in this Symposium show bird species are associated with specific plant communities or plant
structural components. Jerry Via showed the Acadian Flycatcher (Empidonax virescens) on rhododendron hills; Benedict
Pinkowski distinguished between the habitats of three species
of bluebirds; Ross James discussed vireo habitat.
There are several papers showing how forest bird species
are correlated with habitat components (Anderson and Shugart,
1974). However, to understand the role birds have in the community, we must learn why birds are found where they are and
what factors cause them to shift to other habitats.
Several excellent papers addressed these questions in this
Symposium. Jerry Via and Ross James showed where flycatchers
and vireos forage in the forest. Thomas Smith, James Dickson,
Richard Conner, and Montague Whiting discussed temporal and
seasonal shifts in foraging patterns. Thomas Grubb and Shaun
McEllin discussed how some species of forest birds forage.
These data can tell us how birds utilize the habitat in which
they are found.
Hartley (1953) showed that during insect population explo-
sions birds take a larger proportion of insect populations than
at other times. His studies indicated that ecological separation of closely related species (Great Tit, Blue Tit, Coal Tit,
Marsh Tit, and Long-tailed Tit) was broken down during a 1948
caterpillar infestation in England. Birds became more gener-
alists and took the most available food item. Forest birds
therefore appeared to have evolved on the basis of food availability. Competitive forces which maintained habitat separation
broke down during insect population outbreaks.
It is important to realize that habitat separation may
occur only during one period of the year. In this Symposium
Shaun McEllin shows that nuthatches have the greatest dietary
specialization during the breeding season. Williamson (1971)
finds that Red-eyed Vireos (Vireo olivaceus) forage at different heights in a tree at different rates during the breeding
season. Kilham (1965) reports similar results for the Hairy
Woodpecker (Picoides villosus). Morse (1973) shows sexual foraging differences in Yellow Warblers (Dendroica petechia) and
American Redstarts (Setophaga ruticilla).

Tinbergen (1960) shows how birds develop specific search images for insects. In his study he indicates that birds are very selective in their choice of food. Thomas Grubb discusses some of the many factors affecting the development of a search image. This selection is based on the visual stimuli reaching the birds' retina. Such factors as color, size, movement, previous experience, and palatability are involved. Birds search branches, needles, and twigs for insects they have recognized as particularly palatable. The profit birds derive from their search in the form palatable prey is addressed by Edward Garton. Tinbergen's study of tits in England shows that the adoption of these search images seems to be a process of conditioning which takes place soon after development of a new species of prey. There is some reason to believe that they can only have a limited number of specific search images at one time.

What can we say about birds' role in controlling insect populations? Shilova-Krassova (1953) in a Soviet study drew conclusions that appear to be the theme of most studies in the literature. In general, under conditions of low but increasing prey density, the role of the birds in diminishing the insect number is more significant than in areas where there is a very high infestation. Bruns (1960) has a similar conclusion. Birds generally are not able to break down an insect plague, but function more in preventing insect plagues.

IV. MANAGEMENT STRATEGIES

During the Symposium, participants have discussed, in general, methods of increasing the population size of woodpeckers, flycatchers, chickadees, bluebirds, vireos, and cavity nesters. These data are very valuable in planning a management program for these and closely related species. Now, we must view the community of birds in the forest ecosystem. It is most important that management techniques do not overmanage for one species and thereby exclude others. We must be sure to consider migratory birds which appear during the height of insect reproduction.

Leon Folse suggests that species having similar habitat requirements can be identified through multivariate analysis. Such an approach would be very useful in identifying key bird species and key habitat. Once key species or habitat type are identified, managers can maintain the associated group of species.

V. ROLE OF BIRDS IN THE FOREST ECOSYSTEM

McCullough (1970) indicates that our emphasis of bird populations as controlling insects may be neglecting other important roles that these animals play in the ecosystem. He points out that few studies of birds and mammals have attempted to place these animals in an ecosystem context, and particularly in forest ecosystems since the proportion of total energy and nutrients cycling through birds and mammals is relatively small. Birds have a very insignificant role in secondary productivity. However, the importance of birds and mammals should not be evaluated primarily in terms of the number of insects that they are taking since many bird populations as well as mammals are important sources of animal proteins in human diets.

Birds and mammals perform a number of functions in ecosystems due to activities other than feeding. For example, they disturb areas so that air can circulate and other animals can move. They perform a selective feeding role in controlling certain pest populations. While birds do not always control outbreaks, they do maintain some populations of insects at a level which keeps them from exploding into destructive population sizes. Riley McClelland shows this effect in his Pileated Woodpecker (<u>Dryocopus</u> <u>pileatus</u>) study. Birds, because of their mobility, play an important role in transporting material within and between both terrestrial and aquatic communities as well as dispersing seeds.

Wiens (1973) in a paper on grassland birds suggests that birds, which are largely opportunistic, vary their diets depending on the number of different food items available. He feels birds do not have a closely knit interrelationship with other components in the ecosystem. However, he points out that their opportunistic nature may enable birds to exert functional control in the system by flexibly responding to, for example, grasshopper outbreaks at infrequent but important intervals.

Wiens goes on to state that the evolution of structural and functional organization of the ecosystem might be a gradual process in which some components are first able to exist off the system excess without the reciprocal feedback effects that later evolve. He discusses the role of birds as governors or controllers rather than direct participants in the functional pattern of the ecosystem. Avian predators might act to dampen oscillations of herbivorous insects.

Birds therefore have a number of roles in the ecosystem. Imre Otvos states they have a direct and indirect effect on insect population. Wiens and McCullough suggest that they influence stability, transport, and energy flow in the system.

VI. CONCLUSION

The forest ecosystem, as Jerome Jackson stated in his opening remarks, underwent major changes in the last 200 years. We now have forests greatly influenced by human activity. We must learn what role birds have in this system so that the desired management practices can be instituted and undesirable ones discontinued.

Birds do not appear to have a major effect on productivity but appear to exert a density dependent effect on prey. Further, they are responsible for transporting seeds and nutrients, and appear to have a minor role in mineral cycling. Basically birds function in maintaining the stability of the system.

This Symposium points out several areas of investigation that should be followed.

1. We need to learn more about avian habitat requirements as they relate to habitat structure, food availability, and other animal species.

2. We must find out what the direct and indirect effects of avian temporal and seasonal foraging tactics are on insect populations.

3. We need more information on foraging changes that occur as prey populations move, increase and decrease in size, and disappear.

4. Finally we must realize that an important end product of our work is management. Species management has been common. Now management must be concerned with community and ecosystem levels.

This Symposium serves as a valuable tool to bring together experts studying different insectivorous birds and their prey base in the forest ecosystem. The proceedings serve as a stimulus in the effort to evaluate the role of birds in the forest ecosystem. As we look at the results of these data we must place them in the context of other ecosystem data, remembering that the term "ecosystem" is the most important word in the title of the Symposium, "The Role of Insectivorous Birds in Forest Ecosystems."

REFERENCES

Anderson, S., and Shugart, H. (1974). *Ecology 55*, 828.
Bruns, H. (1960). *Bird Study 7*, 193.
George, J., and Mitchell, R. (1948). *J. Forestry 16*, 454.
Hartley, P. (1953). *J. Anim. Ecol. 22*, 261.
Kendeigh, S. (1947). Ontario Dept. of Lands and Forest. Biol. Bull. No. 1.

Keve, A., and Reichart, G. (1960). *Falke 7*, 20.
Kilham, L. (1965). *Wilson Bull. 77*, 134.
Korol'kova, G. (1956). *In* "Ways and Means of Using Birds in Combating Noxious Insects" (L. Poznanin, ed.), p. 55. USSR Ministry of Agriculture, Moscow.
McCullough, D. (1970). *In* "Analysis of Temperate Forest Ecosystems" (D. Reichle, ed.), p. 107.
Morse, D. (1973). *Ecology 54*, 346.
Robbins, C. (1978). For. Serv. Gen. Tech. Rep. SE-14, p. 142. U.S. Dept. Agri., Asheville, North Carolina.
Shilova-Krassova, S. (1953). *Zoologicheski Zhurnal 32*, 955.
Tinbergen, L. (1960). *Arch. Neerl. Zool. 13*, 265.
Tothill, J. (1923). Proc. Acadian Ent. Soc. for 1922, *8*, 172.
Wiens, J. (1973). *Ecol. Monogr. 43*, 237.
Williamson, P. (1971). *Ecol. Monogr. 41*, 129.